SOLID STATE PHYSICS

VOLUME 40

Founding Editors

FREDERICK SEITZ

DAVID TURNBULL

SOLID STATE PHYSICS

Advances in Research and Applications

Editors

HENRY EHRENREICH

DAVID TURNBULL

Division of Applied Sciences
Harvard University, Cambridge, Massachusetts

VOLUME 40

1987

ACADEMIC PRESS, INC.
Harcourt Brace Jovanovich, Publishers

Orlando San Diego New York
Austin Boston London Sydney
Tokyo Toronto

ACADEMIC PRESS, INC.
Orlando, Florida 32887

United Kingdom Edition published by
ACADEMIC PRESS INC. (LONDON) LTD.
24–28 Oval Road, London NW1 7DX

LIBRARY OF CONGRESS CATALOG CARD NUMBER: 55-12200

ISBN 0–12–607740–1 (alk. paper)

PRINTED IN THE UNITED STATES OF AMERICA

87 88 89 90 9 8 7 6 5 4 3 2 1

Contents

Stage Ordering in Intercalation Compounds

S. A. Safran

Experimental Studies of the Structure and Dynamics of Molten Alkali and Alkaline Earth Halides

Robert L. McGreevy

Contributors to Volume 40

Numbers in parentheses indicate the pages on which the authors' contributions begin.

M. Y. CHOU, *Department of Physics, University of California, and Materials and Molecular Research Division, Lawrence Berkeley Laboratory, Berkeley, California 94720* (93)

MARVIN L. COHEN, *Department of Physics, University of California, and Materials and Molecular Research Division, Lawrence Berkeley Laboratory, Berkeley, California 94720* (93)

WALT A. DE HEER, *Department of Physics, University of California, Berkeley, California 94720* (93)

W. D. KNIGHT, *Department of Physics, University of California, Berkeley, California 94720* (93)

ROBERT L. MCGREEVY, *Clarendon Laboratory, University of Oxford, Oxford OX1 3PU, England* (247)

D. G. PETTIFOR, *Department of Mathematics, Imperial College, London SW7 2BZ, England* (43)

S. A. SAFRAN, *Exxon Research and Engineering, Corporate Research Science Laboratories, Annandale, New Jersey 08801* (183)

D. WEAIRE, *Physics Department, Trinity College, Dublin 2, Ireland* (1)

F. WOOTEN, *Department of Applied Science, University of California, Davis/Livermore, California 94550* (1)

Preface

The articles in this volume deal with a wide range of materials including amorphous semiconductors, metallic alloys and clusters, intercalation compounds, and ionic liquids.

Wooten and Weaire review the methods of construction, by handbuilding or computer, and the properties of four-fold coordinated random-network models. They describe the computer generation of a model in which a diamond cubic structure is randomized and then relaxed by a Monte Carlo procedure. Its properties have proved to be in remarkably good agreement with experimental results for amorphous silicon and germanium.

The Miedema Rules for alloy formation, discussed in the article by Pettifor, have been remarkably successful in predicting the heats of formation of many solid alloys in terms of ''macroscopic'' atoms having properties associated with real surfaces, in particular, the work function and surface charge density. Pettifor examines the quantum mechanical origin of the heat of formation of metals, clarifying not only the significance of the Miedema Rules, but in the process providing new insight into the nature of bonding in simple metals and their alloys. While the picture involving ''macroscopic'' atoms, as originally proposed, is not supported by the quantum theory of alloys, the scheme (when applied as Miedema intended it to be) has a great deal of practical utility. The basis for Pettifor's discussion in terms of pseudopotentials is supplied in part by the article of Heine and Weaire in Volume 24 of this series.

Unlike these ''macroscopic'' atoms, the metallic clusters described from both experimental and theoretical viewpoints in the article by de Heer *et al.* come far closer to having bulk properties, because they are ten to one hundred times bigger. The science of clusters is a rapidly growing interdisciplinary field with great promise for the production of new ideas and physical systems. Its ideas are relevant to related problems in the physics of atoms, molecules, condensed matter, and transitions among these systems. Recent developments have brought about an important increase in language compatibility and collaborations among experimentalists, theorists, physicists, and chemists. The review emphasizes the electronic shell model, which is elegant and simple, avoids the complexity of elaborate quantum chemical computer calculations for clusters containing large numbers of atoms, and possesses the power of predictability. Two of the authors, M. L. Cohen and W. D. Knight, have made distinguished contributions to this series previously (cf. Volumes 2, 24, and 31). Readers will find Lang's article concerning the electronic structure of metal surfaces in Volume 28 to supply useful background information.

Intercalation compounds, for example, involving graphite or the transition metal chalcogenides, exhibit a rich variety of novel structural and physical properties. Among these are quasi-two-dimensional solid, liquid, and gas behavior and incommensurate solid phases along with their associated phase transitions. The article by Safran provides a review of the field with emphasis on staging. This term describes the periodicity of the intercalate layer sequences that can exist in these solids. While its point of view is theoretical, this article emphasizes the connections and applications to experiment. It discusses the microscopic origin of the interactions responsible for staging, staging phase transitions, and the kinetics of staging and intercalation. The article's comprehensive albeit brief introduction to intercalation compounds and their electrons and transport properties makes it accessible to the general reader.

While the physics of the ordered crystalline state has been extensively studied, comparable studies of liquids on the atomic scale have only become possible during the last two decades because of the insufficient development of the requisite experimental techniques. McGreevy reviews the information provided by these techniques and its physical interpretation for the molten alkali and alkaline earth ion liquids, which may be considered to be completely ionic and unencumbered by the complexity of molecular or complex ions. The understanding of the structure and dynamics of these liquids provided by neutron and light scattering and computer simulation is discussed in considerable detail. These ideas and techniques evidently have more general applicability to other binary liquids, to monatomic and polyatomic fluids, and to amorphous materials. As the author points out, physical insight into the structure and dynamics of liquids that show interesting properties such as glass-forming ability or fast ion conduction in the solid state will result in a better understanding of such properties.

The articles appearing in this volume were all received by the editors in early 1986. Their publication has suffered an unfortunate delay through no fault of the authors. We apologize for the unusual circumstances that caused this delay and assure them and others who will contribute to this series in the future that steps are being and will continue to be taken to prevent a recurrence of this situation.

HENRY EHRENREICH
DAVID TURNBULL

SOLID STATE PHYSICS, VOLUME **40**

Modeling Tetrahedrally Bonded Random Networks by Computer

F. WOOTEN

Department of Applied Science,
University of California,
Davis/Livermore, California

D. WEAIRE

Physics Department,
Trinity College,
Dublin, Ireland

I. A Short History of Random-Network Models

1. Introduction

Glassmaking has been an important human activity since at least the seventh century B.C.,[1] and a sufficiently subtle process to merit mystery and special ceremony. The Assyrians advised[1]: "Do not allow any stranger to enter the building.... Offer the due libations to the gods daily."

Today's solid-state physicists treat glasses (and the wider class of amorphous solids) with the same admiration and respect in the investigation of their basic properties.

One cannot get very far along the road to understanding those properties without confronting the problem of the nature of the detailed structure of the material in question. To determine this structure is, for many, an end in itself; for many others it is the necessary foundation of theory, upon which they can build.

In principle, the connection between properties and structure may be inverted in almost every case to infer structural information from a given physical measurement, such as that of optical absorption. To do so is tempting but dangerous; it is usually better to stick to those techniques which bear directly on structure. The principal experimental technique, x-ray diffraction, has not changed much since the 1930s, although it is now augmented by a variety of more specialized probes.

In the same way, the attempt to build up a theoretical picture of the structure of an amorphous solid can rest on indirect arguments (from thermodynamics, "common sense," or whatever) or upon the direct testing of *models*, where the latter is interpreted literally. We incline to the latter approach, following Lord Kelvin[2]:

> I advise any of you who wish to study crystallography to contract with a wood-turner or a maker of beads for furniture tassels or for rosaries, for a thousand wooden balls of about half an inch diameter each.

In Fig. 1 we show an early example of such wooden balls, put to good use by Nernst[3] in illustrating the diamond cubic structure, which he suggested for diamond, just in advance of its experimental discovery by Bragg and Bragg.[4] It offers us an appropriate introduction to our theme, since it is

[1] S. Toulmin and J. Goodfield, "The Architecture of Matter." Univ. of Chicago Press, Chicago, Illinois, 1962.

[2] Lord Kelvin, "The Molecular Tactics of a Crystal," p. 9. Oxford Univ. Press (Clarendon), London and New York, 1894.

[3] W. Nernst, "The Theory of the Solid State." Univ. of London Press, London, 1914.

[4] W. H. Bragg and W. L. Bragg, *Proc. R. Soc. London, Ser. A* **89,** 277 (1913).

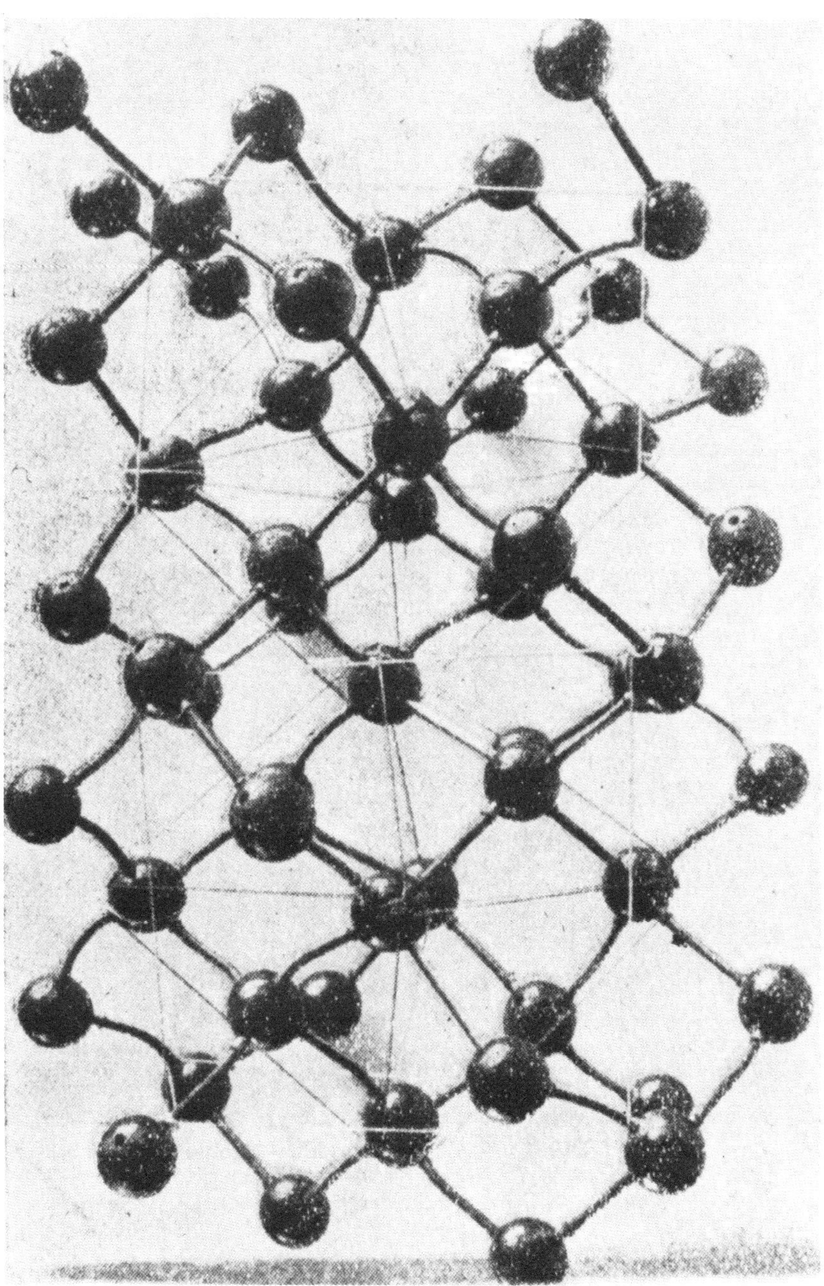

FIG. 1. W. Nernst's hypothetical cubic crystal structure for diamond.[3]

(we think) the first ball–and–stick model for a crystal inspired by a notion of a preferred bonding arrangement, and one of special significance for the world today and this article.

In due course such models became commonplace in both crystal and molecular physics. The work of Crick and Watson is a celebrated example of a case in which "hands-on" experience with a model was an essential guide and constraint to creativity.

But in our subject, that of amorphous solids, the use of actual physical models was delayed until about 1970. The reasons are not obvious; we shall comment upon this again in later sections.

After 1970 many such models were built. Use was made of computers to relax and analyze models and, in a few cases, actually to construct them. There has never been much doubt that the future lay in the direction of computational methods of construction. Building models by hand is often uncomfortably subjective; amusing at first, it soon becomes tedious, and the range of models that can be explored is limited in practice. Computational procedures offer great flexibility and power, once a satisfactory algorithm has been developed. The latter is, however, nontrivial.

In this article we shall review our own program of research in the development of such an algorithm for the case of tetrahedrally bonded amorphous solids, setting it in the context of previous and parallel efforts directed toward the same end.

2. Background: Some Key Questions

Accounts of the early history of crystallography tell of its guiding principle: that the symmetric external form of crystals is due to the internal order of atoms or molecules. The converse of this idea, that amorphous solids have a disordered internal structure, was also current in the 18th and 19th centuries. It was asserted, for example, in one of John Tyndall's public lectures[5]:

> To many persons here present a block of ice may seem of no more interest and beauty than a block of glass; but in reality it bears the same relation to glass that orchestral harmony does to the cries of the marketplace. The ice is music, the glass is noise; the ice is order, the glass is confusion. In the glass, molecular forces constitute an inextricably entangled skein; in ice they are woven to a symmetric texture...

This vivid and poetic image of an "inextricably entangled skein" is particularly appropriate to the theme of the present article, since the computer algorithm with which we shall be concerned creates just such a system.

[5] J. Tyndall, "Heat—A Mode of Motion." Longmans, Green, London, 1863.

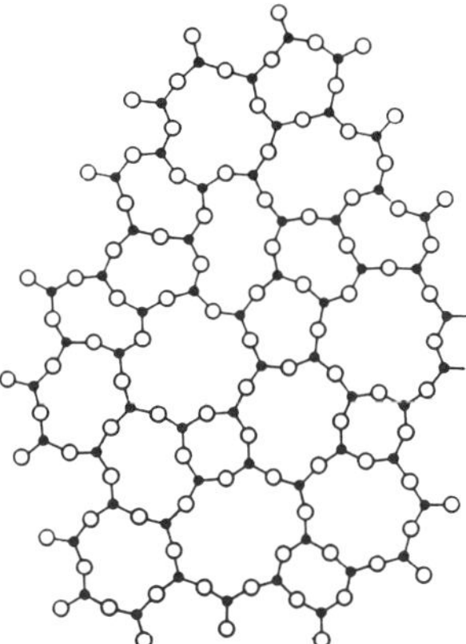

FIG. 2. Zachariasen's schematic diagram illustrates his continuous random network model for amorphous silica.[6]

For *molecular forces* we should today substitute *covalent bonds.* With this modernization Tyndall's picture is clearly that of a *continuous random network.* This concept is commonly attributed to Zachariasen,[6] who came nearly a century later: "...the atomic arrangement in glass is characterized by an extended three-dimensional network which lacks symmetry and periodicity." The actual term *random network* seems to have been first used by Warren.[7]

Zachariasen's celebrated schematic diagram, which illustrates the nature of this model, is shown in Fig. 2. His ideas were exceptionally coherent and timely. They emerged at the end of a period which had witnessed the spectacular succcess of x-ray crystallography in the determination of crystal structures. When the crystallographers turned some of their attention to liquids and amorphous solids, around 1930, they were naturally prejudiced in favor of structural interpretations that took the single crystal as their starting point. Thus the microcrystallite hypothesis, which seeks to explain the structures of amorphous solids, and even liquids, in terms of a very fine polycrystalline structure, was popular. It may be traced back to Lebedev,[8] in particular, in 1921.

It should be realized that a great deal of experimental work on amorphous solids was done in the period in question. As W. L. Bragg said in 1933,[9] "Diffraction by amorphous bodies has been the subject of an immense amount

[6] W. H. Zachariasen, *J. Am. Chem. Soc.* **54,** 3841 (1932).

[7] B. E. Warren, *Phys. Rev.* **45,** 657 (1934).

[8] E. A. Porai-Koshits, *J. Non-Cryst. Solids* **73,** 79 (1985).

[9] W. L. Bragg, *in* "The Crystalline State" (W. H. Bragg and W. L. Bragg, eds.), Vol. 1, p. 195. Bell, London, 1933.

FIG. 3. The hand-built model of Bell and Dean for amorphous silica[11] (Crown copyright).

of research." The attempts to fit the data with microcrystallite models must have been many, and the lack of success of such attempts made Zachariasen's alternative ideas attractive. They were immediately recognized as "a definite step forward"[10] and gradually gained ground, until the random network model became orthodox within the community of crystallographers working on glass. However, this community itself became less prominent; the "sharp end" of structural studies moved elsewhere, and the nature of amorphous solids receded once again into the scientific background. Perhaps this is why no actual physical model based on Zachariasen's principle was built until the late 1960s. At that time a number of such models were built for SiO_2, most notably that of Bell and Dean,[11] for which an impressive degree of agreement with diffraction data was demonstrated.[12] This was the belated fulfillment of the program outlined by Zachariasen (Fig. 3).

At the same time the field widened and deepened to include a broad range of amorphous solids, so that amorphous metals and amorphous elemental

[10] J. T. Randall, "The Diffraction of X-rays and Electrons by Amorphous Solids, Liquids and Gases," Chapman & Hall, London, 1934.

[11] R. J. Bell and P. Dean, *Nature (London)* **212,** 1354 (1966).

[12] R. J. Bell and P. Dean, *Philos. Mag.* [8] **25,** 1381 (1972).

semiconductors, among others, were intensively studied for the first time. The amorphous Group-IV semiconductors, *a*-Si and *a*-Ge, came to be regarded as prototypical covalent amorphous solids, and our own work follows in that relatively new tradition. There is, however, a very close connection betweem amorphous silicon and amorphous silica, at least in principle. A random-network model for one can be converted into a corresponding model for the other by the addition or subtraction of the oxygen atoms which link the silicon atoms.

Phillips[13] has nevertheless placed these two amorphous solids in distinct categories, amorphous silica being a *glass* (quenched from the melt, and undergoing a glass transition), while amorphous silicon cannot be so made: It is said to be "overconstrained." As a consequence, it is suggested that amorphous silicon must necessarily have a discontinuous structure, in which regions of continuous random network are limited in size.[13,14] The underlying notion, that tetrahedral bonds cannot be randomly connected *ad infinitum* in a disordered manner, is intuitive rather than proven. Our own prejudice is to accept the infinite random network as an idealization and regard departures from it (evidenced, for example, by electron spin resonance) as accidental and dependent on its preparation.

Practical applications have recently attracted much attention to *hydrogenated* amorphous silicon, in which sufficient hydrogen is incorporated in the material to radically alter its properties, chiefly by saturating dangling bonds. In many respects the basic structure of this material appears to have much in common with pure amorphous silicon and germanium, but there is now little doubt that it is generally inhomogeneous, containing regions of relatively high and low hydrogen concentration. There are many outstanding questions regarding this material, which we shall not attempt to analyze in this article, but it seems that the random network will be one ingredient in any satisfactory complete picture of its structure.

The first random-network models built specifically for Si and Ge date from the early 1970s and are reviewed in the following section.

3. Model Building for Amorphous Silicon and Germanium

The first model for a tetrahedrally bonded amorphous solid was built by Polk[15] with the guidance of Turnbull, and is shown in Fig. 4. It is a roughly spherical cluster of plastic units. Anyone contemplating it is bound to ask: What rules were used to construct it? The answer is, in part, that remarkably few rules were needed. If no attempt is made to develop a periodic structure,

[13] J. C. Phillips, *J. Non-Cryst. Solids* **34**, 153 (1979).

[14] J. C. Phillips, *J. Non-Cryst. Solids* **43**, 37 (1981).

[15] D. E. Polk, *J. Non-Cryst. Solids* **5**, 365 (1971).

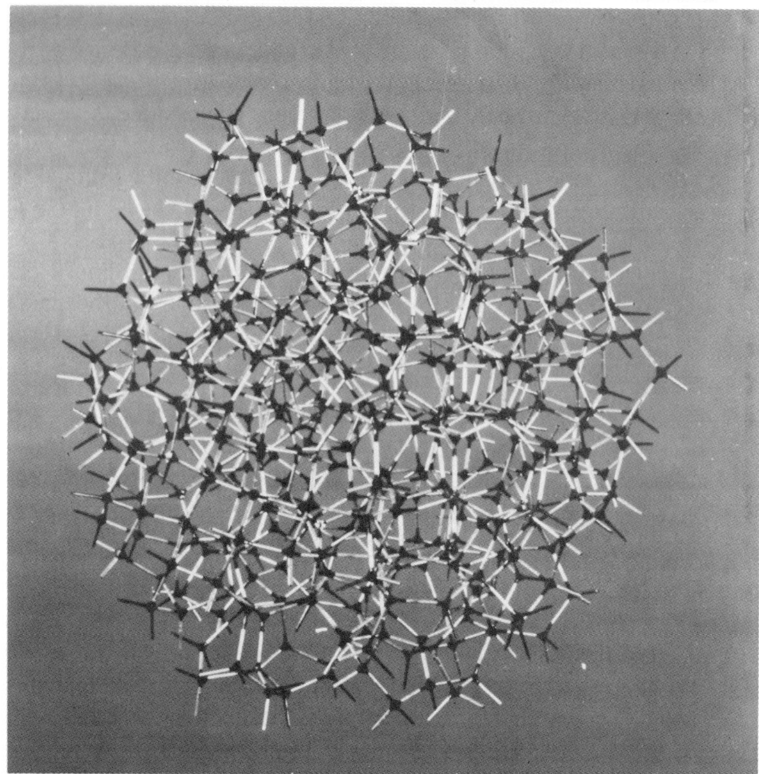

FIG. 4. Polk's original hand-built model for amorphous silicon.[15]

and if fairly rigid units are used, then the only rule that is really necessary is that no dangling bonds are left within the cluster, and it is maintained in a roughly spherical shape. Typically there are only a couple of alternatives for the bonding of an atom in a given locality on the surface: An arbitrary "random" choice is made, just as in the work of Bell and Dean.

This is not to say that a different structure cannot be made if one sets out to do so, with additional rules. Later on, Connell and Temkin[16] did so, eliminating odd rings of bonds. However, such exercises, while interesting, do not improve agreement with experiment in this case. Polk's original comparison with diffraction data is shown in Fig. 5. The significance of the radial distribution function (RDF) and its relation to diffraction experiments is recalled in Appendix A.

[16] G. A. N. Connell and R. J. Temkin, *Phys. Rev. B: Solid State* [3] **9**, 5323 (1974).

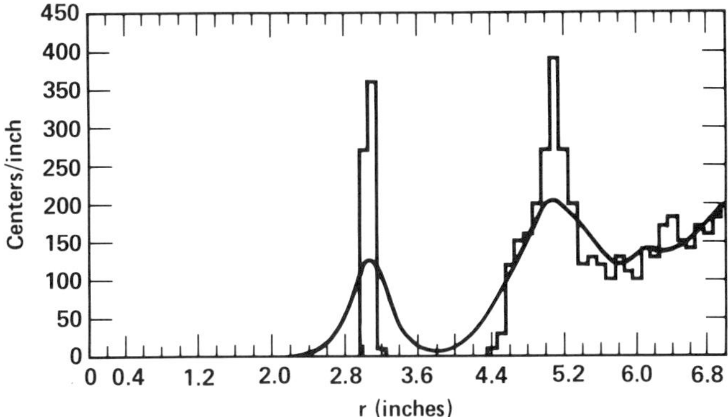

FIG. 5. Comparison with the experimentally determined radial distribution function for Polk's original model.

Most of the models built before 1985 have been critically reviewed by Etherington *et al.*,[17] with emphasis on the comparison with diffraction data. We have one or two minor points of disagreement with that review, but its uniform presentation of radial distribution functions is very useful.

A "back-of-the-envelope" calculation will show that, in a cluster of a few hundred atoms, quite a large fraction lie close to the surface. This complicates the analysis of a model such as Polk's in the investigation of bulk properties, or even calculation of a radial distribution function. This was recognized at an early stage by Henderson, who set out to create a model with periodic boundary conditions. By this we mean that the model is contained in the unit cell of a lattice, in such a way that the periodic structure so defined has the same sort of random tetrahedral structure as the interior of Polk's model; that is, there is no mismatch at the boundaries of the unit cell. Free surfaces are thereby eliminated. By a *tour de force* of hand building, Henderson actually produced such a model,[18] containing 61 atoms. He also experimented with computer algorithms for creating periodic structures, but did not succeed in producing one which gave good agreement with experiment. (It is, however, a feature of the primary literature that "good agreement" is almost always claimed in original publications, and such articles as the present one are necessary to put this into perspective.)

Henderson's early work[19] contained two essential ingredients of our own: periodic boundary conditions and a computer algorithm. Indeed the latter,

[17] G. Etherington, A. C. Wright, J. T. Wenzel, J. C. Dore, J. H. Clarke, and R. N. Sinclair, *J. Non-Cryst. Solids* **48,** 265 (1982).

[18] D. Henderson, *J. Non-Cryst. Solids* **16,** 317 (1974).

[19] D. Henderson and F. Herman, *J. Non-Cryst. Solids* **8–10,** 359 (1972).

which was based on the iterative relaxation of highly random structures, was in somewhat the same spirit as ours, namely that a *simple* algorithm is highly desirable. In contrast to this, the attempts by Shevchik[20] and Polk and Boudreaux[21] to mimic the hand-building process were rather cumbersome, although quite successful in building up the same sort of cluster that Polk had studied.

Thus by 1980 the random-network model had become well established for amorphous silicon and germanium and a range of more or less acceptable structural models existed. Nevertheless, the model-building process still looked clumsy and inflexible. Given the enduring interest in this structure and a number of related problems, a fully satisfactory computer model-building procedure was still desirable.

4. Strategy and Tactics for Computer Modeling

One may distinguish three distinct strategies for building a random network model by computer:

1. Accretion. By this we mean the progressive addition of atoms to a growing cluster. This now seems attractive only when surface properties or the kinetics of the deposition process are of interest.
2. Molecular dynamics. Molecular dynamics has already been used for a number of glasses,[22] and has much to commend it. It is doubtful if satisfactory structures can now be generated for amorphous Si and Ge by this method (see Section IV), but one may anticipate important future developments in this area.
3. Randomization and relaxation. By this we refer to a class of procedures in which a highly random structure is created and then relaxed towards a low-energy structure which is still random. In our own work we have called it colloquially "shake-and-bake," which perhaps captures its essence.

In comparison with molecular dynamics, our third and favored category suffers from the disadvantage that its rules may be rather arbitrary and "unphysical." On the other hand, it can have advantages of extreme simplicity and transparency, without being entirely divorced from the relevant physics. It can use very simple interatomic potentials and restrict covalent bonding to a given type, such as the tetrahedral bonding which we have in mind here. It works within a restricted space of bonding possibilities, in a similar manner to studies based on model spin Hamiltonians in the theory of magnetism.

[20] N. J. Shevchik, *Phys. Status Solidi* B **58,** 111 (1973).

[21] D. E. Polk and D. S. Boudreaux, *Phys. Rev. Lett.* **31,** 92 (1973).

[22] S. A. Brawer and M. J. Weber, *Phys. Rev. Lett.* **45,** 460 (1980).

As we have seen, this strategy was attempted in a primitive form by Henderson.[18] More recently, Guttman[23,24] has also done so. Guttman's approach is quite similar to ours, except that he does not settle on a single, simple algorithm, and his calculations have not been carried as far as ours; hence they do not produce the same degree of agreement with diffraction data.

Our own work has been aimed from the outset at meeting a number of criteria:

1. Compatibility with *periodic boundary conditions* in the sense described above;
2. Practicality for samples of at least *several hundred atoms*;
3. A *simple* computer algorithm; and
4. *Good agreement* with the observed radial distribution function.

It developed in two stages, of which the second was not entirely anticipated.

First, a range of tetrahedrally bonded structures was created by "randomizing" the diamond cubic structure. The latter is chosen simply because it gives us a convenient starting configuration, consisting of tetrahedral bonds and compatible with periodic boundary conditions.

Second, such randomized structures were altered in an attempt to minimize the total energy. In doing so, we were strongly influenced by the widespread adoption of the technique of *simulated annealing*[25] even in fields far afield from physics. We were thus encouraged to use an elementary Monte Carlo method at this point, although its physical significance is not obvious.

The aim of this two-step process of randomization and Monte Carlo relaxation is to create what is, within the given set of rules, an "inextricably entangled skein" of tetrahedral bonds. When it does so, the degree of agreement with experiment is remarkable, as we shall see.

II. Computer Generation of Continuous Random-Network Models with Periodic Boundary Conditions

5. Introduction

It is perhaps surprising that there have been so few attempts to explore the tetrahedrally bonded random network systematically along the lines suggested in the last section. In part this may have been due to a suspicion that periodic boundary conditions are somehow artificial and undesirable.

[23] L. Guttman, *AIP Conf. Proc.* **20,** 224 (1974).
[24] L. Guttman, *AIP Conf. Proc.* **31,** 268 (1976).
[25] S. Kirkpatrick, C. D. Gelatt, Jr., and M. P. Vecchi, *Science* **220,** 671 (1983).

However, in our view such boundary conditions are very much the lesser of two evils, the other being the free surface of cluster models. The influence of the boundary conditions on the structure within the unit cell is an intriguing question, but certainly there is no evidence that it seriously detracts from the value of models such as ours.

The first systematic program of this kind was that of Guttman,[23,24] which we shall review in some detail.

Before embarking on an analysis of the two approaches, it is of interest to note what is probably the most obvious technique to ensure perfect connectivity and at the same time to introduce both topological and geometric disorder, that is, to start from the diamond cubic structure, remove single atoms at random, and rejoin the four dangling bonds in pairs. This procedure introduces five- and sevenfold rings of bonds into the network and destroys six- and eightfold rings. Alben *et al.*[26], starting from a 64-atom supercell, generated a 58-atom model in this way for purposes of comparison with other models, while engaged in the study of vibrational properties. They (and we, in unpublished studies of similar models containing 700–1000 atoms) concluded that such structures are rather unsatisfactory as representations of *a*-Si. The structures have large bond-angle distortions, and their RDFs are in poor agreement with experiment. One can devise modifications of this scheme, but only at the cost of a loss in simplicity and with no significant improvement in the model.

6. Guttman's Method

The first approach to the computer generation of periodic random networks to be extensively developed was that of Guttman.[23,24]

He considered identical atoms with fourfold coordination. His aim was to construct an ideal random network with no dangling bonds, with application to *a*-Ge.

Starting from a finite region of a perfect crystal chosen as a unit cell, containing many unit cells of the crystal and subject to periodic boundary conditions, Guttman's algorithm provides the following prescription: A set of n "neighbors," $n > 4$, is assigned to each atom. Then, by a stochastic process, four neighbors are selected to which bonds are made. This process is repeated until the desired network is constructed or the boundary conditions prohibit completion. Note that there is no guarantee of success. Guttman found that the probability of successfully constructing a network with no dangling bonds

[26] R. Alben, D. Weaire, J. E. Smith, Jr., and M. H. Brodsky, *Phys. Rev. B: Solid State* [3] **11,** 2271 (1975).

decreases with increasing cell size, varying from 30% success with a 64-atom cell to 0.1% for a 512-atom cell.

Guttman originally chose to start with the diamond structure and assign to each atom the set of four first- and twelve second-nearest neighbors, from which four were to be selected for bonding. If these bonding assignments are made purely at random, too much disorder is introduced and the density after relaxation is excessive. The latter is to be expected: 75% of the bonds ($\frac{12}{16}$) in the unrelaxed structure would have bonds 66% too long, corresponding to the second-neighbor distance. Thus the structure would be under strong compressive stress. To reduce this effect, Guttman chose arbitrarily to weight the group of four nearest neighbors by a factor of 100.

Two 512-atom networks generated by this scheme were chosen for relaxation using a potential with a simple Hooke's-law bond-stretching term and a similar angle-bending term. The cell size was varied to find the density at which the potential energy of the relaxed structure was a minimum. These densities averaged 15% greater than the crystal density (the experimental value being within a few percent of that for the crystal). Moreover, the rms angular deviation from perfect tetrahedral bonding was about 17°, much greater than the experimental value (roughly 10° [17]).

This first attempt by Guttman suffered from the same deficiencies as did the first work of Wooten and Weaire[27] and other early work mentioned above: too much emphasis on introducing topological randomness and not enough on the energies involved. Guttmann recognized the flaw and, presumably recognizing also that hand-built models of the day were better than his computer-generated models, suggested that "...it should be possible to guide the computer in whatever direction is necessary to imitate the human model-builder." Of course, what one really would like to do is mimic the actual physical processes that lead to amorphous structures in nature, meaning in this case the laboratory.

Guttman next used his algorithm as a starting point, in order to generate a random network, which was then modified by alteration of the bonding pattern on a trial basis.

It is useful to consider the kind of topological rearrangements used by Guttman. Two examples are shown in Fig. 6, where atoms 1, 2, 3, and 4 define a path of alternating bonds and nonbonds. Interchanging the bonds and nonbonds along the path produces a topological rearrangement that preserves fourfold coordination when the path is constrained to terminate on the starting atom or pass through its image in an infinite succession of periodic cells. *Any* topological rearrangement that preserves the coordination number

[27] F. Wooten and D. Weaire, *J. Non-Cryst. Solids* **64**, 325 (1984).

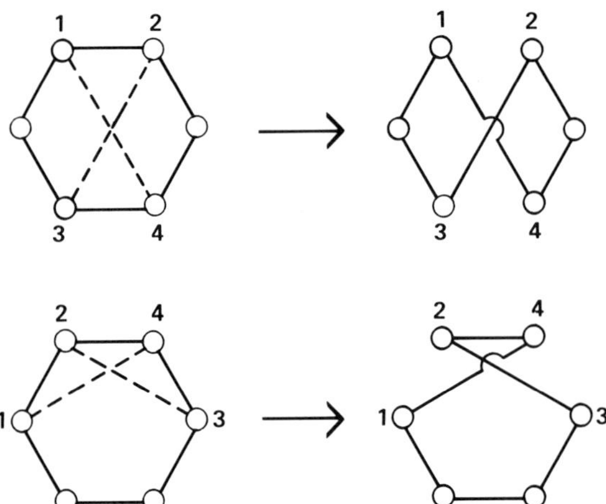

FIG. 6. Left: Two examples of a set of four atoms selected such that a path can be defined ($1 \rightarrow 2 \rightarrow 3 \rightarrow 4 \rightarrow 1$) whose steps are alternately between bonded atoms and nonbonded atoms. Right: Bonds and nonbonds along the chosen paths have been interchanged.

for *all* atoms can be described in terms of such a transformation along a path of alternating bonds and nonbonds.

Guttman's method is to program the computer to search for paths of some maximum number of steps, starting from the shortest nonbond. Having found a path, the bonds and nonbonds are interchanged on a trial basis. If, after relaxation to its minimum energy configuration, the new structure has a lower energy, it is accepted; if the energy has increased, the change is rejected. The process is continued until the strain energy has been lowered as much as seems reasonably possible.

Guttman first applied his refined method to the construction of 64-atom models, achieving some improvement in density and RDF, but falling short of the desired agreement between models and experiment. In later work[28,29] he used 54-atom unit cells based on the body-centered cubic structure as a starting point.

An ensemble of models was generated, from which he selected a few for detailed examination. He exploited one of the distinct advantages of computer-generated models, namely, the ease of making and studying a large number of models, something that is impossible in practice with hand-built models. However, in this case the overall effect was to reinject the element of subjective judgment for which the hand-builders had been criticized.

[28] W. Y. Ching, C. C. Lin, and L. Guttman, *Phys. Rev. B: Solid State* [3] **16,** 5488 (1977).
[29] L. Guttman, W. Y. Ching, and J. Rath, *Phys. Rev. Lett.* **44,** 1513 (1980).

7. Bond Transpositions

As we have seen, one needs to introduce rearrangements that do not produce excessive bond distortion. A particularly simple bond rearrangement is illustrated in Figs. 7 and 8. It exerts a torque on the bond connecting atoms 2 and 5, which can move to relieve much of the strain imposed. It does not produce gross distortions of geometry as do the ring-twisting actions illustrated in Fig. 6. The use of this simple type of rearrangement was inspired by an analogous process described by Weaire and Lambert in studies of two-dimensional networks such as soap froths, in a body of work reviewed by Weaire and River.[30] It is characterized by switching two second-neighbor bonds that are parallel to each other in the perfect diamond structure (or approximately so in other structures). Its salient feature is that, of the obvious local rearrangements, this one introduces the minimum strain into the otherwise perfect diamond structure. In particular, the strain energies associated with introducing the rearrangements shown in Figs. 6a,b and 7 (or 8) are, roughly, 19, 12, and 4.5 eV, respectively, for structures relaxed using the Keating potential (Section 8,b). Of course, in an already randomized lattice, the values for strain energy introduced by a particular type of *topologial* rearrangement will show a spectrum of values depending on the accompanying *geometrical* rearrangement. However, the trends will be the same, and it is important to exclude topological rearrangements that would generally introduce very large strain energies. A schematic representation of the energy barrier that must be overcome in making the bond transposition is shown in Fig. 9.

The key idea in our work is that *only* this one simple type of rearrangement is used throughout the two-stage process which creates the model. It appears to be sufficient for its purpose, and is both efficient and unprejudiced. Rather as transpositions may be used to build up an arbitrary permutation, so this one

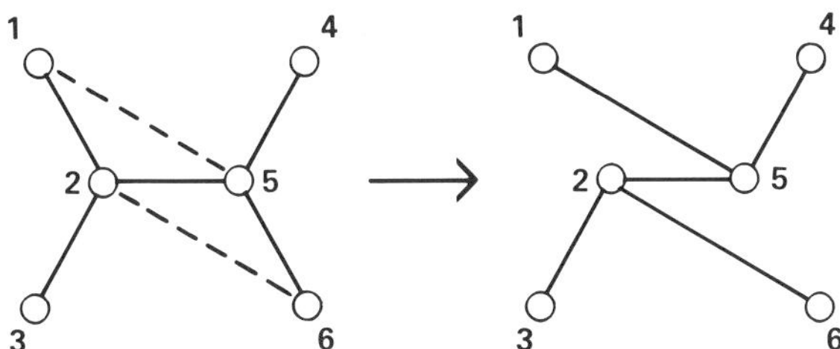

Fig. 7. A simple bond rearrangement used in studies of two-dimensional networks such as soap froths.[30]

[30] D. Weaire and N. Rivier, *Contemp. Phys.* **25,** 59 (1984).

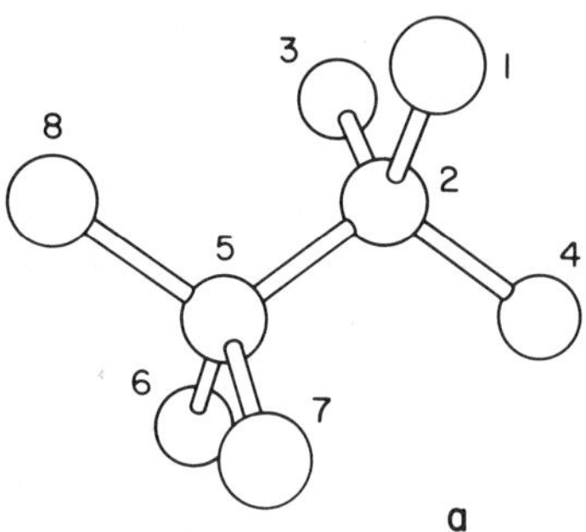

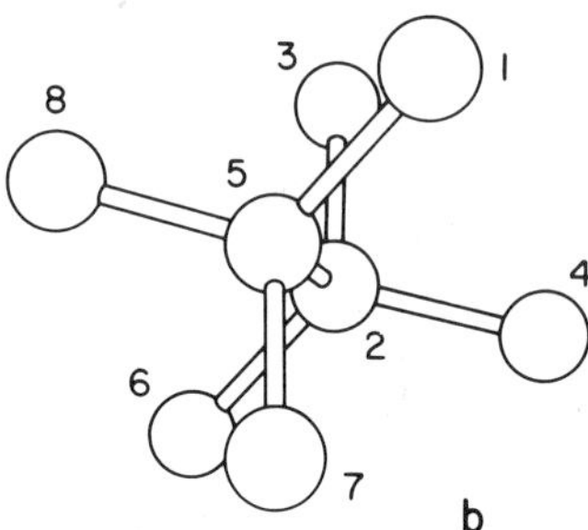

FIG. 8. Bond rearrangement or "bond switch" involving the exchange of two parallel bonds. This is the three-dimensional analog of the rearrangement shown in Fig. 7. The atoms are embedded in a 216-atom unit cell. (a) Configuration of bonds in the diamond cubic structure. (b) Relaxed configuration after switching two bonds.

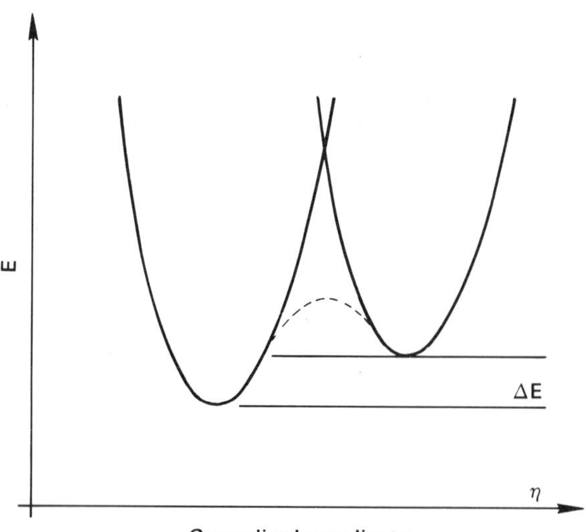

FIG. 9. Schematic illustration of the variation of potential energy around two alternative configurations that are related by a bond transposition. In a more accurate description, the energy would show a smooth variation as indicated by (– – –), defining the barrier. This can only be very crudely estimated in the present model, which gives the energies indicated by (———).

elementary process is used to explore a wide range of random structures. Indeed, this may be regarded as a transposition in the table of nearest neighbors.

Originally we referred to such rearrangements as "defects," but quickly grew to regret this inappropriate term. Here we propose to call them "bond transpositions" or "bond switches." When combined with prescriptions for the calculation of strain energies and a few subsidiary rules, the specification of this one allowed rearrangement defines a model material which approximates silicon and germanium for some purposes. In the spirit of jellium and Lennard–Jonesium, we may call this model material *sillium.*

8. Definition of Sillium

Briefly, the sillium model is defined by the following rules:

1. Each atom is bonded to *four* neighbors.
2. The total energy is the sum of bond-bending and stretching terms, as given by the Keating potential. No vibrational motion is considered.
3. The only degrees of freedom allowed consist of the bond transpositions discussed in the previous section, and defined more precisely below.

In principle this might be regarded as a (rather artificial) thermodynamic system, leading to the specification of a phase diagram, etc. In practice, the model "just grew," and we have not adopted such an attitude up to this point. In future such quantities as the melting temperature (if this term is appropriate for the order–disorder transformation in this case) may well be of interest.

Added to the above rules are some more technical considerations, which complete the recipe for investigation of the model for most of our work:

1. Periodic boundary conditions; and
2. Monte Carlo rules for structural rearrangements at finite temperature T, based on energy differences of equilibrium structures. Using this terminology, our *modus operandi* is as shown in Fig. 10.

Each of these points will be given more detailed explanation below.

a. Tetrahedral Bonding

The tetrahedral bond, realized in its most perfect symmetric form by the elements which form the diamond cubic structure and, with some distortion, in various other crystal structures, is the only bonding configuration considered in this work. Thus, for example, varieties of amorphous carbon which involve sp^2 bonding are excluded, as is the metallic bonding found in liquid Si and Ge.

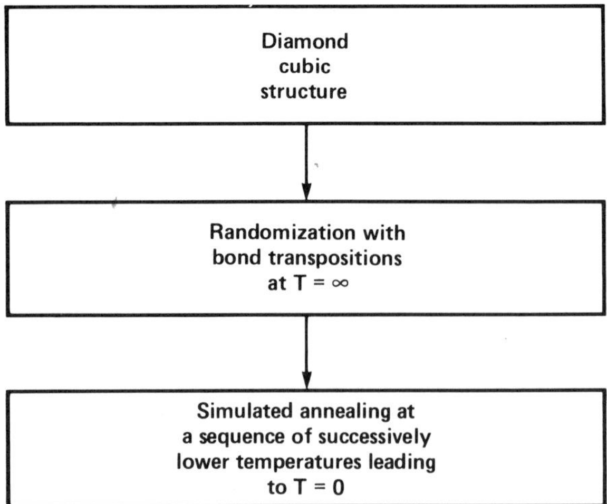

FIG. 10. Procedure used here to generate random networks (see Note Added in Proof).

A surprising range of topologically distinct structures is achievable with slightly distorted tetrahedral bonds. These include the diamond cubic and related structures (wurtzite, etc.) and the silicon and germanium polymorphs (Si III, Ge III, clathrates).[31] To these may be added further structures found in SiO_2 and H_2O which are essentially tetrahedral in character. It follows that a wide range of disordered structures must also be possible. From this point of view, it is surprising that the structure of amorphous Si and Ge seems to be rather well defined and that a variety of rather arbitrary model-building procedures have approximated it quite well.

b. The Keating Potential

The Keating potential, which we adopt for the description of bond-bending and stretching forces, arose in the context of attempts to fit the elastic and vibrational properties of Group-IV elements.[32] It is one of various alternative potentials which offer a semiempirical description of bonding forces, and involve only a few parameters. They may be contrasted with the earlier tradition in lattice dynamics, due to Born,[33,34] in which the total energy is

[31] F. C. Weinstein, *in* "Fifth International Conference on Amorphous and Liquid Semiconductors" (J. Stuke and W. Brenig, eds.) Taylor & Francis, London, 1974.

[32] P. N. Keating, *Phys. Rev.* **145,** 637 (1966).

[33] M. Born, *Ann. Phys. (Leipzig)* [4] **44,** 605 (1914).

[34] M. Born and K. Huang, "Dynamical Theory of Crystal Lattices." Oxford Univ. Press, London and New York, 1954.

simply expanded for a given crystal in terms of atomic displacements, and force constants emerge as the coefficients, to be adjusted to fit experiments. This frankly empirical approach, which at first seems logical and unprejudiced, ran into great difficulties. If taken much beyond nearest neighbors, the fitted forces are underdetermined and dependent on truncation; moreover, the transfer of the forces to a different structure poses problems. Hence the Born scheme has been largely supplanted by others in which physical or chemical insight prescribes a suitable form for the total energy, with few parameters. In our case these include the valence force field model[35,36] and the Keating potential,[32] which are roughly equivalent, and the more elaborate Weber bond charge model.[37]

The Keating potential consists of two terms, as indicated in Fig. 11a and detailed in Eq. (8.1).

$$V = \frac{3}{16}\frac{\alpha}{d^2}\sum_{l,i}(\mathbf{r}_{li}\cdot\mathbf{r}_{li} - d^2)^2 + \frac{3}{8}\frac{\beta}{d^2}\sum_{l\{i,i'\}}(\mathbf{r}_{li}\cdot\mathbf{r}_{li'} + \tfrac{1}{3}d^2)^2 \tag{8.1}$$

where α and β are the bond-stretching and bond-bending force constants, respectively, and $d = 2.35$ Å is the Si–Si strain-free equilibrium bond length in the diamond structure. The first sum in the expression is on all atoms l and their four neighbors specified by i; the second sum is on all atoms and pairs of distinct neighbors; and $\mathbf{r}_{li}$ is the vector from atom l to its ith neighbor. The bond-stretching force constant, $\alpha = 4.75 \times 10^4$ dyn/cm, is from Alben *et al.*[38] Various values of β have been used, and they are mentioned at appropriate places.

Central bond-stretching forces alone (the α term) do not suffice to stabilize a tetrahedrally bonded structure. The constraints imposed by maintaining constant bond lengths are outnumbered by the degrees of freedom. Hence the bond-bending term is essential. For many purposes, however, including the understanding of vibrational spectra[26,39] or the structures considered here, it is important to remember that the bond-stretching forces are much larger in practice. The value of β/α is to some extent arbitrary, in view of the approximate nature of the model, but must fall somewhere in the range 0.1–0.3 for Si/Ge. We have followed Martin,[36] taking $\beta/\alpha = 0.285$ (when building the model), but the precise value is not crucial in our work. This is due to the dominance of the bond-stretching forces. What generally happens, when an acceptable model for *a*-Si or *a*-Ge is relaxed to minimize energy, is as follows.

[35] M. J. P. Musgrave and J. A. Pople, *Proc. R. Soc. London, Ser. A* **268,** 474 (1962).

[36] R. M. Martin, *Phys. Rev. B: Solid State* [3] **1,** 4005 (1970).

[37] W. Weber, *Phys. Rev. B: Solid State* [3] **15,** 4789 (1977).

[38] R. Alben, J. E. Smith, Jr., M. H. Brodsky, and D. Weaire, *Phys. Rev. Lett.* **30,** 1141 (1973).

[39] D. Weaire and R. Alben, *Phys. Rev. Lett.* **29,** 1505 (1972).

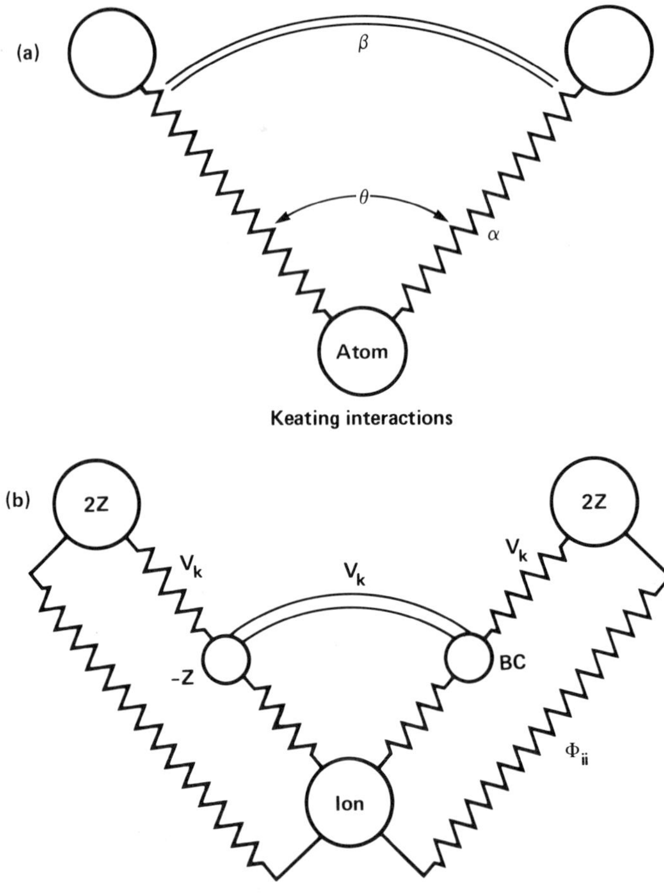

FIG. 11. Keating and Weber interactions. (a) The Keating potential consists of bond-stretching and bond-bending terms with force constants α and β, respectively. (b) Weber's adiabatic bond-charge model includes four interactions: a Coulomb interaction between all ions and bond charges, a Keating interaction coupling two bond charges to the common ion, an ion–ion central interaction, and a Keating-type interaction between each bond charge and its associated ions.

The bonds relax to their ideal unstretched length (to within a few percent). As we have seen, this underdetermines the structure, and the β term is responsible for the choice of a structure to minimize the bond-bending energy, within the subspace defined by constant bond lengths. Hence, to a rough approximation the total energy simply scales with β, and the relaxed structure is independent of β/α.

Both the Keating and valence force field models are capable of further extension with additional force constants, but one soon runs into the same problems which bedevilled the Born formalism. A better solution, which is in some sense an extension of the Keating model, but with a vital extra ingredient borrowed and adapted from the shell model,[40] is the Weber bond charge model.[37] In this, extra degrees of freedom are introduced, which represent bonding electrons as classical charges on the bonds. With a few more parameters this achieves very accurate agreement with vibrational spectroscopy for crystalline Si and Ge. We have thus been tempted to explore its use in our own model. This has been done,[41–43] but the results, on the whole, do not deviate much from the Keating model, and will not be discussed in much detail below.

We should note in passing that recent first-principles calculations have achieved good results for the relative energies of tetrahedrally bonded crystal structures.[44–46] Compared with these, our own prescription is very primitive, but the marriage of the sophistication of the local-density-functional method to the full complexity of the structures considered here still lies in the future.

c. Rules for Bond Transpositions

The entire process of randomization and annealing is achieved using the simple type of rearrangement illustrated in Fig. 8, but subsidiary rules and procedures have also been used, as explained below.

Each atom is numbered, its coordinates are listed in a coordinate table, and a neighbor table is constructed.

Four atoms are chosen at random, subject to the restriction that they form a chain such as that given by atoms 1, 2, 5, and 6 in Fig. 8. In order to minimize bond distortion, the two bonds to be switched must be as nearly parallel as possible. Bonds 1–2 and 5–6 meet this criterion; bonds 1–2 and 5–7 do not. The latter two bonds belong to a common sixfold ring in the diamond structure. Switching them would correspond to the rearrangement illustrated in Fig. 6b, which introduces excessively large strain energy. To ensure that the bonds to be switched are roughly parallel (which is the best that can be done in the random network), it is required that they *not* be members of the same 5-, 6-, or 7-fold ring.

Bonds must not be excessively long. Without this restriction, a sequence of bond switches can result in a bond extending the length of the unit cell,

[40] W. Cochran, *Proc. R. Soc. London, Ser. A* **253,** 260 (1959).

[41] F. Wooten, K. Winer, and D. Weaire, *Phys. Rev. Lett.* **54,** 1392 (1985).

[42] K. Winer and F. Wooten, *Comput. Phys. Commun.* **34,** 67 (1984).

[43] B. J. Hickey, G. J. Morgan, D. L. Weaire, and F. Wooten, *J. Non-Cryst. Solids* **77/78,** 67 (1985).

[44] M. T. Yin and M. L. Cohen, *Phys. Rev. Lett.* **45,** 1004 (1980).

[45] M. T. Yin and M. L. Cohen, *Phys. Rev. B: Condens. Matter* [3] **26,** 5668 (1982).

[46] R. Biswas, R. M. Martin, R. J. Needs, and O. H. Neilsen, *Phys. Rev. B: Condens. Matter* [3] **30,** 3210 (1984).

creating a Gordian knot. In practice, bond switches are rejected that would produce an unrelaxed bond of length greater than 1.7 times the ideal bond length. This is a distance slightly greater than the normal second-neighbor distance. If the factor were 1.6, no bond switches would be permitted in the starting diamond structure. Experience shows that increasing it to 1.8, say, gains nothing but does increase computing time.

No 4-fold rings are allowed. Such rings necessarily involve large bond distortions. They could be allowed in the randomization process in the expectation that they would be removed during the annealing process, but experience shows it would be at a high cost in computer time. Our subsidiary rules are really a matter of practical computing convenience rather than necessity, which is why they were not included in our earlier definition of sillium.

d. The Initial Process of Randomization

The word "random," when applied to random network models, raises deep mathematical issues which are unlikely to be resolved by any precise prescription—a degree of common sense is required. Certainly when a model is generated from an initial configuration that is crystalline, one should demand the elimination of all recognizable features associated with the crystalline order which are not also associated (so far as we know) with the amorphous state.

The most stringent test of this relates to the structure factor $|S(q)|^2$ associated with those reciprocal lattice vectors that are labeled (111) for the diamond cubic structure.[47]

In the starting structure, which is a supercell with the perfect diamond structure, the structure factor is of zero intensity for most values of the reciprocal lattice vector appropriate to the chosen supercell. It is nonzero only for some of the q values which relate to the smaller unit cell of the diamond structure. The introduction of disorder by random bond switching progressively redistributes the structure factor among *all* of the reciprocal lattice vectors of the supercell which contains the model. We should therefore require that the structure factor be of roughly equal intensity for all q values (see Section 9).

e. Geometrical and Topological Relaxation

In this section we shall explain the various procedures schematically indicated in Fig. 10. At all stages, the structure is to be relaxed to the geometrical configuration which minimizes its energy, as given by the Keating

[47] F. Wooten and D. Weaire, *J. Phys. C* **19,** L411 (1986).

potential. We assume this to be unique; we have no evidence that it is not for the structures in question. This is an interesting question in the context of the study of low-temperature thermal properties, and we shall have cause to return to it later (Section 14).

Geometric relaxation is accomplished by following the prescription of Steinhardt *et al.*[48] The force (vector) and force gradient (tensor) are calculated for each atom in turn, the atom being moved to its position of equilibrium under the bond-stretching and bond-bending forces due to its nearest and next-nearest neighbors, all other atoms being fixed.[48]

The vector distance through which each atom must be moved can be found from a Taylor series expansion for the (zero) force about the minimum energy position for the atom, yielding

$$\nabla V_{\mathbf{r}} = -(\nabla^2 V_{\mathbf{r}})\Delta\mathbf{r} \tag{8.2}$$

which thus requires a matrix inversion of the gradient tensor to find the distance $\Delta\mathbf{r}$.

Each atom is repositioned in turn, the process being repeated over enough cycles that convergence to equilibrium is achieved for the entire structure, with sufficient accuracy for the purposes at hand.

From starting coordinates that yield a distribution of bond lengths that are too long by 60% on average, bond lengths converge to within 3% of their final values in 5 cycles and to 1 part in 10^3 after 25 cycles. This is the same as found by Steinhardt *et al.*[48] All final results reported here are for 100 cycles of relaxation, at which point convergence to 1 part in 10^4 has been reached for bond lengths, angles, and strain energies.

The accuracy required for geometric relaxation varies throughout the model-building process. During the initial randomization the purpose of geometric relaxation is to ensure accuracy in bond lengths of a few percent so that it is possible to apply the criterion that bonds not be too long. Four cycles of relaxation after each pair of bond switches is sufficient. In a very large cell (4096 atoms) this can be restricted to local relaxation in the neighborhood of the site at which the bonds have been switched.

As indicated in Fig. 10, a highly randomized model is first created by the random incorporation of many bond transpositions. In the language of the Monte Carlo method, this corresponds to $T = \infty$, but we have not tried to take this process to any limit; rather, we have stopped when it was considered that the structure was thoroughly randomized, say, having $\Delta\theta = 22^\circ$ for the rms bond-angle deviation (see Note Added in Proof).

Following this, the structure is held for some time at a sequence of finite temperatures leading to $T = 0$, and in this way a random structure of

[48] P. Steinhardt, R. Alben, and D. Weaire, *J. Non-Cryst. Solids* **15,** 199 (1974).

particularly low energy may be found. The essential process which takes place at each finite temperature may be called *topological* relaxation. In practice two bonds are switched at random on a trial basis, and the energies of the two different structures are compared. If the new structure has a lower energy, it is accepted; if the energy is higher, it is accepted with a probability $\exp(-\Delta E/kT)$.

We do not know of any ideal prescription for the sequence of temperatures to be used in this process. In practice we used two or three steps, and do not think that much further lowering of energy is achievable by adding to this sequence.

This approach is much like the Metropolis algorithm[49] applied to optimization by simulated annealing.[25,50,51] It is a technique that provides a powerful practical solution of a wide range of problems, ranging from statistical mechanics to wiring problems and the design of computers.

III. Results

In the sections that follow we present and discuss results obtained from applying the algorithm for generating tetrahedrally bonded random networks.

The first section provides a reminder that realistic structures cannot be produced simply by a process of randomization. The discussion emphasizes the necessity of including a Boltzmann factor in deciding the acceptability of bond switches during simulated annealing.

A presentation and discussion of some properties of the models and comparisons with experiment are included.

9. Initial Randomization

If one of our bond rearrangements is introduced into the diamond cubic structure, its effects upon the topology of the structure are as follows:

1. *Four* fivefold rings are introduced (resulting from the shortening of sixfold rings);
2. *Twelve* sixfold rings are removed (converted into five- and and sevenfold rings).
3. *Sixteen* sevenfold rings are created and *twenty-four* eightfold rings are removed.

[49] N. Metropolis, A. Rosenbluth, M. Rosenbluth, A. Teller, and E. Teller, *J. Chem. Phys.* **21,** 1087 (1953).
[50] V. Cerny, *J. Optimization Theory Appl.* **45,** 41 (1985).
[51] D. Vanderbilt and S. G. Louie, *J. Comput. Phys.* **56,** 259 (1984).

Our definition of a ring is any closed nonreturning path of bonds, as in Ref.[48], *not* that used by Etherington *et al.*[17] The latter has the advantage of yielding a distribution function for *n*-fold rings which is zero at high *n*, but it is less closely tied to physical properties.

The way in which the ring statistics evolve with increasing concentration of bond switches can be understood by a simple rate-equation model,[27] the results of which are shown in Fig. 12.

As the model becomes increasingly randomized, the RDF becomes gradually more featureless, as shown in Fig. 13. We have followed Etherington *et al.*[17] in representing the radial distribution function $g(r)$ by the correlation function

$$t(r) = r^{-1}g(r) \tag{9.1}$$

We have included a Gaussian line broadening, corresponding to a full width of 0.23 Å at half-height, in calculating the correlation functions shown below, in order to facilitate comparison with experiment.[17] The use of a Gaussian to smooth and broaden the correlation function is convenient, but is only an approximation to the peak function $P(r)$ as given by Etherington *et al.*[17] The true peak function, with its satellite features on either side of the central maximum, introduces extra detail in the experimental correlation function, which is not significant at this point (see also Appendix A).

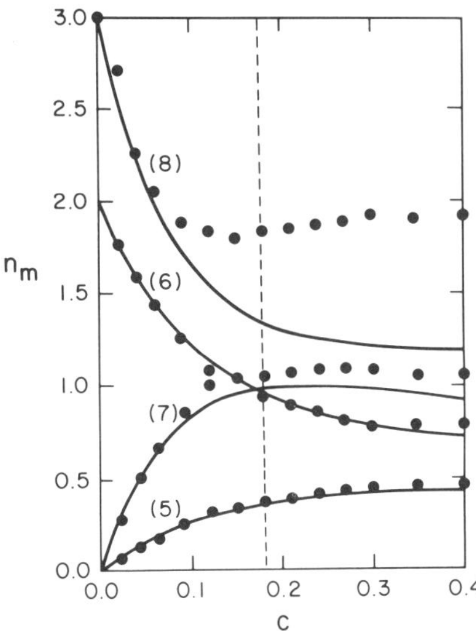

FIG. 12. Number of *m*-fold rings per atom as a function of *c* (the number of bond-pair switches per atom), for $m = 5–8$. (———) are given by the theory described in Ref. 27.

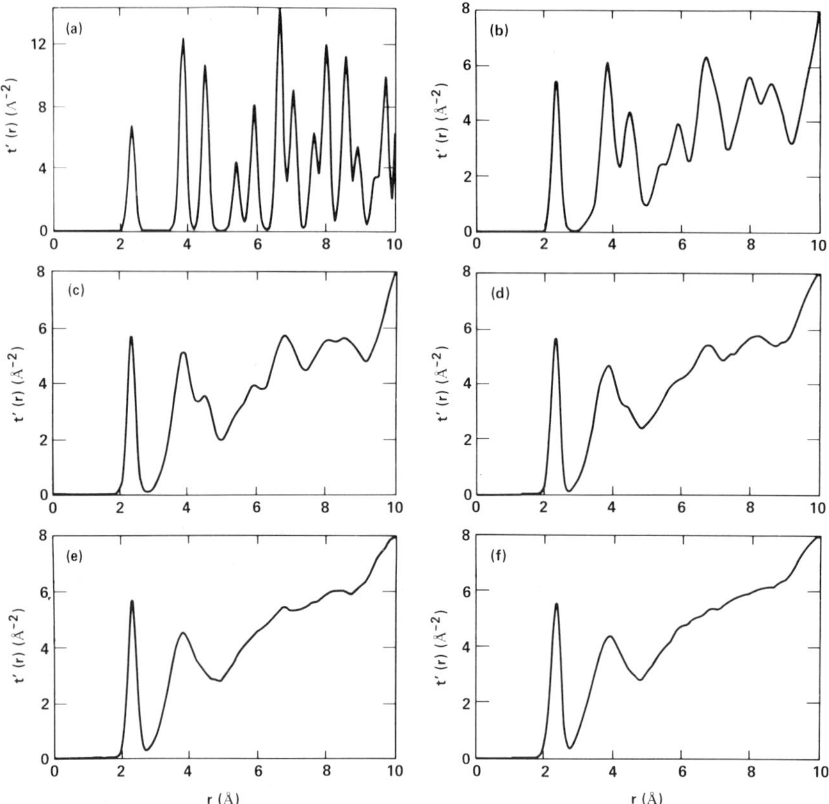

FIG. 13. Correlation function for different amounts of randomization. (a) The perfect diamond cubic structure; (b)–(f), concentrations of bond-pair switches per atom are $c = 0.06$, 0.12, 0.18, 0.24, and 0.30, respectively.

Also included for comparison in Fig. 14 is the radial distribution function of *a*-Ge, after Etherington *et al.*[17]

In the case of ring statistics, both the Polk/Boudreaux and Steinhardt models have $n_5 \approx 0.4$, $n_6 \approx 0.9$ (rings per atom), after correction for surface effects,[48] as does the Henderson periodic model.[18] This corresponds roughly to models with $c = 0.18$, as is clear from Fig. 12. Thus, to generate ring statistics similar to those of the previous models, a concentration of bond switches greater then 0.18 may be used. There is very little change in ring statistics beyond $c = 0.18$, asymptotic values being reached by $c = 0.3$. Furthermore, by the time $c = 0.3$ has been reached, the RDF has become fairly featureless, apart from the first- and second-neighbor peaks, which are characteristic of tetrahedral bonding. One might conclude that the model is

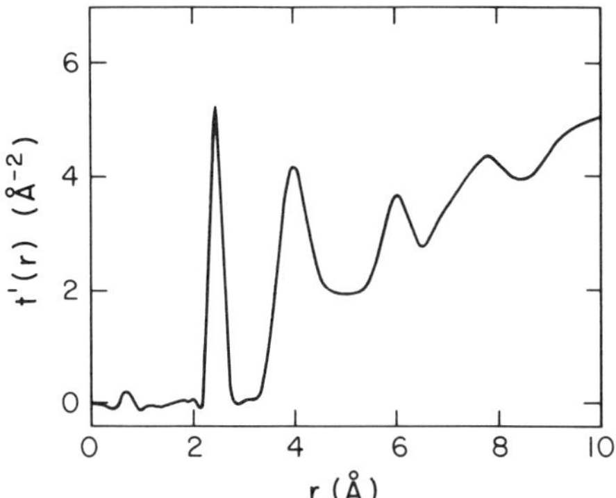

FIG. 14. Experimental correlation function for amorphous Ge, after Etherington *et al.*,[17] for comparison with Fig. 13.

"truly random" at this point; however, the structure factor imposes a stricter criterion than either the ring statistics or the RDF, if no features characteristic of the diamond structure are to be retained.[47]

In Fig. 15 we show how the structure factor $|S(q)|^2$ varies with q for the diamond structure and three randomized models.[47] The introduction of disorder by random bond switching progressively reduces the values of the structure factor associated with the original reciprocal lattice vectors.

It is now evident that the initial randomization is insufficient even at $c = 0.3$, if this criterion is applied. It is necessary to go much further in order to satisfy it, as Fig. 15 shows. Roughly speaking, $c = 1.5$ is sufficient. This empirical estimate can be rationalized as follows.

The structure factor in question, here denoted by I, is well approximated by the Debye–Waller expression

$$I = I_0 \exp(-\tfrac{1}{3}q^2\langle\delta r^2\rangle) \tag{9.2}$$

where $\langle\delta r^2\rangle$ is the mean-square displacement of atoms from their original positions and q is a reciprocal lattice vector. Our calculations show that this is, in turn, approximately related to the number of rearrangements by

$$\langle\delta r^2\rangle = kc \tag{9.3}$$

where $k \simeq 5$ Å^2. This is of the same order as the square of the nearest-neighbor distance r, which has the value $r = 2.35$ Å appropriate to Si in the diamond structure.

For very low c, such that the exponent in Eq. (9.2) is much less than unity, the Debye–Waller expression is justified by the fact that it is exact in first order, according to a familiar proof. However, it remains accurate for somewhat higher values, because the combination of random displacements due to

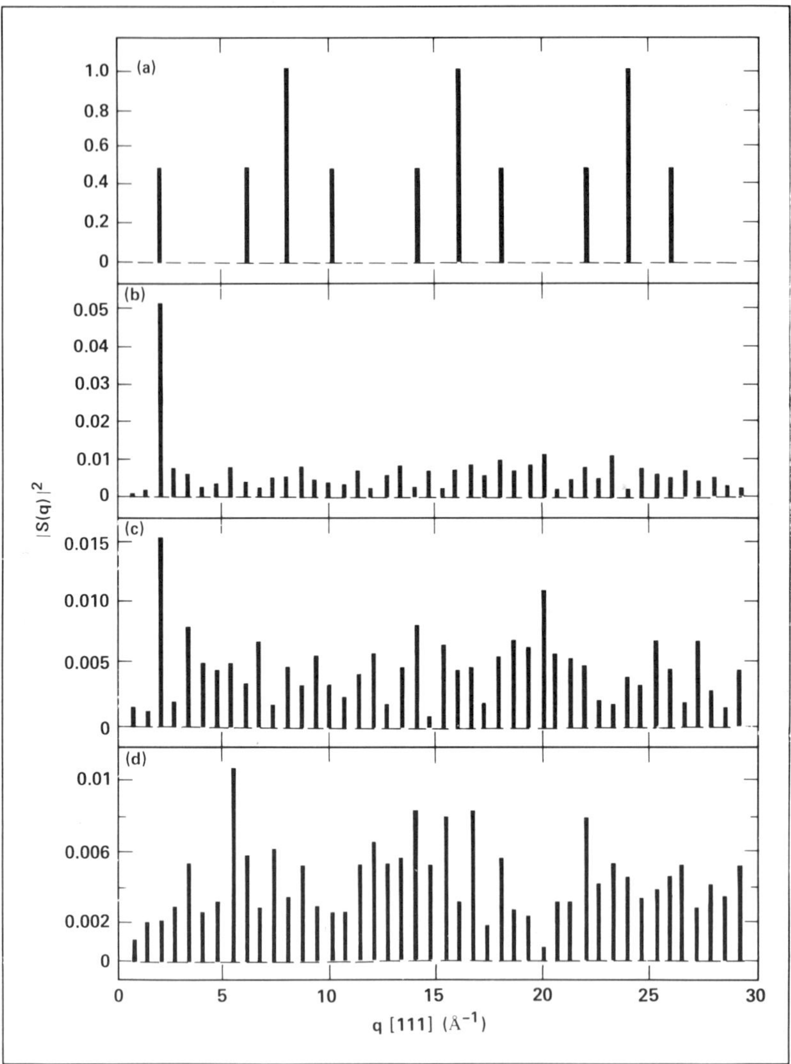

FIG. 15. Values of $|S(q)|^2$ averaged over the four (111) directions of the 216-atom periodic unit cell. (a) The perfect diamond cubic structure. (b)–(d) Randomized structures, with number of bond transpositions per atom denoted by c, which is equal to 0.50, 1.00, and 1.50, respectively.

multiple bond rearrangements causes the distribution of displacements to approach a Gaussian form, which is the necessary condition for the validity of the Debye–Waller expression (see, e.g., Cowley[52]).

In practice the Debye–Waller expression is remarkably accurate, fitting the calculated (111) structure factor to within about 1% up to $c = 0.06$, at which point $I = 0.65\, I_0$.

Using Eqs. (9.2) and (9.3) we can estimate the value of c at which the (111) peak disappears into the background for a 216-atom model, as follows. Any deviation from the ideal diamond structure introduces new reciprocal lattice vectors at all q values appropriate to the chosen supercell, as we have seen. The density of these in reciprocal space is proportional to the volume of the supercell and hence to the number of atoms in the supercell. The effect of randomizing the structure is to spread the structure factor over all these peaks with roughly equal intensity. Hence the required value of $\langle \delta r^2 \rangle$ is given by

$$\exp(-\tfrac{1}{3} q^2 \langle \delta r^2 \rangle) \lesssim 216^{-1} \tag{9.4}$$

When Eq. 9.3 is used to express this condition in terms of c for the peak in question, this requires c to be of order unity. In practice, we have found that c must be about 1.5 for a 216-atom model. Another way of putting this is to say that the rms displacement of each atom must be of the order of the nearest-neighbor distance. However, it should be noted that this criterion depends (albeit only logarithmically) on the sample size.

Thus we see that, in order to truly randomize the structure, we need a concentration c of bond switches which is of order unity—yet it is evident from Fig. 13 that even $c = 0.3$ creates far too much broadening of the second-nearest-neighbor peak. This is due to the large distortion of bond angles. It is the role of the subsequent annealing process to reduce this angular distortion.

10. Annealing the Randomized Model

Having produced a randomized but highly strained model, it is necessary to continue with random bond switches in a search for topological rearrangements that will lower the energy. There is concern at this point that, with annealing, some characteristics of the diamond structure might return.

Refining the model proceeds rapidly at first. Indeed, if the initial randomization corresponds to $c \approx 0.3$, the process leads directly back to the original diamond structure in almost the same number of rearrangements as was used in the original randomization process. With $c > 0.6$, this does not happen, and

[52] J. M. Cowley, "Diffraction Physics," p. 147. North-Holland Publ., Amsterdam, 1975.

after a long annealing process a random model of comparatively low energy is the result.

The annealing process soon reaches a stage where it is difficult to find rearrangements that will further lower the energy. It will often happen that the structure will be trapped in a metastable state such that none of the allowed bond transpositions will lead to a lower energy. The inclusion of the Boltzmann factor is thus an essential ingredient to lift the structure out of its metastable state and into a slightly more strained state from which it might find another path to lower-energy states. We emphasize that no fully satisfactory models can be made without the inclusion of the Boltzmann factor.

The temperature in the Boltzmann factor must be chosen such that the total energy of the model will on average decrease with an increase in time, but such that the structure will be lifted out of a metastable state in a reasonable number of tries. For the values of the parameters we have used in the Keating potential, we have found $kT = 0.4$ eV to be a good choice until the structure is approaching agreement with experiment, as measured by the rms angular deviation being, say, about 13°. We then lower kT to 0.2 eV, which seems to define the lowest useful nonzero temperature.

Figure 16 shows the correlation function at three points in the long annealing process that finally leads to a good model. Even when the rms angular deviation is as much as 15.6° the correlation function is clearly recognizable as being in qualitative agreement with experiment. When it finally reaches 12.6° the agreement with experiment is quite remarkable. What is also remarkable, and argues strongly for the validity of the continuous random network model, is that, when the starting structure is diamond cubic, the final structure is *always* either the original diamond structure or a random structure in very close agreement with experiment.

One of the striking differences between the RDFs for crystalline and amorphous silicon is that in the amorphous structure the third (crystalline) peak is missing. The RDF for the amorphous material is not just a broadened version of that for the crystal.

11. Characteristics of the Model

The simplest and most physically appealing measure of agreement between models and experiment is a comparison of RDFs. Our calculated correlation function for the PRL model[41] is shown in Fig. 17, in comparison with an experimental curve.[17] Note that the essential difference between *a*-Si and *a*-Ge is a simple scaling in proportion to the bond lengths: For Si, $d = 2.35$ Å; for Ge, $d = 2.46$ Å. Apart from this scaling, experimental data are similar for the two cases.

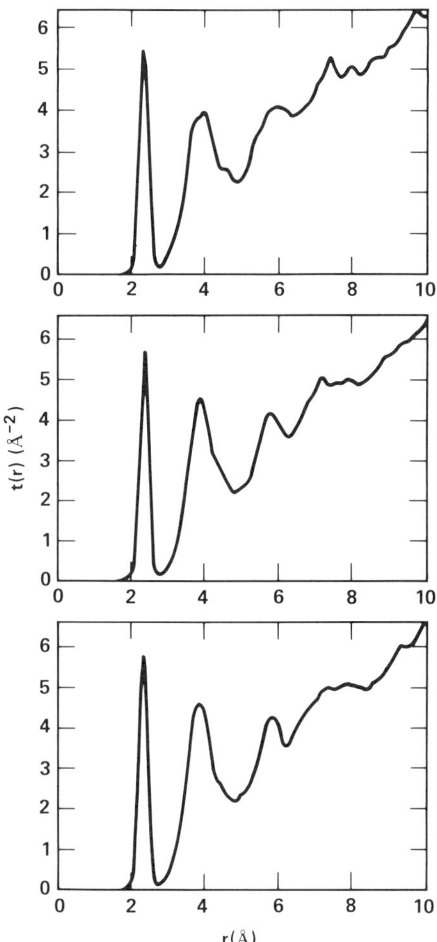

FIG. 16. The correlation function at three points in the annealing process. The rms angles, from top to bottom, are 15.6°, 13.2°, and 12.6°, respectively.

This model[17] has a slight remnant of the structure factor characteristic of the original diamond cubic structure.[47] Nevertheless, its correlation function shows no serious discrepancy with experiment, apart from the systematic deviation at high *r*. *All* models produced with the algorithm we have described have essentially the same RDF and are in remarkable agreement with experiment whether or not they show any diamond cubic memory effects in the structure factor. The latter, if present, arises from an insufficient initial randomization and is not a consequence of the annealing process.[47]

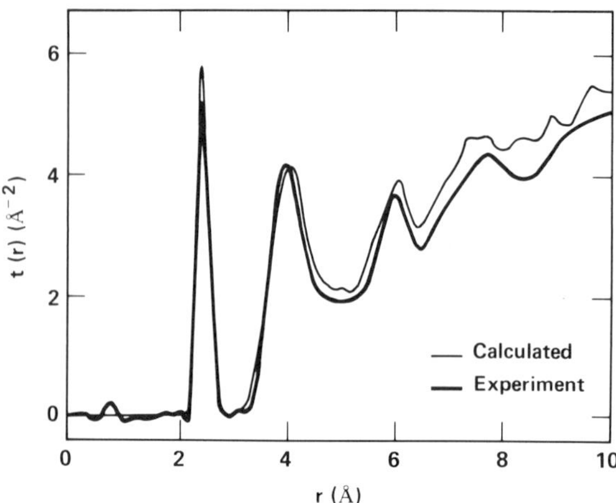

FIG. 17. Comparison of correlation functions for *a*-Ge (experimental, from Ref. 17) and *a*-Si (model scaled to Ge).

We believe the discrepancy in $t(r)$ at high r is related to inhomogeneities in real *a*-Si films,[53–55] particularly voids on a scale of, say, 100 Å. These voids result in the measured macroscopic density of real *a*-Si films being of the order of 90% the density of crystalline Si.[17] However, in the homogeneous region thought to be typical of the bulk material, the density is close to that of crystalline Si. Brodsky *et al.*[56] reported the density of *a*-Si to be 1.01 ± 0.02 times that for the crystal, a value in agreement with random network models.

Some characteristics of the above model, which are typical of all the models we have produced, are listed in Tables I and II.

Table I also includes results for the case in which the model was relaxed using a generalization[42] of Weber's potential[37] (see Section 8,b). Although it is convenient to include long-range interactions with periodic models, it requires much more computer time than do simple short-range interactions like the

[53] J. J. Hauser, *Phys. Rev. B: Solid State* [3] **8,** 3817 (1973).

[54] J. C. Phillips, *Phys. Rev. Lett.* **42,** 1151 (1979).

[55] J. C. Phillips, *Comments Solid State Phys.* **9,** 191 (1980).

[56] M. H. Brodsky, D. Kaplaz and J. F. Zieglev, *Appl. Phys. Lett.* **21,** 305 (1972).

FIG. 18. (a) The 216-atom diamond cubic unit cell. (b) One bond-pair transposition introduced into the unit cell. The view is in the (110) direction.

FIG. 19. (a) 27 bond-pair switches ($c = 0.125$) introduced into the unit cell. The view is in the (110) direction. (b) A fully randomized unit cell, for which $c = 1.50$, viewed in the (100) direction.

FIG. 20. An annealed model with no memory of the original diamond cubic structure.

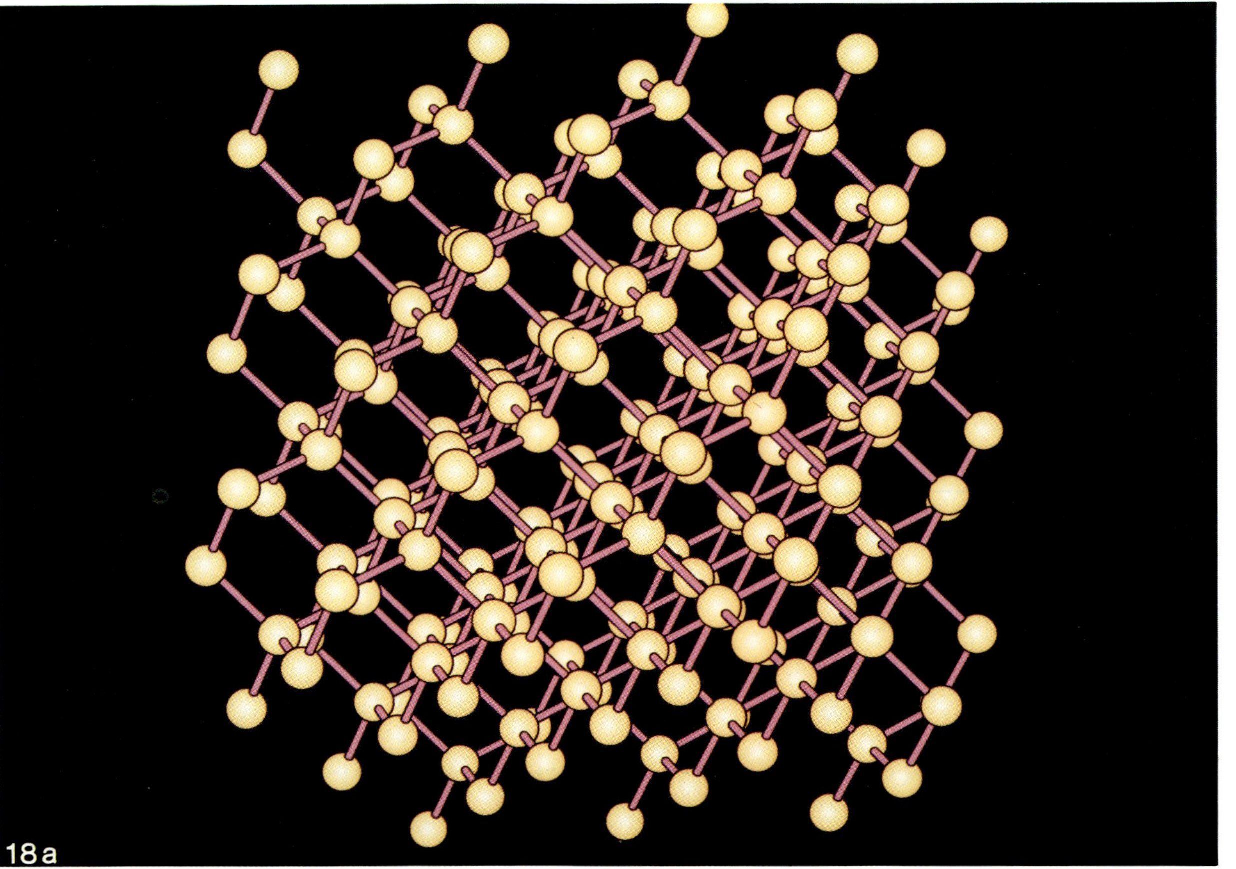

18a

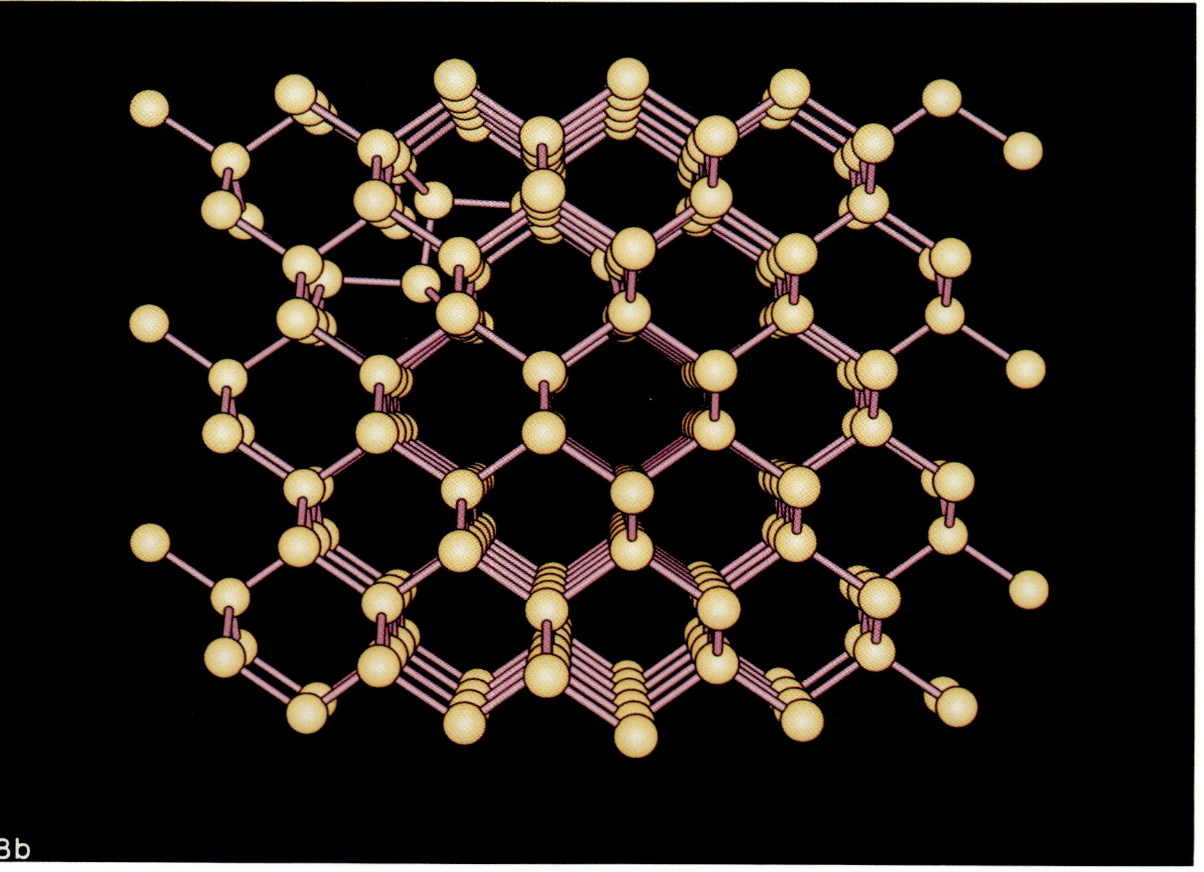
18b

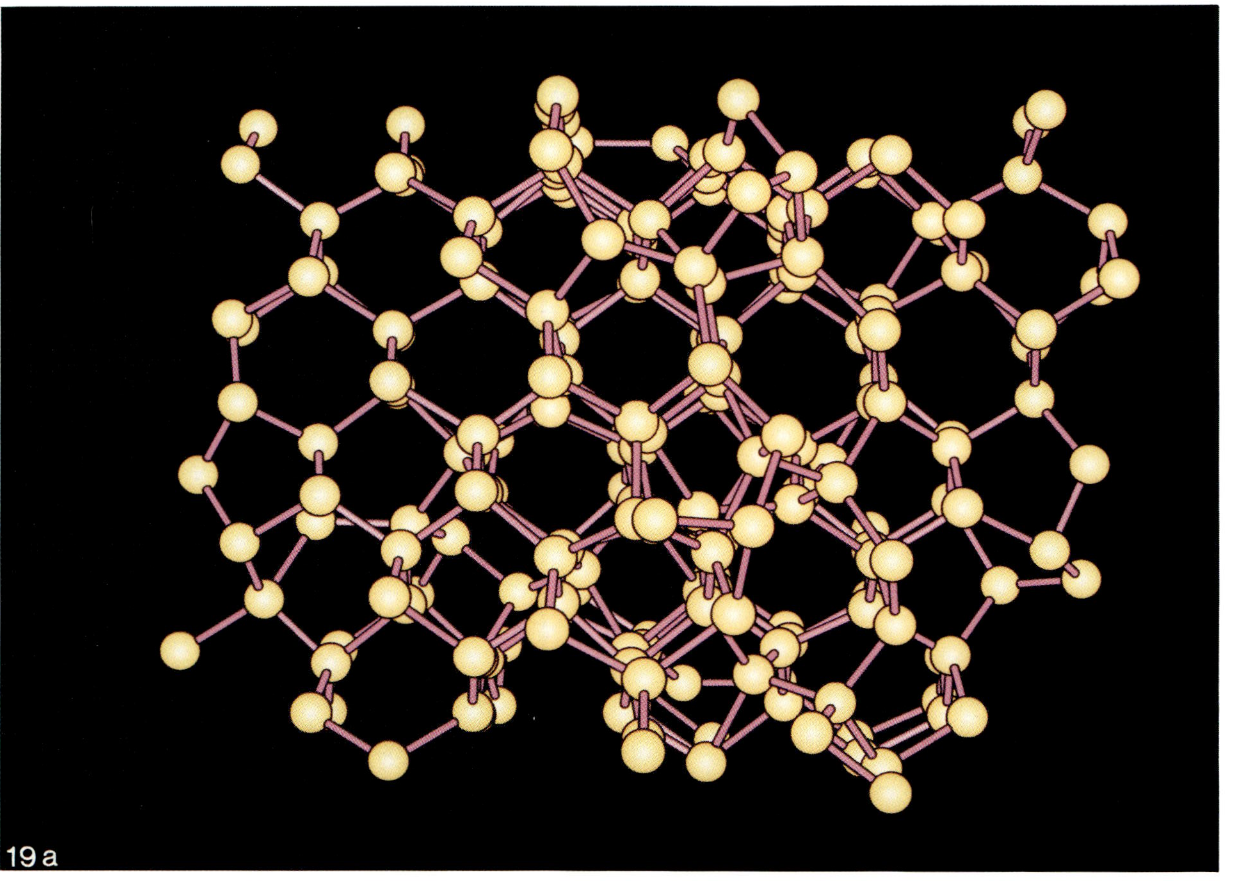
19a

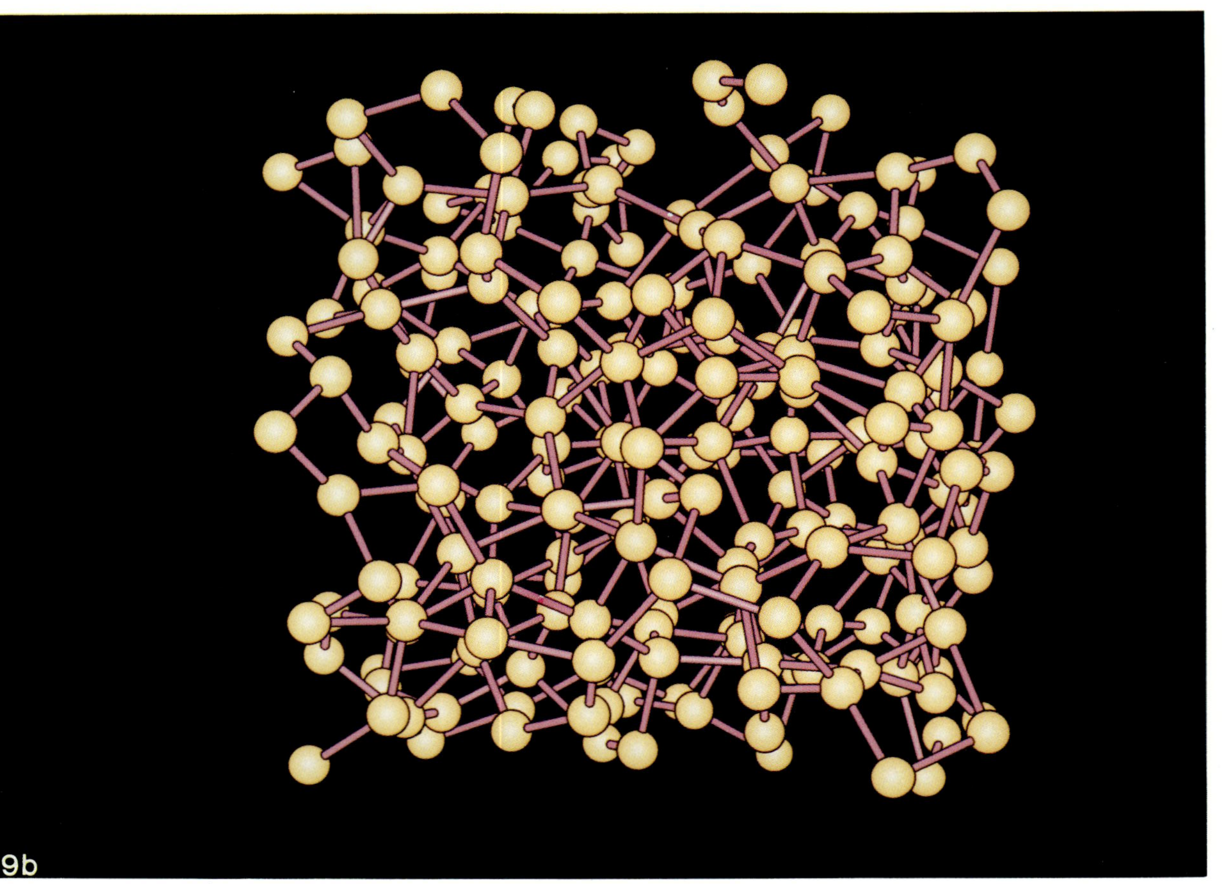

19b

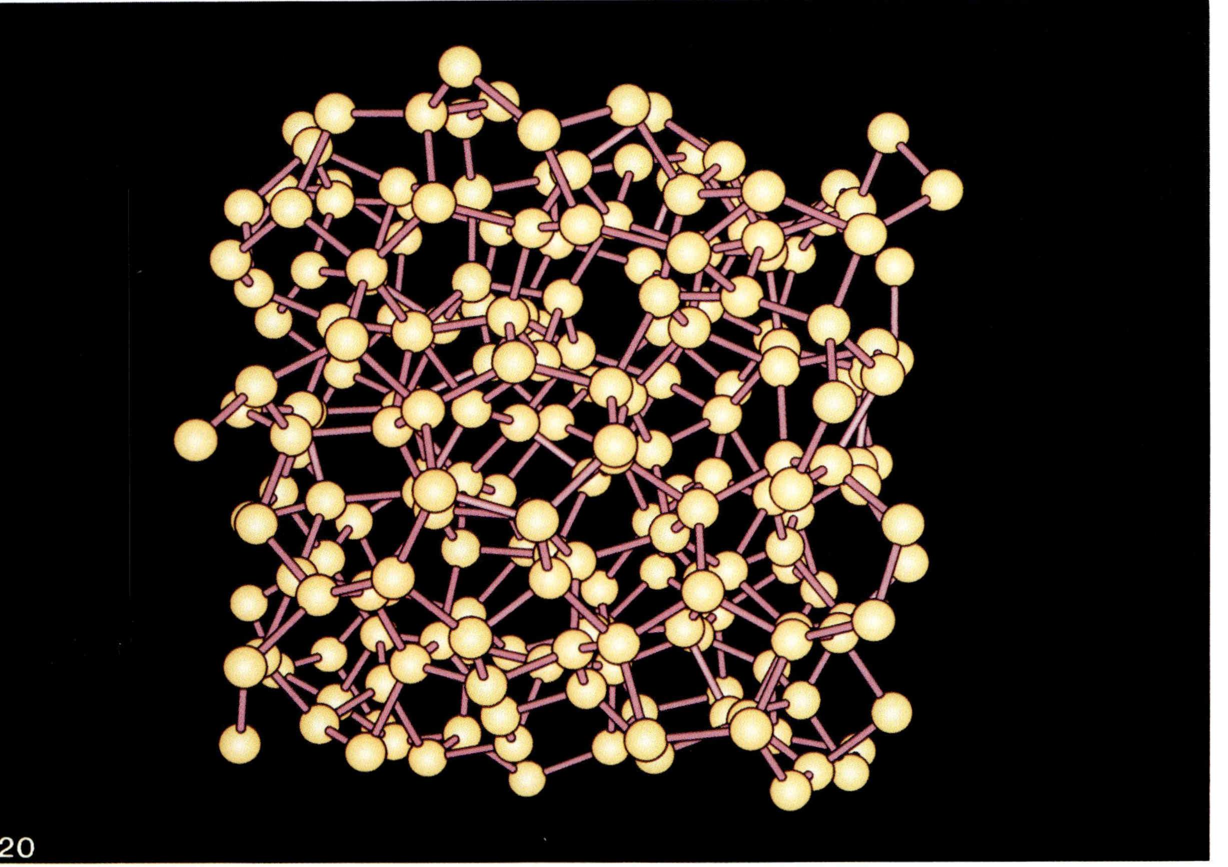

20

TABLE I. RING STATISTICS (RINGS/ATOM)

Structure	5-atom	6-atom	7-atom	8-atom
Diamond cubic	0	2	0	3
PRL model[41]	0.46	0.89	1.05	1.96

TABLE II. COMPARISON OF MODEL CHARACTERISTICS[41]

	Keating potential	Weber bond-charge potential
Density relative to crystalline silicon	1.04	1.03
rms angle deviation from tetrahedral	10.9	11.4
rms bond-length deviation from crystalline value	2.7%	1.9%

Keating potential. The benefits are generally of marginal value. Calculations of the electronic density of states for *a*-Si using the PRL model showed that the effects of a reduced rms bond-length deviation with the inclusion of Coulomb forces are just offset by the increased rms angular deviation.[43]

12. VIEWING THE MODEL: COMPUTER GRAPHICS

It is now possible to display structures on a CRT screen and to obtain high-quality photographs of them.[57] Figures 18–20 provide a beautiful example. Of course, one can always construct a physical model of a computer-generated model, but computer graphics permits rapid viewing of many models.

Here we make use of computer graphics to illustrate the modeling process. Figures 18 and 19 show the starting diamond cubic structure followed by three stages of the randomization process. The last stage is for $c = 1.50$, at which point the structure has no memory of the initial diamond cubic structure. Figure 20 shows an annealed model for which the correlation function is in agreement with experiment and the intensity of the (111) peak in the structure factor has remained low.

[57] K. Winer and F. Wooten, *J. Mol. Graphics* **3**, 76 (1985).

13. Relative Energies

The method which we have described was intended to model the structure itself; it was recognized from the outset that the actual energies which were calculated in the model-building process might be unrealistic. Earlier we made the point that these scale with the bond-bending force constant β, which is not uniquely determined, and so could be adjusted without much changing the nature of the structures generated. Nevertheless, the magnitude of the various energies involved is bound to be of some interest.

We speak here always of total energy, relative to that of diamond cubic. For the Keating potential, this cannot be a negative quantity, and this is in accord with the fact that diamond cubic is the most stable structure for Si and Ge. The relative total energy of wurtzite and other polytype structures is necessarily *zero* in this model: No potential terms which discriminate between staggered and eclipsed bond configurations are included. (Inasmuch as the model-building process is successful, such terms are shown to be insignificant in determining the structure of the amorphous solid.)

Any structure for which the tetrahedral bond is distorted will have a positive relative total energy. This includes our random networks and various polymorphs (Si III, Ge III, clathrates, ...) which have been investigated in recent years.[31,58]

Typically the relative energy in our calculations (for $\beta/\alpha = 0.2$ and α chosen for Si) ranges from 1.0 eV/atom for the unrealistic randomized structures, down to 0.26 eV/atom for the final annealed structure (0.23 eV/atom when relaxed with the modified Weber potential[42]). Donovan *et al.*[59] measured the heat of crystallization of amorphous Si, obtaining 0.123 ± 0.007 eV/atom, which may be compared with the above.

Briefly, some other related calculations are as follows. Weaire and Williams[60] were the first to make estimates in this spirit, for the Si III and Ge III structures, obtaining 0.15 and 0.29 eV/atom, respectively (scaled to Si), but with the valence force field model. Our own estimates for these quantities were 0.21 eV/atom for Si III (BC-8) using the Weber potential. As for amorphous silicon, Steinhardt *et al.*[48] obtained 0.08 eV/atom for their finite cluster of Ge. If rescaled to Si, with $\beta/\alpha = 0.20$, this becomes 0.1 eV/atom. That this is somewhat less than the experimental value is consistent with the fact that the radial distribution function has too much sharp structure, due to the bond angles being insufficiently distorted. But again we would counsel caution in making these very rough comparisons.

Lastly, mention should be made of the first-principles calculations men-

[58] J. D. Joannopoulous and M. L. Cohen, *Solid State Phys.* **31,** 71 (1976).

[59] E. P. Donovan, F. Spaepen, D. Turnbull, J. M. Poate, and D. C. Jacobson, *Appl. Phys. Lett.* **42,** 698 (1983).

[60] D. Weaire and A. R. Williams, *Phys. Status Solidi* B **49,** 619 (1972).

tioned in Section 8,b. The calculation of 0.13 eV/atom by Biswas *et al.*[46] for the relative energy of BC-8 is nicely consistent with the estimate of Weaire and Williams quoted above.

14. The Spectrum of Bond-Switch Energies

As the random network asymptotically approaches agreement with experiment during the annealing process, the amount of computer time required for further improvement in the model becomes progressively greater. This aspect of the annealing process bears a close relationship to the metastability of the amorphous state itself. The random-network structure is stable not just against small distortions, but also against bond rearrangements that can take place within reasonable times.

One way of describing the modeling process is through the spectrum of values of ΔE (one for each bond switch of the type illustrated in Fig. 8). This varies, as shown in Fig. 21, during the building process. The final structure, annealed at $T = 0$, always obeys $n(\Delta E) = 0$, $\Delta E < 0$.

Figure 22 shows a calculated final spectrum for one model. Note that $n(\Delta E)$ tends to a finite value at $\Delta E = 0$. There is considerable uncertainty in extrapolating the density of states to zero energy, but a rough estimate is that $N_0 \approx 8 \times 10^{21}\ \mathrm{eV^{-1}\ cm^{-3}}$ ($\sim 5 \times 10^{46}\ \mathrm{J^{-1}\ m^{-3}}$) for this model.

The spectrum of bond switch energies is also of interest in another context. The details of the bond-switching process for $\Delta E \approx 0$ are worthy of study for their possible application to an understanding of the anomalous thermal and acoustic properties of amorphous solids at very low temperatures.

It is now recognized that a quasilinear term in the specific heat is a very general property of all amorphous materials, possibly including *a*-Si and *a*-Ge below a few degrees kelvin. This has been successfully interpreted phenomenologically in terms of a tunneling model,[61–63] in which atoms, or groups of atoms, tunnel between the two lowest-energy states in a double potential well. What is lacking is detailed microscopic models for the tunneling process.

It now seems clear that no single model will suffice to provide a microscopic description of all tunneling states. But it would be useful to have a well-understood model for any material. Amorphous silicon and germanium are thus of particular interest because of the simplicity of their structures, and, although it once seemed unlikely that the rigidity of fully coordinated tetrahedral networks offered the possibility for tunneling states to exist,[61] the experimental evidence is now quite strong.[63,64]

[61] W. A. Phillips, *J. Low Temp. Phys.* **7,** 351 (1972).

[62] P. W. Anderson, B. I. Halperin, and C. M. Varma, *Philos. Mag.* [8] **25,** 1 (1972).

[63] W. A. Phillips, *J. Non-Cryst. Solids,* **77/78,** 1329 (1985).

[64] J. Y. Duquesne and G. Bellessa, *J. Phys. C* **16,** L65 (1983).

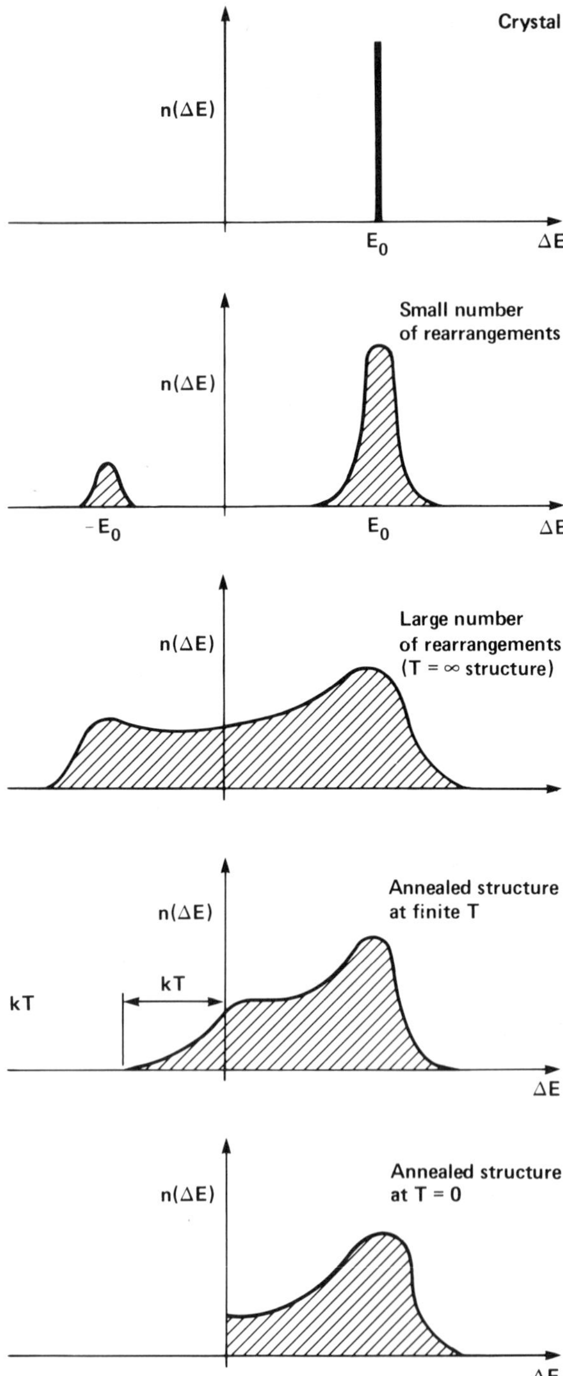

FIG. 21. Schematic picture of the evolution of the spectrum $n(\Delta E)$ at various stages of the building process. This is the spectrum of energies associated with the bond transpositions defined in the text and illustrated in Fig. 8 for the case $\Delta E = E_0$.

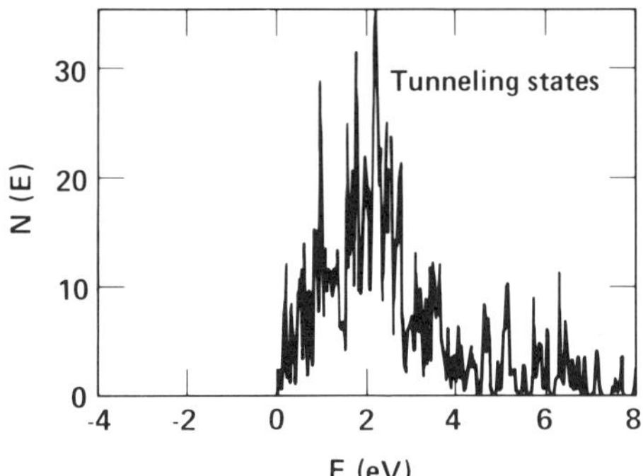

FIG. 22. Calculated spectrum of values of ΔE for bond transpositions in a typical model.

The density of tunneling states for annealed *a*-Ge derived from the low-temperature heat capacity is $N_0 \approx 2 \times 10^{45}\ \mathrm{J}^{-1}\ \mathrm{m}^{-3}$, which is more than an order of magnitude less than the estimated density of bond-switching states in our random network model for $\Delta E = 0$. Clearly the simple bonding rearrangement used in the random-network modeling process provides more than enough two-level systems. But the answer is not so simple.

Only two-level systems of sufficiently small barrier height will contribute to specific heat measurements on any finite time scale, and a calculation of the tunneling probability requires a description of the easiest path between two potential minima. One cannot calculate the barrier except very roughly, using simple Keating forces (see Fig. 9). These are only reliable close to ideal tetrahedral coordination, and, even if the forces were known, the route taken in phase space would remain to be determined.

What then can be said about possible mechanisms for tunneling modes in fully coordinated *a*-Si or *a*-Ge? Clearly any mechanism must involve the rearrangement of at least two atoms if no dangling bonds are to be permitted. The simplest such mechanism must be something akin to the bond-switching mechanism we have used for generating amorphous structures. This mechanism is probably insufficiently subtle to be a good candidate for a tunneling mode—rough estimates indicate barrier heights at least an order of magnitude too high, although we have not fully explored the possibility of exceptional cases.

IV. Conclusion

Admiration of the results of the sillium model for tetrahedrally bonded random networks should be tempered with an awareness of its deficiencies and limitations. As has been noted from time to time, these are several: absence of dangling bonds, surfaces, and voids; restriction to tetrahedral bonding; elementary semiempirical potential; arbitrary restriction to one rearrangement process; no consideration of the role of barriers to this rearrangement. The presence of the barrier between the two metastable states that are considered deserves particular attention. We noted in the last section that some attempt has been made to estimate the barrier height in a few cases, with the low-temperature specific heat in mind. If physical significance is attributed to the sillium model, the barriers must be recognized as playing an important role more generally. If they were all of the same height, they would merely change the true scale associated with the process considered, by the appropriate Boltzmann factor—but this can hardly be realistic. The barrier heights will have a wide spectrum, and it may be that only the lower ones are accessible on an experimental time scale for a given temperature. We have not considered this kind of complication, although it might, in principle, be *roughly* incorporated in our type of calculation. This would mean that, in simulated annealing, allowed bond switches would be restricted according to their barrier heights.

The impracticability of the molecular-dynamics method in this case is partly due to the role of the barriers, which ensure that significant rearrangements take place on time scales much longer than that of the vibrations about (metastable) equilibrium. Hence molecular dynamics is unlikely to produce a low-energy random structure within practical computing times. This appears to be borne out by the calculations of Stillinger and Weber,[65] which provide an interesting comparison with our own work. These authors have used a semiempirical potential intended to represent both tetrahedral and metallic bonding, at least roughly. Melting therefore takes place from the diamond cubic crystal structure to a metallic type of structure, as is indeed the case with Si and Ge. Although the melting process is artificial in this type of calculation (being inhomogeneous in reality), it was obviously of interest to see in what manner the crystal structure became disordered. In fact, after much analysis, what was discovered was the proliferation and interaction of precisely the bond switch which lies at the heart of our own model!

On the other hand, upon cooling the simulated liquid, a much more distorted structure was obtained than in our own work, for the reasons outlined above.

Perhaps a hybrid of the two approaches may be useful in future. In any case, the sillium model itself now seems capable of application (with cautious interpretation) in a variety of new contexts, related to current experiments. These include the incorporation of other types of covalent bonds, the study of

[65] F. H. Stillinger and T. A. Weber, *Phys. Rev. B: Condens. Matter* [3] **31,** 5262 (1985).

recrystallization,[66] and the modeling of various interfaces (amorphous/amorphous, crystal/amorphous etc.). It offers an essential simplicity and transparency which may be of lasting value in constructing and categorizing realistic random networks.

Appendix A The Radial Distribution Function

Scattering of an incident monoenergetic collimated beam from an amorphous material, whether the beam is of x rays, neutrons, or electrons, produces only a concentric sequence of diffuse rings. The absence of sharp structure in the scattering pattern is the principal experimental evidence that the material is amorphous, for even microcrystalline material produces a sequence of sharp circles if the microcrystallites are sufficiently large that the distinction between crystallite and continuous random models is meaningful (i.e., about 15 Å in diameter).

In the case of amorphous materials, the variation in intensity in the diffraction pattern cannot be used to infer a well-defined structure. The most that one can determine is the distribution of the separations of atoms in the form of a pair distribution function. There is an ensemble of possible structures, yet the structure is not simply random. There are physical constraints that must be satisfied—covalent tetrahedral bonding and short-range order, among others. These constraints manifest themselves in a pair distribution function that exhibits structure and is experimentally reproducible.

How, then, is one to interpret scattering patterns? Apart from the most elementary features, the answer must lie in the construction of models, a topic which has been the central theme of this article. One sets out to find a model or, better still, an algorithm to create an ensemble of models that agrees with experiment.

The simplest and most appealing comparison between models and experiment is through the radial distribution function (RDF), which is here denoted by $g(r)$ and defined by the relationship

$$g(r) = 4\pi r^2 \rho(r).$$

where $\rho(r)$ is the local number density of atoms at a distance r from an atom, averaged with respect to the choice of atom.

The appeal of the RDF is that it gives a rather direct physical picture of the spatially averaged structure, albeit only a one-dimensional representation.

[66] F. Wooten, G. A. Fuller, K. Winer, and D. Weaire, *J. Non-Cryst. Solids* **75**, 45 (1985).

The area under the first peak

$$\int g(r)dr = \int 4\pi r^2 \rho(r)dr = Z$$

is the number of atoms in the first coordination shell surrounding an average atom, assuming no overlap with the second peak.

Beyond the first peak, which defines the mean nearest-neighbor distance and mean coordination number, and the second peak, which defines the rms angular deviation, nothing more can be learned directly and reliably from the RDF.[67] Models must be built.

Figure 23 provides a two-dimensional representation of a random network having some of the characteristics of *a*-Si. Note that in the sequence of peaks in the RDF, atoms a, b, c, and d contribute to the third, fourth, fifth, and sixth peaks, but they are topologically third-, fourth-, third-, and fourth-nearest neighbors of the origin atom—there is no correspondence between peak and neighbor number. These peaks are the exact counterparts of peaks in the RDF for the diamond structure that arise from the same topological connections: atoms at opposite vertices of a 6-fold ring, etc. In *a*-Si, however, the introduction of 5- and 7-fold rings and the destruction and distortion of many 6- and 8-fold rings result in a dramatic change in the RDF of *a*-Si compared with Si in the diamond structure: The third peak, which is prominent in the crystalline phase, is missing in *a*-Si!

While the RDFs hold the greatest appeal, it is a closely related function, the correlation function $t(r)$, that should be used for quantitative comparisons.[1] It is related to $g(r)$ by the expression

$$t(r) = g(r)/r$$

The correlation function has two advantages. On a practical level, it permits structural data to be reasonably presented graphically over a greater range of r. [Note that $g(r)$ increases on average as r^2, but $t(r)$ increases only as r.] What is more important and more fundamental is that it is in $t(r)$ and not $g(r)$ that experimental broadening is symmetric and r independent. Thus the best direct comparison between models and experiment is through $t(r)$, with experimental broadening being included in $t(r)$ for the model.

It has sometimes been argued that it is better not to impose a Fourier transform, with its inevitable complications, upon the experimentally observed interference function, but rather to transform the radial distribution function of the model. In principle this may seem preferable, but in practice the interference function is dominated by the pronounced oscillation which

[67] R. J. Temkin, W. Paul, and G. A. N. Connell, *Adv. Phys.* **22,** 581 (1973).

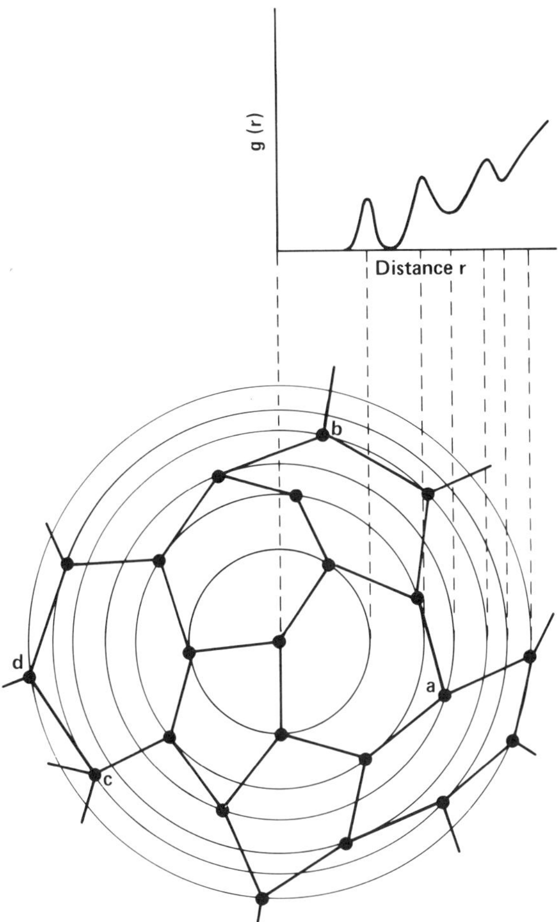

FIG. 23. Schematic illustration of the origin of structural features in the radial distribution function for an amorphous solid having some of the characteristics of *a*-Si. Atoms are shown as lying on sharply defined rings, for simplicity. Broadening is incorporated in $g(r)$.

represents the Fourier transform of the nearest-neighbor peak, and the subtle features beyond the nearest-neighbor peak, which discriminate between rival models, are much less evident than in real space.

ACKNOWLEDGMENTS

Research support by N.B.S.T. (Ireland) and Livermore National Laboratory is acknowledged. Fred Wooten wishes to thank Trinity College, Dublin, for hospitality during the period in which this article was written.

NOTE ADDED IN PROOF

In earlier work,[14] it was pointed out that one could start with the perfect crystal and immediately accept bond transpositions with a probability $\exp(-\Delta E/kT)$, in a process that we likened to melting; but it was presumed that initial randomization at $T = \infty$ followed by annealing is more efficient. Recent studies show that the latter is decidedly not the case.

One should first "melt" the crystal at a temperature just above the melting point. Here, the melting point is defined as that temperature for which the mean square displacement of the atoms from their positions in the diamond structure increases linearly with time and the structure factor, $|S(q)|^2$, approaches a value of order $1/N$. This differs from the true melting behavior of Si and Ge in that tetrahedral bonding is preserved. For $\beta/\alpha = 0.285$, the melting point is $kT = 1.0$ eV.

After the crystal has been randomized by this melting procedure, the temperature is reduced in small steps, establishing thermal equilibrium at each temperature. This procedure greatly decreases the probability that the structure will be trapped in a deep metastable state from which it cannot escape.

It is also now clear that allowing 4-fold rings greatly increases the escape rate from deep metastable states. The 4-fold rings tend to disappear as the structure is annealed, but it is not yet known if, in general, they all disappear during the final stages of annealing.

A Quantum-Mechanical Critique of the Miedema Rules for Alloy Formation

D. G. PETTIFOR

Department of Mathematics,
Imperial College,
London, England

I. The Miedema Scheme

1. INTRODUCTION

During the 1970s Miedema and his colleagues developed an extremely simple scheme for predicting the heats of formation of metallic alloys.[1–8] They considered atoms in the metallic state as being characterized by two

[1] A. R. Miedema, F. R. de Boer, and P. F. de Châtel, *J. Phys. F* **3**, 1558 (1973).
[2] A. R. Miedema, *J. Less-Common Met.* **32**, 117 (1973).
[3] A. R. Miedema, R. Boom, and F. R. de Boer, *J. Less-Common Met.* **41**, 283 (1975).
[4] A. R. Miedema, *J. Less-Common Met.* **46**, 67 (1976).
[5] A. R. Miedema and P. F. de Châtel, *Metall. Soc. AIME Proc.* (1980).
[6] A. R. Miedema, P. F. de Châtel, and F. R. de Boer, *Physica (Amsterdam)* **100**, 1 (1980).
[7] A. R. Miedema and A. K. Niessen, CALPHAD: *Comput. Coupling Phase Diagrams Thermochem.* **7**, 27 (1983).
[8] A. K. Niessen, F. R. de Boer, R. Boom, P. F. de Châtel, W. C. M. Mattens, and A. R. Miedema, CALPHAD: *Comput. Coupling Phase Diagrams Thermochem.* **7**, 51 (1983).

coordinates, namely the work function φ^* and the cube root of the electronic charge density at the metallic Wigner–Seitz sphere radius $\rho_M^{1/3}$, respectively. An *AB* alloy would then be formed by cutting out the Wigner–Seitz cells from the pure metals *A* and *B* and packing them together to create the binary system. By treating the atoms as macroscopic pieces of metal, they argued that there were two contributions to the heat of formation. The first was attractive and arose from the flow of charge from one atom to another due to the difference in the work function $\Delta\varphi^*$. The second was repulsive and arose from the removal of the discontinuity in the charge density $\Delta\rho_M^{1/3}$ across the interface between neighboring *A* and *B* Wigner–Seitz cells. By judicious choice of the parameters φ^* and $\rho_M^{1/3}$, Miedema and his colleagues were able to predict successfully not only the sign of the heat of formation but also its magnitude.

The success of this empirical "macroscopic atom" model has led to a belief that the underlying physical concepts upon which it is based are correct.[9] In particular, the attractive contribution is assumed to be *ionic* in character.[10,11] Unfortunately, however, this point of view is at variance with first-principles quantum-mechanical calculations for *metallic* systems.[12–14] These calculations, which are based on the local-density-functional approximation,[15] allow a detailed breakdown of the various contributions to the metallic bond, both for the elemental metals[16–18] and for the binary systems.[12,13] We shall see that the metallic bond in binary alloys is not well described by the ionic model. Although the Miedema model provides very simple and useful *rules* for alloy formation, the physical *concepts* upon which it is based find no justification in quantum mechanics.

This article examines the quantum-mechanical origin of the heat of formation of metallic alloys. I will begin by discussing the Miedema model in Section I,2 and the failure of the ionic model in Section I,3. The nature of the metallic bond in simple metals and their alloys is presented in Sections II,4 and

[9] See, for example, J. R. Chelikowsky, *Phys. Rev. B: Condens. Matter* [3] **25,** 6506 (1982).

[10] J. A. Alonso and L. A. Girifalco, *J. Phys. F* **12,** 2455 (1978).

[11] J. A. Alonso and L. A. Girifalco, *J. Phys. Chem. Solids* **38,** 869 (1977); **39,** 79 (1978).

[12] A. R. Williams, C. D. Gelatt, and V. L. Moruzzi, *Phys. Rev. Lett.* **44,** 429 (1980).

[13] A. R. Williams, C. D. Gelatt, and V. L. Moruzzi, *Phys. Rev. B: Condens. Matter* [3] **25,** 6509 (1982).

[14] C. D. Gelatt, A. R. Williams, and V. L. Moruzzi, *Phys. Rev. B: Condens. Matter* [3] **27,** 2005 (1983).

[15] P. Hohenberg and W. Kohn, *Phys. Rev.* **136,** B864 (1964); W. Kohn and L. J. Sham, *ibid.* **140,** A1133 (1965).

[16] D. G. Pettifor, *J. Phys. F* **8,** 219 (1978).

[17] A. R. Williams, C. D. Gelatt, and V. L. Moruzzi, *Metall. Soc. AIME Proc.* (1980).

[18] O. K. Andersen, O. Jepsen, and D. Glötzel, *in* "Highlights of Condensed Matter Theory," LXXXIX Corso. Soc. Ital. Fis., Bologna, Italy, 1985.

II,5 within the nearly-free-electron approximation. We shall see that, although there is a rearrangement of the electronic charge on the formation of an alloy, each atom remains perfectly screened within its new metallic environment, so that the ionic model is not appropriate. In Sections II,6 and II,7 the nature of the metallic bond in transition metal alloys is discussed within the tight-binding approximation. The attractive contribution to the heat of formation will be identified with the change in bond order which accompanies alloy formation. In Section III,8 we discuss the significance of the Miedema parameters φ^* and $\rho_M^{1/3}$ and present conclusions.

2. The Miedema Scheme

The energy change involved in the formation of an alloy is very small compared to the cohesive energy which is required to separate the atoms infinitely far apart. Miedema *et al.*,[1–8] therefore, chose coordinates which characterized an atom in its *metallic* rather than free-atom state. They assumed that each atom within its Wigner–Seitz cell behaves like a macroscopic piece of metal. The AB alloy may be considered as a packing together of the Wigner–Seitz cells of the elemental metals A and B, respectively. As shown in Fig. 1, this is accompanied by only a small change in the shape of the cells, their volumes being preserved.

The heat of formation arises from the fact that the electronic charge density within the elemental Wigner–Seitz cells no longer satisfies the boundary conditions appropriate to that of the alloy. There will be a redistribution of the electrons within and between the cells in order to achieve the equilibrium alloy density. Miedema *et al.*[1–8] argued that this was accompanied by two distinct contributions to the heat of formation ΔH. The first is due to the discontinuity in the density of electrons $\Delta\rho_M^{1/3}$ at the interface between dissimilar Wigner–Seitz cells.[19] As the charge densities had originally been in their equilibrium state in the Wigner–Seitz cells of the elemental metals, the removal of this discontinuity will raise the energy. This positive contribution to ΔH is

FIG 1. The "macroscopic atom" model for alloy formation.

[19] The original version in Refs. 1 and 2 had the repulsive contribution proportional to $(\Delta\rho_M)^2$. Later versions replaced this by $(\Delta\rho_M^{1/3})^2$ which leads to much better quantitative predictions (see Ref. 3).

expected to be proportional to $(\Delta\rho_{\mathrm{M}}^{1/3})^2$. The second contribution is due to the difference in the work function of the elemental metals $\Delta\varphi^*$. If the atoms are regarded as macroscopic pieces of metal, the contact potential difference leads to a flow of charge from one atom to the other and the setting up of a dipole layer across the $A-B$ interface. This lowers the energy by an amount proportional to $(\Delta\varphi^*)^2$.

The heat of formation ΔH of a *disordered* binary alloy is, therefore, written within the Miedema scheme[4–8] as

$$\Delta H = 2c_A c_B g(c_A, c_B)\lambda_{\mathrm{M}}^{-1}\,\Delta\hat{H} \tag{2.1}$$

where

$$g(c_A, c_B) = V_A^{2/3}V_B^{2/3}/(c_A V_A^{2/3} + c_B V_B^{2/3}) \tag{2.2}$$

$$\lambda_{\mathrm{M}} = (\rho_{\mathrm{M}}^{A})^{-1/3} + (\rho_{\mathrm{M}}^{B})^{-1/3} \tag{2.3}$$

and

$$\Delta\hat{H} = -P(\Delta\varphi^*)^2 + Q(\Delta\rho_{\mathrm{M}}^{1/3})^2 \tag{2.4}$$

The atomic concentration and volume of the A and B constituents are represented by c_A, c_B and V_A, V_B, respectively. The prefactor $g(c_A, c_B)$ gives the deviation in ΔH from regular solution behavior due to the difference in the sizes of the atoms V_A and V_B. The heat of formation within the Miedema model is clearly proportional to the surface area that is shared by dissimilar cells. This accounts for the factor $V_A^{2/3}V_B^{2/3}$ in Eq. (2.2). The prefactor $\lambda_{\mathrm{M}}^{-1}$ is argued[4] to reflect the influence of the electronic screening length on the width of the dipole layer at the $A-B$ interface.

The magnitude of the heat of formation for an *ordered* compound is larger than that for the disordered alloy. This is due to more unlike atoms being nearest neighbors and hence increasing the contact area between dissimilar cells. Miedema[2] has treated this effect empirically by writing

$$\Delta H_{\mathrm{ord}} = [1 + 8(c_A^{\mathrm{S}} c_B^{\mathrm{S}})^2]\,\Delta H_{\mathrm{dis}} \tag{2.5}$$

where ΔH_{dis} is given by Eq. (2.1). c_A^{S} and c_B^{S} are the surface-area concentrations which can be expressed in terms of the atomic concentrations c_A and c_B by

$$c_{A,B}^{\mathrm{S}} = c_{A,B} V_{A,B}^{2/3}/(c_A V_A^{2/3} + c_B V_B^{2/3}) \tag{2.6}$$

For constituents with the same atomic volume, we see that an alloy with equiatomic composition will have a 50% larger heat of formation in the ordered than in the disordered state. The ordering energy, therefore, comprises one-third of the ordered heat of formation.

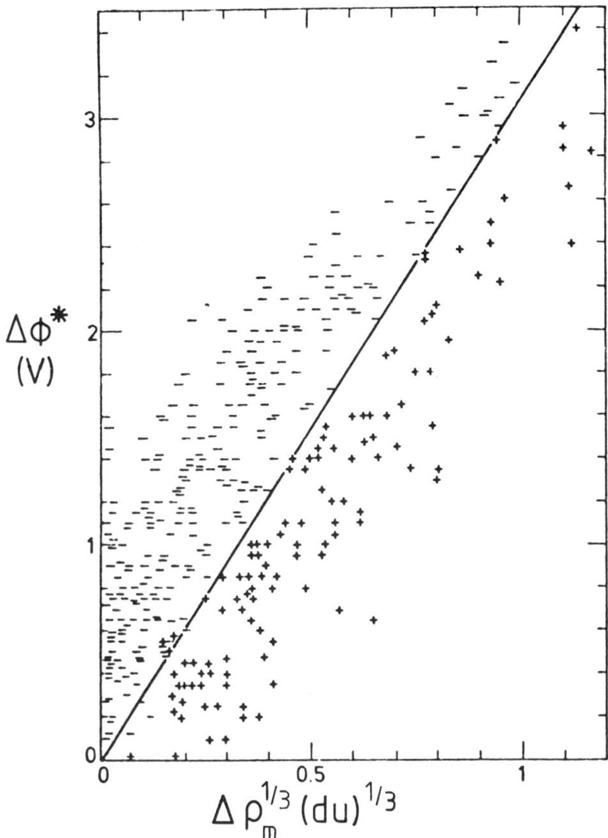

FIG. 2. Demonstration that the Miedema scheme reproduces the sign of ΔH for solid binary alloys consisting of a transition metal and a transition, noble, alkali, or alkaline earth metal[6] (see text for details).

It follows from Eqs. (2.1) and (2.4) that the *sign* of the heat of formation is fully determined by the ratio of $\Delta\varphi^*$ to $\Delta\rho_M^{1/3}$. In particular,

$$\Delta H \lesseqgtr 0 \qquad \text{for } \Delta\varphi^*/\Delta\rho_M^{1/3} \lesseqgtr \sqrt{Q/P} \tag{2.7}$$

We see from Fig. 2 that the Miedema scheme reproduces the sign of ΔH for all binary alloys of two transition metals and of transition metals alloyed with noble, alkali, and alkaline earth metals. Each binary system is characterized by some point ($\Delta\rho_M^{1/3}$, $\Delta\varphi^*$) on the figure. If the binary system contains stable intermetallic compounds, it is represented by a minus sign, whereas if no compounds exist and there is only a limited range of solid solubility (<10 at. %

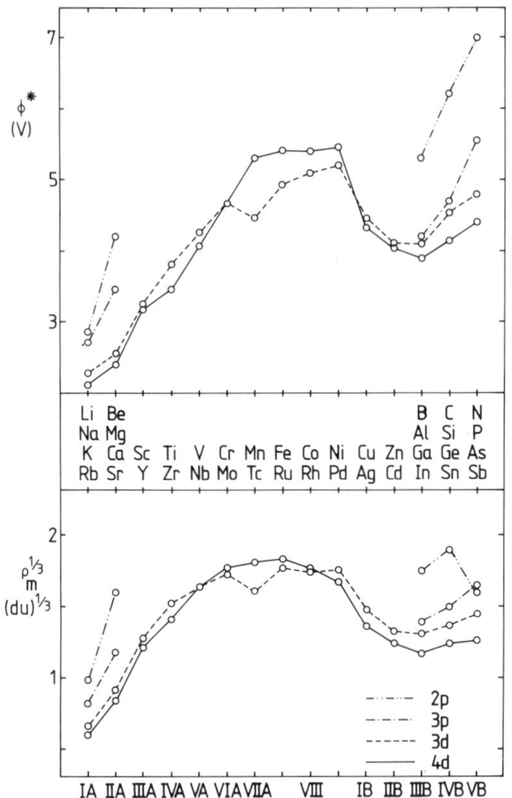

FIG. 3. The Miedema parameters[8] φ^* and $\rho_M^{1/3}$. (The values for the 5d row have not been included for clarity.)

at both sides), then it is represented by a plus sign. The plusses and minuses in Fig. 2 are separated nearly perfectly by the straight line $\Delta H = 0$ corresponding to $Q/P = 9.4\ V^2/(\text{d.u.})^{2/3}$.

The parameters[8] φ^* and $\rho_M^{1/3}$, which lead to this separation of alloys with positive and negative heats of formation, are shown in Fig. 3. φ^* is measured in volts, whereas ρ_M is given in density units (d.u.) such that $\rho_M(\text{Li}) \simeq 1$ d.u. The asterisk was added to the coordinate φ^* as a reminder that it is not the empirical work function φ which enters the scheme, but a suitably adjusted parameter which leads to the separation displayed in Fig. 3. Although φ and φ^* behave in a similar fashion across the series, values for individual elements can vary by as much as 1 V. Since it is the *difference* $\Delta\varphi^*$ which enters the expression for ΔH, this distinction between φ^* and the work function φ is, therefore, crucial for the success of the Miedema scheme. We will return to a discussion of the significance of the parameters φ^* and $\rho_M^{1/3}$ in Section III,8.

We should note, however, that the coordinate ρ_M also deviates from the predicted theoretical values of the electronic charge density at the Wigner–Seitz sphere radius ρ_{WS}. These deviations, although usually small on the scale of the charge density, can be important in determining the sign of ΔH since this involves the difference in charge density.

Miedema found that alloys of transition metals with *sp* valence elements to their right in the Periodic Table are not separated by a simple straight line through the origin as in Fig. 2. We see from Fig. 3 that these *sp*-bonded elements comprise the Group IIB elements Zn, Cd, and Hg, the Group IIIB elements B, Al, Ga, In, and Tl, the Group IVB elements C, Si, Ge, Sn, and Pb, and the Group VB elements N, P, As, Sb, and Bi. A plot of $\Delta\varphi^*$ versus $\Delta\rho_M^{1/3}$ for alloys of transition metals with elements from a particular *sp* valence group showed that the regions of positive and negative heats of formation were separated by a hyperbolic demarcation line. Moreover, the intercept of this demarcation line with the $\Delta\rho_M^{1/3}$ axis moved progressively further away from the origin as the group number of the *sp*-bonded element increased.

Thus, the heat of formation of these alloys of transition metals with non-transition metals could be modeled by adding an extra attractive contribution to $\Delta\hat{H}$ in Eq. (2.4), namely

$$\Delta\hat{H}_{pd} = -P(\Delta\varphi^*)^2 + Q(\Delta\rho_M^{1/3})^2 - R_{pd} \tag{2.8}$$

R_{pd} was argued to reflect the additional bonding arising from the hybridization between the valence d electrons of the transition element and the p electrons of the Group B element. Its observed dependence on the group number of the *sp*-bonded element was incorporated by writing $R_{pd} = R_p R_d$, where R_d takes the value unity for all transition elements except Sc and Y, for which $R_d = 0.7$. R_p/P takes the value[20] 1.4 for Group IIB, 1.9 for Group IIIB, 2.1 for Group IVB, and 2.3 for Group VB, respectively. This third term in Eq. (2.8) is sizeable, contributing about 0.5 eV/atom to the heat of formation of an alloy with equiatomic composition.

A value of P can be found by comparing the predictions of the model with known experimental data. Using Eq. (2.1) with atomic volumes in cm^3, φ^* in volts, ρ_M in density units (d.u.), and ΔH in kJ/mole, P takes the values[6] of 14.1, 10.6, and 12.3 for alloys of two transition metals, two non-transition metals, and a transition metal with a non-transition metal, respectively (1 kJ/mole = 0.01 eV/atom).

This very simple scheme has been successful not only in predicting the heats of formation of binary alloys, but also in treating other related problems such as, for example, surface segregation,[5] heats of adsorption,[21] the dissociation

[20] R_{pd} is reduced by a factor of 0.73 for liquid alloys.

[21] A. R. Miedema and J. W. F. Dorleijn, *Surf. Sci.* **95**, 447 (1980).

energy of diatomic molecules,[22] and the interatomic energy of impurities in ternary alloys.[23]

3. The Ionic Model

The success of the Miedema scheme is often judged to validate the physical concepts upon which the model is based.[9] In particular, the attractive term is assumed to be ionic in character. In this section we examine this assumption by considering the ionic model of Alonso and Girifalco,[10,11,24–26] which was proposed in order to account for the attractive and repulsive terms in the Miedema scheme. The theory is a modification of the earlier work of Hodges and Stott.[27]

The electronic ground state of a binary alloy can be reached in a two-step process. First, the Wigner–Seitz cells of the two dissimilar metals are "prepared" by either expanding or contracting them until they have identical charge density at their Wigner–Seitz boundaries, as illustrated schematically in Fig. 4.[28]

This preparation of the Wigner–Seitz cells of the elemental metals is performed under the constraint that the *total* volume of the alloy remains unchanged. This step, which removes the charge-density discontinuity at the interface between dissimilar cells,[29] leads to a positive contribution to the heat of formation. It is given for an AB alloy (in energy units per atom) by

$$\Delta H_{\text{prep}} = \tfrac{1}{4}[(\mathscr{B}_A/V_A)(V_A^{\text{prep}} - V_A)^2 + (\mathscr{B}_B/V_B)(V_B^{\text{prep}} - V_B)^2] \tag{3.1}$$

where $\mathscr{B}_A$ and $\mathscr{B}_B$ are the bulk moduli of the elemental metals A and B, respectively, V_A^{prep} and V_B^{prep} being their prepared volumes. Thus, the repulsive contribution to the Miedema heat of formation is identified with the elastic energy required to prepare the elemental cells so that the charge-density discontinuity across the Wigner–Seitz interface $\Delta\rho_{\text{WS}}$ is removed.

The second step comprises the relaxation of the electronic charge of the prepared atoms into the ground state of the binary alloy. The ionic model assumes that this redistribution of the charge is accompanied by a change in

[22] A. R. Miedema and K. A. Gingerich, *J. Phys. B* **12,** 2081 (1979).

[23] J. A. Alonso, T. E. Cranshaw, and N. H. March, *J. Phys. Chem. Solids* **46,** 1147 (1985).

[24] J. A. Alonso and L. A. Girifalco, *Phys. Rev. B: Condens Matter* [3] **19,** 3889 (1979).

[25] J. A. Alonso, D. J. Gonzalez, and M. P. Iniguez, *Solid State Commun.* **31,** 9 (1979).

[26] L. A. Girifalco and J. A. Alonso, *Metall. Soc. AIME Proc.* (1980).

[27] C. H. Hodges and M. J. Stott, *Philos. Mag.* [8] **26,** 375 (1972).

[28] A. R. Williams, C. D. Gelatt, and V. L. Moruzzi, unpublished (1980).

[29] It is clear from Fig. 1 that the charge density discontinuity in the alloy can only be removed if the shape of the cell is also allowed to change. In the spirit of the Miedema model this explicit structural contribution is neglected.

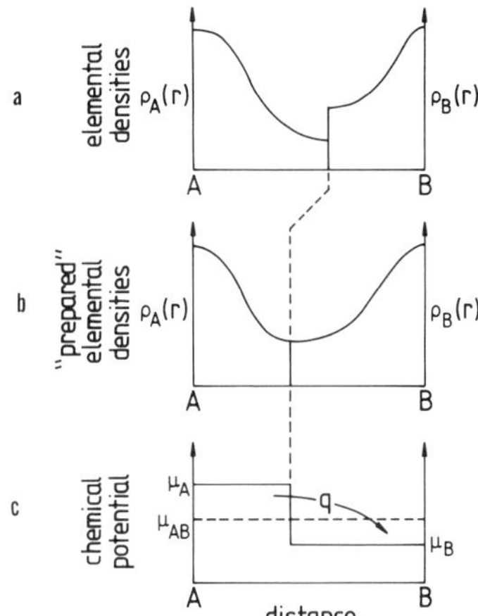

FIG. 4. Schematic representation of the two steps constituting the ionic model of alloy formation. In the first step from (a) to (b) the constituent possessing the smaller (larger) Wigner–Seitz electron density is compressed (expanded) until the boundary-density mismatch is eliminated. In the second step, which is represented by (c), the discontinuity in the chemical potential between the elemental Wigner–Seitz cells is removed by the transfer of q electrons to the cell possessing the lower chemical potential (after Williams *et al.*[28]).

energy, which is given by[27]

$$\Delta H_{\text{ionic}} = -\tfrac{1}{4}q(\mu_A - \mu_B) \tag{3.2}$$

where μ_A and μ_B are the chemical potentials associated with the prepared A and B Wigner–Seitz cells. qe is the amount of charge which flows from one atomic cell to another in order to equilibrate the chemical potentials, as shown in Fig. 4. Assuming[24,27] that the charge transfer qe (where e is the electronic charge) is proportional to $(\mu_A - \mu_B)$, this ionic contribution is negative and proportional to $(\mu_A - \mu_B)^2$. Thus the attractive contribution to the Miedema heat of formation is identified with the ionic energy [Eq. (3.2)], which results from the flow of charge from one atom to another on alloy formation.

The validity of the ionic model has been tested by Williams *et al.*[28] for the case of binary AB transition metal alloys. The heats of formation of the ordered 4d transition metal compounds with respect to the bcc lattice were computed from first principles within the local-density-functional (LDF) approximation.[15] This approximation is known[30–32] to predict accurately

[30] V. L. Moruzzi, J. F. Janak, and A. R. Williams, "Calculated Electronic Properties of Metals." Pergamon, Oxford, 1978.

[31] S. Lundquist and N. H. March, eds., "Theory of the Inhomogeneous Electron Gas." Plenum, New York, 1983.

[32] J. Callaway and N. H. March, *Solid State Phys.* **38,** 135 (1984).

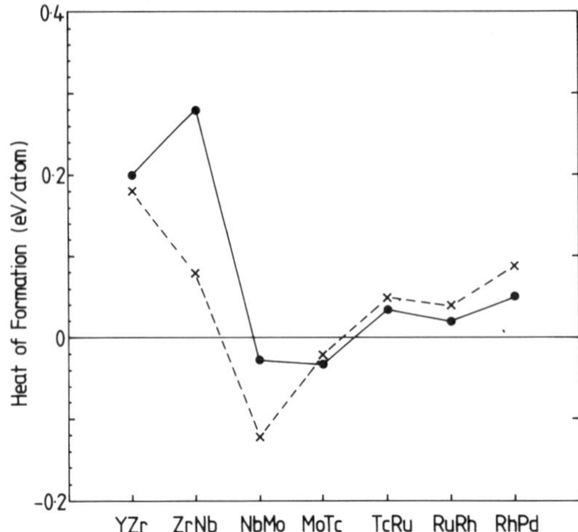

FIG. 5. (×) give the predicted LDF heats of formation of ordered transition metal compounds with respect to the bcc lattice, which are formed from constituents in neighboring groups of the 4*d* series. (●) show that the error associated with the ionic model is as large or larger than the heats of formation itself (after Williams *et al.*[28]).

the bulk properties of solids. The crosses in Fig. 5 show the variation of ΔH_{LDF} for elements in neighboring groups across the 4*d* transition metal series. We see that the heat of formation is positive near the beginning and end of the series, and negative near the middle. This behavior is consistent with Miedema's predictions and is discussed in detail in Section II,7.

The error within the ionic model is due solely to the assumption that the energy of the last step in Fig. 4 is given by Eq. (3.2). The LDF calculations can compute this error directly since it is given by

$$\text{error} = (\Delta H_{\mathrm{LDF}} - \Delta H_{\mathrm{prep}}) - \Delta H_{\mathrm{ionic}} \tag{3.3}$$

where the preparation energy is unambiguously defined by the first step in Fig. 4. The ionic energy is also obtainable, since both the chemical potential difference $(\mu_A - \mu_B)$ and the charge transfer from one prepared cell to another may be evaluated explicitly.[33] The circles in Fig. 5 show that the error of the ionic model can be as large as or larger than the heat of formation itself. Moreover, the sign of the error oscillates across the series, being positive at the edges but negative in the middle. The origin of the oscillation in the error will be discussed in Section II,7.

[33] The LDF calculations were performed using the Augumented Spherical Wave method of A. R. Williams, J. Kübler, and C. D. Gelatt, *Phys. Rev. B: Condens Matter* [3] **19,** 6094 (1979).

Pettifor and Varma[34] have shown that the Hodges–Stott relation [Eq. (3.2)] does not correctly describe the change in boundary conditions which the electronic eigenstates experience when going from the prepared cells to the binary alloy. The ionic model fails to describe the quantum-mechanical metallic bond, whose nature we explore in the following sections.

II. The Nature of the Metallic Bond

4. Bonding in Simple Metals

Wigner and Seitz[35] were the first to apply quantum mechanics to a study of the metallic bond. In 1933 they solved the Schrödinger equation for the sodium atom subject to the bonding boundary condition that the derivative of the wave function vanishes over the Wigner–Seitz sphere. The resulting eigenvalue located the bottom of the conduction band, Γ_1, since this corresponds to the most highly bonded eigenstate. By adding to this the average kinetic energy per electron of a free-electron gas, they obtained the total binding energy per atom, as shown in Fig. 6. They found values of the cohesive energy, equilibrium atomic volume, and bulk modulus that were within 10% of experiment.

Figure 6 shows that, as the atoms are brought together from infinity, the bonding state becomes more and more bonding until about 3 a.u., when Γ_1 turns upwards and rapidly loses its binding energy. This behavior is well described at metallic densities by the Fröhlich–Bardeen expression[36]

$$\Gamma_1 = -3/S + r_c^2/S^3 \tag{4.1}$$

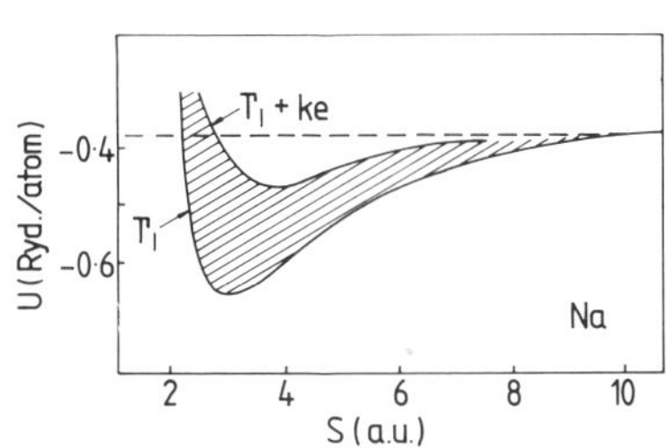

FIG. 6. The binding energy U as a function of the Wigner–Seitz radius S for sodium (after Wigner and Seitz[35]). The bottom of the conduction band Γ_1 is given by the lower curve, to which is added the average kinetic energy per electron (the shaded region).

[34] D. G. Pettifor and C. M. Varma, *J. Phys. C* **12,** L253 (1979).

[35] E. P. Wigner and F. Seitz, *Phys. Rev.* **43,** 804 (1933).

[36] J. Callaway, "Energy Band Theory." Academic Press, London, 1964.

where S is the Wigner–Seitz radius and r_c a constant. [Unless otherwise stated, expressions throughout the remainder of this article will be in atomic units such that the unit of energy is the rydberg (1 Ry = 13.6 eV) and the unit of length is the atomic unit (1 a.u. = 0.529 Å). In these units $\hbar^2/2m = 1$ and $e^2 = 2$. 1 mRy/atom = 1.32 kJ/mole.] The negative contribution to Eq. (4.1) gives the potential energy of a free electron in the Coulomb field of a point-ion core. The positive contribution is associated with the kinetic energy which arises from the *valence* state Γ_1 having to be orthogonal to the *core* states of the sodium ion. This orthogonality constraint introduces the well-known nodes in the outer-shell radial wave functions (see, for example, Fig. 6 of Ref. 37). As the volume of the Wigner–Seitz sphere decreases, the valence electron spends more time in the core region, and its kinetic energy increases, due to the rapid oscillation of the wave function between the nodes.

These oscillations in the valence eigenfunctions may be removed by a transformation of the Schrödinger equation that replaces the true potential by a pseudopotential while leaving the eigenvalues unaltered.[38,39] Pseudopotential theory explains why simple metals such as sodium, magnesium, and aluminum behave like nearly-free-electron (NFE) systems: The relevant matrix elements of the pseudopotential are small due to electrons scattering only weakly from the ion cores. Thus, the bulk properties of simple metals may be obtained by perturbing a free-electron gas by the presence of the ion-core pseudopotential lattice.[37–39]

To *first* order in the pseudopotential, the total binding energy per atom may be written[37–41]

$$U^{(1)} = U_{\mathrm{eg}} + U_{\mathrm{ion}} \tag{4.2}$$

The first term is the electron-gas energy, which may be split into kinetic and exchange-correlation contributions, respectively

$$U_{\mathrm{eg}} = U_{\mathrm{ke}} + U_{\mathrm{xc}} \tag{4.3}$$

The average kinetic energy per electron is given by

$$U_{\mathrm{ke}}/Z = 2.21/r_s^2 \tag{4.4}$$

whereas the exchange-correlation energy per electron is

$$U_{\mathrm{xc}}/Z = -0.916/r_s - (0.115 - 0.0313 \ln r_s) \tag{4.5}$$

[37] D. G. Pettifor, *in* "Physical Metallurgy" (R. W. Cahn and P. Haasen, eds.), Chapter 3. North-Holland Publ., Amsterdam, 1983.

[38] W. A. Harrison, "Pseudopotentials in the Theory of Metals." Benjamin, New York, 1966.

[39] V. Heine and D. Weaire, *Solid State Phys.* **24,** 250 (1970).

[40] L. A. Girifalco, *Acta Metall.* **24,** 759 (1976).

[41] W. A. Harrison, "Electronic Structure and the Properties of Solids." Freeman, San Francisco, California, 1980.

Z is the number of valence electrons per atom and r_s is the radius of the sphere which contains one electron so that $r_s = S/Z^{1/3}$. The density of the electron gas ρ is given by $\rho = (\frac{4}{3}\pi r_s^3)^{-1}$.

The second term in Eq. (4.2) is the change in energy to first order due to the presence of the lattice of ion cores. This may also be split into two contributions, namely

$$U_{\text{ion}} = U_{\text{es}} + U_{\text{core}} \tag{4.6}$$

Replacing the Wigner–Seitz cell by a sphere and neglecting intercell interactions, the electrostatic contribution is given by

$$U_{\text{es}} = -3Z^2/S + 1.2Z^2/S = -1.8Z^2/S \tag{4.7}$$

The first and second terms are the potential energies of the free-electron gas interacting with a point ion and with itself, respectively. The core contribution in Eq. (4.6) depends on the difference in the zeroth Fourier component of the pseudo- and Coulomb potentials. For the simple ionic pseudopotential[39]

$$\begin{aligned} v_{\text{ion}}(r) &= -2Z/r_c, &\quad &\text{for } r < r_c \\ &= -2Z/r, &\quad &\text{for } r > r_c \end{aligned} \tag{4.8}$$

it takes the form

$$U_{\text{core}} = Z^2 r_c^2/S^3 \tag{4.9}$$

We see, therefore, that the constant r_c in the Fröhlich–Bardeen expression [Eq. (4.1)] may be associated with the core radius of this model potential. In addition, we should note that the Fröhlich–Bardeen potential energy contribution $-3/S$ is very similar to that of the first-order expression, Eq. (4.2), for the case of alkali metals with $Z = 1$. The self-energy of the electron gas, $1.2/S$ in Eq. (4.7), is almost exactly cancelled by the exchange-correlation energy, Eq. (4.5).

In order to discuss the bonding in binary simple metal alloys, it is necessary to go beyond the first-order approximation and allow the gas of free electrons to respond and screen the ion cores. The total binding energy per atom for elemental simple metals is given to *second* order within the real-space representation[42,43] by

$$U^{(2)} = U_{\text{eg}} - \tfrac{1}{2}V\kappa_{\text{eg}}^{-1} + \tfrac{1}{2}\varphi(\mathbf{R}=0) + \tfrac{1}{2}\sum_{\mathbf{R}\neq 0}\varphi(\mathbf{R}) \tag{4.10}$$

where V is the atomic volume and κ_{eg} is the compressibility $(V\,d^2U_{\text{eg}}/dV^2)^{-1}$. The kinetic energy and exchange-correlation contributions of the first two

[42] E. G. Brovman, Y. Kagan, and A. Kholas, *Sov. Phys.—JETP (Engl. Transl.)* **30,** 883 (1970).
[43] M. W. Finnis, *J. Phys. F* **4,** 1645 (1974).

electron-gas terms may be combined to write

$$U_{\mathrm{eg}} - \tfrac{1}{2}V\kappa_{\mathrm{eg}}^{-1} = U_{\mathrm{ke}}^{(2)} + U_{\mathrm{xc}}^{(2)} \tag{4.11}$$

where

$$U_{\mathrm{ke}}^{(2)}/Z = 0.982/r_s^2 \tag{4.12}$$

and

$$U_{\mathrm{xc}}^{(2)}/Z = -0.712/r_s - (0.110 - 0.313 \ln r_s) \tag{4.13}$$

The term $\frac{1}{2}\varphi(\mathbf{R} = 0)$ in Eq. (4.10) is one-half of the electrostatic interaction between an ion and its own screening cloud of electrons. This represents the binding energy of the screened pseudoatom. It is consistent with the virial theorem, where the equilibrium binding energy of *all* the electrons in a free atom is just one-half of the total potential energy. The screening clouds around sodium, potassium, magnesium, and aluminum ions in free-electron environments of the appropriate equilibrium metallic densities are shown in Fig. 7.[44,45] We see that the ions are perfectly screened, with only about 30% of the electronic charge lying beyond the equilibrium Wigner–Seitz radii of the metals. This compares with approximately 80% for the case of free-atom

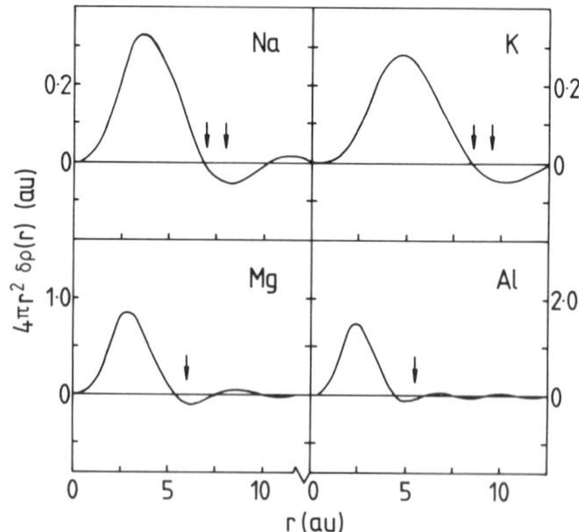

FIG. 7. The radial charge distribution of the screening clouds around sodium, potassium, magnesium, and aluminum ions in free-electron environments of the appropriate equilibrium metallic densities. The arrows mark the positions of the first-nearest neighbors for the close-packed lattices of Mg and Al, and the first- and second-nearest neighbors for the bcc lattices of Na and K (after Rasolt and Taylor[44] and Dagens *et al.*[45]).

[44] M. Rasolt and R. Taylor, *Phys. Rev. B: Solid State* [3] **11,** 2717 (1975).
[45] L. Dagens, M. Rasolt, and R. Taylor, *Phys. Rev. B: Solid State* [3] **11,** 2726 (1975).

charge densities.[46] The quantum-mechanical nature of the screening process is indicated by the oscillations in the tail of the screening clouds.

The interatomic pair potential $\varphi(R \neq 0)$ in Eq. (4.10) represents the electrostatic interaction between an ion and a second ion and its screening cloud some distance R away. For local pseudopotentials it is given by

$$\varphi(R \neq 0) = \frac{2Z^2}{R}\left(1 - \frac{2}{\pi}\int_0^\infty \chi(q, \rho)\hat{v}^2_{\text{ion}}(q)\frac{\sin qR}{q}\,dq\right) \tag{4.14}$$

where $\hat{v}_{\text{ion}}(q)$ is the normalized pseudopotential matrix element $(Vq^2/8\pi)v_{\text{ion}}(q)$ and $\chi(q, \rho)$ is the response function of a free-electron gas of density ρ. The first term is the direct ion–ion Coulomb repulsion, the second the attractive ion-screening cloud interaction. Since the response function may be written in terms of the dielectric function as $1 - \varepsilon^{-1}(q, \rho)$, Eq. (4.14) reduces to the single term

$$\varphi(R \neq 0) = \frac{4Z^2}{\pi R}\int_0^\infty \hat{v}^2_{\text{ion}}(q)\varepsilon^{-1}(q, \rho)\frac{\sin qR}{q}\,dq \tag{4.15}$$

The dielectric function of a free-electron gas may be written[47] in the following simple, yet accurate, form

$$\varepsilon(q, \rho) = 1 + (\kappa_{\text{TF}}/q)^2 f_0(q/2k_{\text{F}})/[1 - af_0(q/2k_{\text{F}})] \tag{4.16}$$

where κ_{TF} and $2k_{\text{F}}$ are the Thomas–Fermi and spanning Fermi wave factors, respectively. $f_0(q/2k_{\text{F}})$ is the Lindhard function,[48] which is given by

$$f_0(x) = \tfrac{1}{2} + [(1 - x^2)/4x]\ln[(1 + x)/(1 - x)] \tag{4.17}$$

The coefficient a in Eq. (4.16) is chosen[47] to satisfy the compressibility sum rule, which leads to the value $a = (0.166 + 0.004r_s)r_s$. In the limit of high metallic densities, $r_s \to 0$, $a \to 0$, and $f_0(q/2k_{\text{F}}) \to 1$, so that the Thomas–Fermi approximation $\varepsilon = 1 + (\kappa_{\text{TF}}/q)^2$ is recovered. The Lindhard or random-phase approximation corresponds to setting $a = 0$ in Eq. (4.16).

An analytic expression for the pair potential $\varphi(R \neq 0)$ can be derived[49] by writing the inverse dielectric function in Eq. (4.15) as the sum of partial fractions, namely

$$\varepsilon^{-1}(q, \rho) = \sum_n \frac{D_n q^2}{q^2 - q_n^2} \tag{4.18}$$

[46] F. Herman and S. Skillman, "Atomic Structure Calculations." Prentice-Hall, Englewood Cliffs, New Jersey, 1963.

[47] R. Taylor, *J. Phys. F* **8**, 1699 (1978).

[48] J. Lindhard, *Mat.-Fys. Medd.—K. Dan. Vidensk. Selsk.* **28**, 8 (1954).

[49] D. G. Pettifor, *Phys. Scr.* T **1**, 26 (1982).

so that the integral in Eq. (4.15) may be evaluated[50] by contour integration.

The weights $D_n = d_n \exp(i\delta_n)$ and poles $q_n = k_n + i\kappa_n$ may be found by approximating the Lindhard function in Eq. (4.16) by a rational polynomial[49] which has the correct low- and high-q behavior and is such that $f_0(1) = 0.5$. Sufficient accuracy for predicting the structural behavior of the simple metals Na, Mg, and Al is achieved[51] with a rational polynomial which leads to six terms in Eq. (4.18). The corresponding poles are complex for $r_s > 1.66$ a.u., so that we may assign $q_4^2 = (q_1^2)^*$, $q_5^2 = (q_2^2)^*$, and $q_6^2 = (q_3^2)^*$. The interatomic potential for $r_s > 1.66$ a.u. may, therefore, be written as the sum of *three* damped oscillatory terms, namely

$$\varphi(R \neq 0) = \frac{2Z^2}{R} \sum_{n=1}^{3} A_n \cos(k_n R + \alpha_n) e^{-\kappa_n R} \tag{4.19}$$

where the amplitude A_n is given by

$$A_n = 2d_n |\hat{v}_{\text{ion}}(q_n)|^2 \tag{4.20}$$

and the phase α_n is given by

$$\alpha_n = \delta_n + 2 \arg \hat{v}_{\text{ion}}(q_n) \tag{4.21}$$

We note that the wavevector k_n and the screening length κ_n^{-1} depend only on the density of the free-electron gas ρ, whereas the amplitude A_n and the phase α_n depend also on the nature of the ion core.

Figure 8 shows the analytic pair potentials for Na, Mg, and Al at their observed equilibrium atomic volumes, which were obtained using the local Ashcroft empty-core pseudopotential.[52] The solid curves are very similar to the pair potentials resulting from the interaction of the screening clouds in Fig. 7 with neighboring ion cores, which were computed using nonlocal pseudopotential theory.[44,45] The solid curves in Fig. 8 are the sum of three contributions: a short-range repulsive term, a medium-range attractive near-neighbor term, and a long-range oscillatory term, which is primarily responsible for which structure the lattice adopts.[51,53] Typical structural energy differences between fcc, bcc, and hcp simple metals are, therefore, expected to be less than or of the order of 1 mRy or 0.01 eV per atom.

Table I gives the empirical binding energies and r_s values for magnesium, aluminum, and the alkali metals.[43,54] The pair-potential contribution is the only *structurally* dependent term in the second-order binding-energy ex-

[50] For the case where $\hat{v}_{\text{ion}}(q)$ has no poles and is an even function of q.

[51] D. G. Pettifor and M. A. Ward, *Solid State Commun.* **49,** 291 (1984).

[52] N. W. Ashcroft, *Phys. Lett.* **23,** 48 (1966).

[53] A. K. McMahan and J. A. Moriarty, *Phys. Rev. B: Condens. Matter* [3] **27,** 3235 (1983).

[54] N. W. Ashcroft and D. C. Langreth, *Phys. Rev.* **155,** 682 (1967).

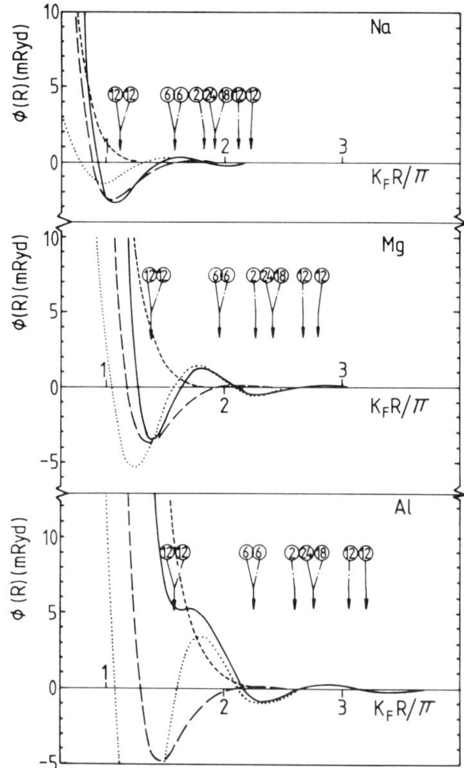

FIG. 8. The analytic pair potentials (———) for Na, Mg, and Al.[51] The short-range (---), medium-range (— — —), and long-range (· · ·) oscillatory terms are also given. The full (dot-dashed) arrows mark the positions of the first four (five) nearest-neighbor shells in fcc (ideal hcp).

TABLE I. THE EMPIRICAL BINDING ENERGIES AND r_s VALUES FOR MAGNESIUM, ALUMINUM, AND THE ALKALI METALS; r_c IS THE MODEL POTENTIAL CORE RADIUS WHICH IS FITTED TO REPRODUCE THE BINDING ENERGY

	Li	Na	K	Rb	Cs	Mg	Al
r_s (a.u.)	3.25	3.93	4.86	5.20	5.65	2.65	2.07
U (eV/atom)	−7.0	−6.3	−5.3	−5.0	−4.7	−24.3	−56.3
r_c (a.u.)	2.64	3.07	3.94	4.30	4.64	2.39	2.04

pression in Eq. (4.10). However, as can be deduced from Fig. 8, it contributes at the equilibrium atomic volume only about 0.3 eV/atom, which is small compared to the total binding energy. The binding energy is, therefore, determined primarily by the structure-independent terms in Eq. (4.10). The dominant central-site contribution $\frac{1}{2}\varphi(\mathbf{R} = 0)$ may be expressed analytically for local pseudopotentials by writing the inverse dielectric function as the sum of partial fractions as in Eq. (4.18). For the simple ionic pseudopotential of Eq. (4.8) we have for $r_s > 1.66$ a.u.

$$\varphi(\mathbf{R} = 0) = -\frac{2Z^2}{r_c}\left(1 - \sum_{n=1}^{3} \mathscr{A}_n d_n r_c^{-1}[\cos(\varphi_n + \delta_n) - e^{-2\kappa_n r_c}\cos(2k_n r_c + \varphi_n + \delta_n)]\right) \tag{4.22}$$

where the amplitude $\mathscr{A}_n$ is given by

$$\mathscr{A}_n = (k_n^2 + \kappa_n^2)^{-1/2} \tag{4.23}$$

and the phase φ_n is given by

$$\varphi_n = \tan^{-1}(k_n/\kappa_n) \tag{4.24}$$

Figure 9 shows the behavior of $\frac{1}{2}\varphi(\mathbf{R} = 0)$ as a function of the density of the electron gas for the case of sodium with $r_c = 3.035$ a.u. when $-Z^2/r_c = -4.48$ eV. The oscillations in the solid curve at low densities or larger r_s values reflect the density dependence of the oscillations in the screening cloud about the sodium ion illustrated in Fig. 7. The dashed curve in Fig. 9 results from neglecting the exponentially damped term in Eq. (4.22) and may be denoted by $\frac{1}{2}\varphi_0(\mathbf{R} = 0)$. We see that the difference in energy between the solid and dashed curves is at most 0.02 eV/atom. We may, therefore, approximate the binding

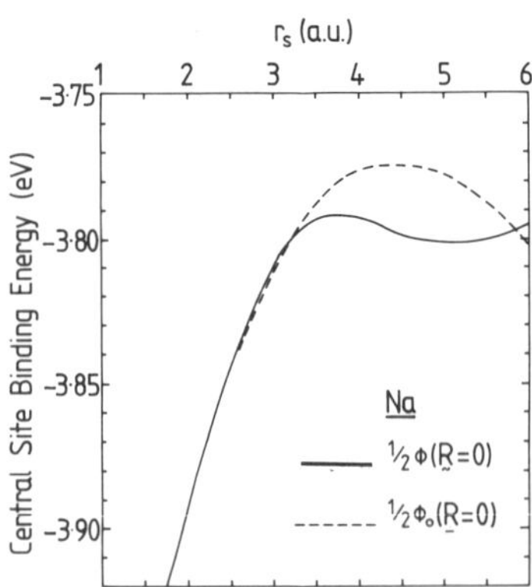

FIG. 9. The pseudoatom binding energy of sodium in a free-electron gas with density specified by r_s. The sodium ion is characterized by a model potential of radius $r_c = 3.035$ a.u. (---) results from neglecting the exponentially damped term in Eq. (4.22).

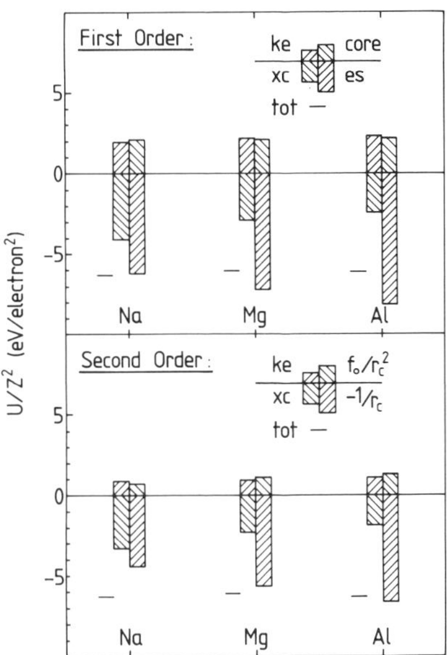

FIG. 10. The contributions to the normalized binding energy U/Z^2 of sodium, magnesium, and aluminum within first- and second-order perturbation theory, respectively.

energy by the dashed curve, which is given by

$$\tfrac{1}{2}\varphi_0(\mathbf{R} = 0) = -Z^2/r_c + (Z/r_c)^2 f_0(\rho) \tag{4.25}$$

where $f_0(\rho)$ is fitted to better than 0.5% accuracy for $r_s < 6$ a.u. by

$$f_0(\rho) = \tfrac{1}{2}(1 - r_s^{1.18}/20)/\kappa_{TF} \tag{4.26}$$

with $\kappa_{TF} = 1.563/\sqrt{r_s}$ and $r_s = 0.620\rho^{-1/3}$. The density-independent contribution $-Z^2/r_c$ was shown by Finnis[43] to reflect the variation in the binding energy among different simple metals using values of r_c from the model potentials of Appapillai and Heine.[55] The density-dependent factor $f_0(\rho)$ reduces to the Thomas–Fermi expression $1/2\kappa_{TF}$ in the limit of high densities.

The different contributions to the binding energy may now be examined directly within first- and second-order perturbation theories by using Eqs. (4.3), (4.4), (4.7), (4.9), (4.12), (4.13), and (4.25). The value of the pseudopotential core radius r_c was chosen so that the second-order structure-independent terms reproduce the empirical binding energy (c.f. Table I). We see from Fig. 10 that, although the individual first- and second-order expressions and contributions look very different, the final total binding energies are very similar. The first- and second-order values are -6.3 and -6.3 for Na,

[55] M. Appapilai and V. Heine, TCM Tech. Rep. No. 5. Cavendish Laboratory, Cambridge, England, 1972.

-24.1 and -24.3 for Mg, and -55.4 and -56.3 eV/atom for Al, respectively. The dominant contribution within second-order theory is the central-site density-independent[56] term $-Z^2/r_c$, although the exchange-correlation contribution also plays an important role in the bonding of simple metals. This is consistent with the fact that the Hartree approximation predicts a cohesive energy in aluminum, for example, which is at least two orders of magnitude too small.[57]

We have seen in Fig. 6 that the metallic bond in Wigner and Seitz's treatment[35] arises from the creation of the bonding eigenstate at the bottom of the conduction band. For *elemental* metals this approach is unambiguous as there exists a well-defined Wigner–Seitz cell across which the Bloch boundary conditions must be satisfied. This approach has been very fruitful. It leads directly to the so-called force theorem,[34,58,59] which gives explicitly the first-order change in total energy of the Wigner–Seitz cells when their boundary conditions are altered, for example by expansion of the lattice. However, for *binary* alloys the choice of the unit cell about each type of atom is ambiguous, so that the partitioning of the binding energy can become arbitrary and unphysical. We will see in the next section that the very different perspective of second-order perturbation theory with its screened, neutral pseudoatoms[60] provides a natural description of the bonding in simple metal alloys.

5. Simple Metal Heats of Formation

The alloys of simple metals, which lie in the same or neighboring groups of the periodic table, are also well described by the nearly-free-electron approximation. Figure 11 shows the density of states of the ordered CsCl binary compounds NaMg, NaAl, and MgAl, which were computed[61] within the LDF scheme. We see that NaMg and MgAl with $\Delta Z = 1$ both have densities of states which are only weakly perturbed from the free-electron values. NaAl with $\Delta Z = 2$ is more strongly perturbed, as expected. Second-

[56] Note that second order perturbation theory is valid only at metallic densities where the scattering from the ionic pseudopotentials is weak. The expressions must not be extrapolated to the free atom limit so that the theory can not be used to predict cohesive energies. It can predict heats of formation, however, because these involve the energy changes between metallic systems.

[57] P. Mohn, private communication.

[58] D. G. Pettifor, *Commun Phys.* **1,** 141 (1976); *J. Chem. Phys.* **69,** 2930 (1978).

[59] A. R. Mackintosh and O. K. Andersen, *in* "Electrons at the Fermi Surface" (M. Springford, ed.), §5.3. Cambridge Univ. Press, London and New York, 1980.

[60] J. M. Ziman, *Adv. Phys.* **13,** 89 (1964).

[61] C. D. Gelatt, V. L. Moruzzi, and A. R. Williams, unpublished (1980).

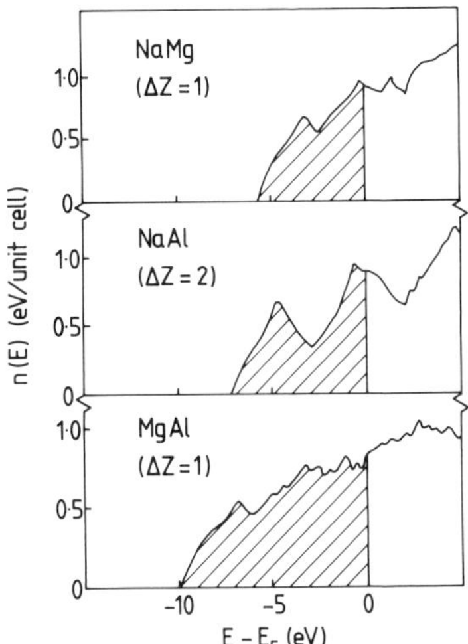

FIG. 11. The density of states of the ordered CsCl binary compounds NaMg, NaAl, and MgAl, which were computed within the LDF approximation by Gelatt *et al.*[61]

order perturbation theory may, therefore, be used for discussing the cohesion of these binary systems.[62–66]

The heat of formation of an equiatomic *AB* alloy may be written within the real-space representation as

$$\Delta H = \Delta H_{\text{eg}} + \Delta H_{\text{pa}} + \Delta H_{\text{struct}} \tag{5.1}$$

where the electron-gas, pseudoatom, and structure-dependent contributions correspond to the different terms resulting from Eq. (4.10). Assuming zero volume of formation by taking $V_{AB} = \frac{1}{2}(V_A + V_B)$, the electron-gas contribution can be expanded as a Taylor series in powers of the difference in the cube root of the elemental electron densities $\Delta\rho^{1/3}$. To second order it is given[66] (in eV/atom) by

$$\Delta H_{\text{eg}} = \bar{Z} f_{\text{eg}}(\overline{\rho^{1/3}})(\Delta\rho^{1/3})^2 \tag{5.2}$$

where

$$f_{\text{eg}}(x) = -43.40 + 7.80/x + 0.16/x^2 \tag{5.3}$$

[62] T. M. Hayes, H. Brooks, and A. R. Bienenstock, *Phys. Rev.* **175,** 699 (1968).

[63] J. E. Inglesfield, *J. Phys. C* **2,** 1285 (1969).

[64] J. Hafner, *J. Phys. F* **6,** 1243 (1976).

[65] R. Taylor *in* "Electrons in Disordered Metals and at Metallic Surfaces" (P. Phariseau, B. L. Gyorffy, and L. Scheire, eds.), p. 473. Plenum, New York, 1979.

[66] D. G. Pettifor and C. D. Gelatt, *in* "Atomistics of Fracture" (R. Latanision, ed.), p. 296. Plenum, New York, 1983.

The three terms in the prefactor, which depend only on the average of the cube root of the electron density, are the kinetic, exchange, and correlation contributions, respectively. The electron gas with the average electron density of the alloy has a lower kinetic energy but higher exchange-correlation energy than the average of these energies for the elemental metals. This follows directly from Eqs. (4.12) and (4.13), since the kinetic energy is concave upwards as a function of the volume per electron, whereas the exchange and correlation energies are concave downwards. The prefactor, and hence ΔH_{eg}, is negative for $\rho^{1/3} > 0.20$ or $r_s < 3.1$ a.u. when the kinetic energy contribution dominates the exchange and correlation.

Figure 12 shows the different contributions to the heat of formation of equiatomic binary alloys of alkali metals. The open circles correspond to the Miedema values for the ordered state, whereas the crosses correspond to the first-principles LDF values with respect to the bcc lattice.[61] The electron-gas contribution ΔH_{eg}, which is the sum of the kinetic and exchange-correlation values in Fig. 12, is positive for alkali alloys since $\bar{r}_s > 3.1$ a.u. The pseudoatom contribution ΔH_{pa} gives the change in the binding energy of the screened pseudoatoms $\frac{1}{2}\varphi(\mathbf{R} = 0)$ due to the difference in the electron-gas density between the alloy and the elemental metal. It reflects the change in binding energy that accompanies the change in shape of the screening clouds in metallic environments of different electron density. It is obtained directly from the second term in Eq. (4.25). We see that ΔH_{pa} is positive and negligibly small for alkali alloys since the pseudoatom binding energy shows only a weak

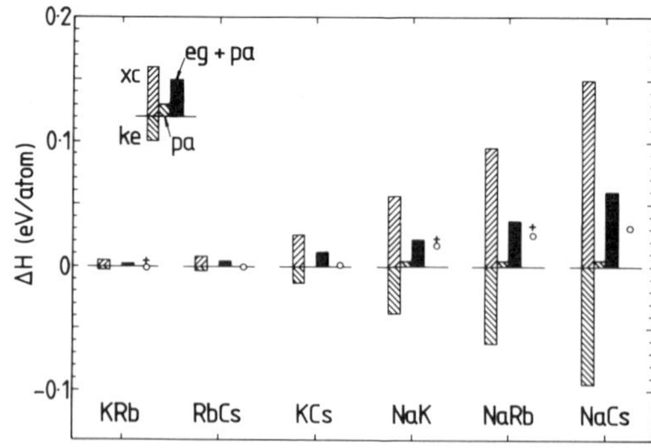

FIG. 12. The contributions to the heat of formation of equiatomic binary alloys of alkali metals. The solid histograms correspond to the sum of the electron-gas (eg) and pseudoatom (pa) contributions. The former is given by the sum of the kinetic energy (ke) and exchange-correlation (xc) contributions. (○) represent the Miedema values for the ordered alloy, whereas (+) represent the first-principles LDF values with respect to the bcc lattice.[61]

density dependence at these metallic densities (see, for example, Fig. 9). The sum of the volume-dependent terms $\Delta H_{eg} + \Delta H_{pa}$ clearly mirrors the behavior of the LDF calculations across the bcc binary alloys in Fig. 12.

Hafner[67] has demonstrated that the structurally dependent contribution ΔH_{struct} can be important in determining the sign of ΔH. He found that, whereas Na_2K, K_2Cs, and Rb_2Cs have positive heats of formation with respect to the bcc lattice, they have small negative heats of formation with respect to the ordered Laves structure due to the arrangement of the nearest-neighbor atoms around the minimum in the pair potential. Thus, provided the volume-dependent contribution to ΔH is not too large and positive, the structural contribution due to the pair potential can stabilize the binary phase. This sensitivity of the heat of formation to structure effects was also found by Miedema *et al.* when modeling the energetics of simple metal alloys (see Section 2,3,2 of Ref. 6).

Figure 13 shows the electron-gas and pseudoatom contributions to the renormalized heat of formation $\Delta H/(\Delta Z)^2$ for the equiatomic binary alloys of Na, Mg, and Al. The open circles again correspond to the Miedema values for the ordered state, whereas the crosses correspond to the LDF values with respect to the bcc lattice.[61] The pseudoatom contribution ΔH_{pa} is large and positive in these alloys[67] due to the sensitivity of the binding energy on the electron density over the range $r_s = 2$ a.u. (corresponding to elemental

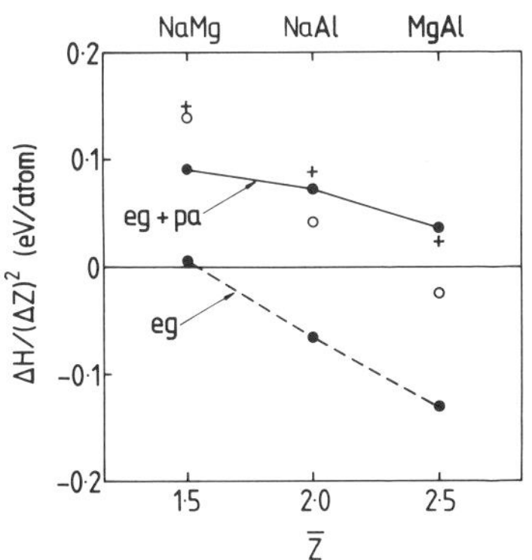

FIG. 13. The electron-gas (eg) and pseudoatom (pa) contributions to the normalized heat of formation $\Delta H/(\Delta Z)^2$ for the equiatomic binary alloys of Na, Mg, and Al. (○) and (+) are as for Fig. 12.

[67] But note that 0.1 eV/atom is still very small compared to the binding energy of 56.3 eV/atom for aluminium.

aluminum) to $r_s = 4$ a.u. (corresponding to elemental sodium) (c.f. Fig. 9). The electron-gas contribution changes sign around $\bar{Z} = 1.5$. MgAl is predicted by the structure-independent contributions and by the bcc LDF calculations to have a positive heat of formation. This is consistent with experiment as no equiatomic equilibrium phase MgAl has been found. However, the structure-dependent contribution to the energy is sufficient to drive ΔH negative for the 2:3 and 17:12 stoichiometries since the intermetallic phases Mg_2Al_3 and $Mg_{17}Al_{12}$ exist.

Second-order perturbation theory, therefore, leads to a description of the bonding in simple metal alloys which is very different from the ionic model. The ionic picture is essentially *classical*, the atoms being regarded as macroscopic pieces of metal which are characterized by the work function or chemical potential of the bulk materials. The bonding in the alloy arises from the flow of *charge* from one atom to another in order to equilibrate these local chemical potentials. The lowering in energy results from the alloy density of states being assumed implicitly to be the simple superposition of the densities of states of the elemental metals, the so-called *rigid*-band approximation. The charge flows "downhill" from one band to another, as illustrated schematically in Fig. 4.[28]

On the other hand, the second-order theory is *quantum mechanical*, the ions being screened by the response function of the *common* band of free electrons which is created when the atoms are brought together to form the solid. Although there is a redistribution of the electronic charge on alloy formation, each individual ion remains perfectly screened, the alloy charge density being given by the superposition of *neutral* pseudoatoms. The attractive contribution to the heat of formation arises from the free-electron-gas kinetic energy contribution being lowered on alloy formation by an amount proportional to $\bar{Z}(\Delta\rho^{1/3})^2$.

Although quantum mechanics appears to offer no justification for the physical basis of Miedema's scheme for simple metal alloys, we have seen in Figs. 12 and 13 that it is very successful as a predictive tool for the sign and magnitude of the heats of formation. On the other hand, although the simple pseudopotential model presented here clarifies the concepts underlying the bonding in simple metal alloys, it does not have the predictive capability of the Miedema scheme.

The single-parameter pseudopotential given by Eq. (4.8) cannot reproduce simultaneously the binding energy, equilibrium atomic volume, and crystal structure of the elemental simple metals. A natural extension of this local potential is the two-parameter parabolic model potential, which is obtained by joining a parabola smoothly to the Coulomb potential tail outside the core so that both the magnitude and derivative of the potential are continuous at r_c. It

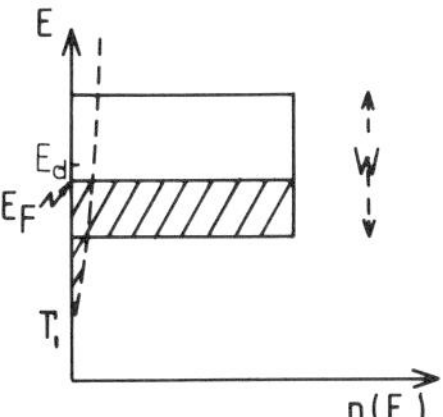

FIG. 14. A schematic representation of the transition metal *sp* (---) and *d* (———) densities of states when *sp*–*d* hybridization is neglected.

has the form[68]

$$v_{\text{ion}}(r)/2Z = [(2 - A)/r_c^3]r^2 + [(2A - 3)/r_c^2]r - A/r_c, \qquad \text{for } r < r_c \quad (5.4)$$
$$= -1/r \qquad \text{for } r > r_c$$

so that $v_{\text{ion}}(r = 0) = -2ZA/r_c$. This potential has the advantage that the third-order perturbative contribution[69,70] to the heat of formation will be small because $v_{\text{ion}}(q)$ falls off rapidly as q^{-4} rather than as q^{-3} for the optimized model potential[71] [Eq. (4.8)] or as q^{-2} for the Ashcroft empty-core pseudopotential.[52] However, the analytic expression for $\varphi(\mathbf{R} = 0)$ is too complicated[68] for the simple discussion attempted in this section. We have ignored nonlocal effects for the same reason, although they are known to be important in the alkaline earths Ca, Sr, and Ba and the Group IIB elements Zn, Cd, and Hg.[72,73]

6. Bonding in Transition Metals

Transition metals are characterized[37] by a fairly tightly bound valence *d* band that overlaps and hybridizes with a broader nearly-free-electron *sp* band, as illustrated in Fig. 14. The different roles played by the valence *sp* and *d* electrons in the bonding of the elemental transition metals has been analyzed in detail,[16,18,74] both within the renormalized-atom approximation[74] and the atomic-sphere approximation.[75] Figure 15 shows the different renormalized-atom contributions[74] to the equilibrium cohesive energy. It clearly demonstrates

[68] M. A. Ward and D. G. Pettifor, to be published; M. A. Ward, Ph.D. Thesis, Imperial College, London (1985).

[69] E. G. Brovman, Y. Kagan, and A. Kholas, *Sov. Phys.—JETP (Engl. Transl.)* **34,** 394 (1972).

[70] M. Hasegawa, *J. Phys. F* **6,** 649 (1976).

[71] R. W. Shaw, *Phys. Rev.* **174,** 769 (1968).

[72] J. A. Moriarty, *Phys. Rev. B: Condens. Matter* [3] **26,** 1754 (1982).

[73] H. L. Skriver, *Phys. Rev. Lett.* **49,** 1768 (1982).

[74] C. D. Gelatt, H. Ehrenreich, and R. E. Watson, *Phys. Rev. B: Solid State* [3] **15,** 1613 (1977).

[75] O. K. Andersen, *Solid State Commun.* **13,** 133 (1973).

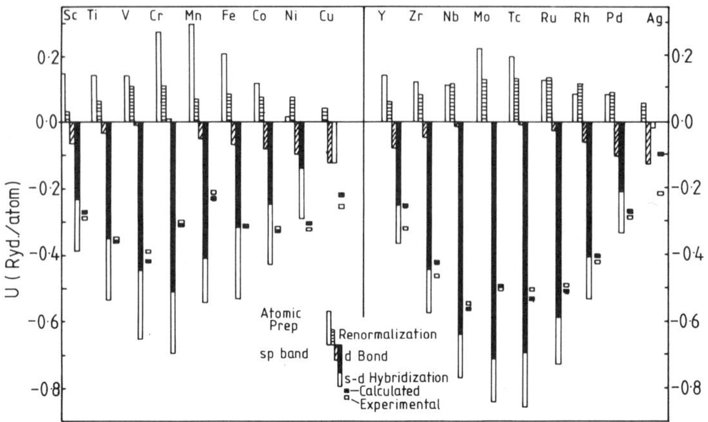

FIG. 15. The contributions to the cohesive energy of 3*d* and 4*d* transition metals within the renormalized atom approximation (after Gelatt *et. al.*[74]).

that the parabolic trend in the cohesion across the nonmagnetic 4*d* series is due to the bonding of the valence *d* electrons, as first suggested by Friedel.[76] If the density of states $n_d(E)$ is assumed to be rectangular, as drawn in Fig. 14, then the *d* bond energy per atom may be written

$$U^d_{\text{bond}} = \int^{E_F} (E - E_d) n_d(E)\, dE = -\frac{W}{20} N_d(10 - N_d) \tag{6.1}$$

where E_d and W are the center of gravity and width of the *d* band, respectively, and N_d is the number of valence *d* electrons.

The simplest model for describing the cohesion and structure of transition metals is, therefore, given by approximating the binding energy of the solid with respect to that of the free atoms as

$$U = U_{\text{rep}} + U_{\text{bond}} \tag{6.2}$$

where U_{rep} is assumed[77] to be given by the sum of central pair potentials. U_{bond} is the bond energy which results from evaluating the density of states $n_d(E)$ within the two-center, orthogonal tight-binding (TB) approximation.[37,78,79]

This TB bond model can be justified within density-functional theory by approximating the exact ground-state charge density which enters the Schrödinger equation by the sum of overlapping atomic charge densi-

[76] J. Friedel, *in* "The Physics of Metals" (J. M. Ziman, ed.), p. 494. Cambridge Univ. Press, London and New York, 1969.

[77] F. Ducastelle, *J. Phys. (Orsay, Fr.)* **31,** 1055 (1970).

[78] J. C. Slater and G. F. Koster, *Phys. Rev.* **94,** 1498 (1954).

[79] D. W. Bullett, R. Haydock, V. Heine, and M. J. Kelly, *Solid State Phys.* **35** (1980).

ties.[80–82] Harris[80] has shown that the error involved is second order in the difference between the overlapping free atom and the exact ground-state charge densities. The errors are less than 15% for the equilibrium separation and vibrational frequency of diatomic molecules such as C_2, N_2, and Cu_2.[80] By identifying[82] the second contribution in Eq. (6.2) with the bond energy rather than the band energy which enters the density-functional scheme, the first contribution is simply the sum of the Coulomb, exchange, and correlation interactions between individual pairs of atoms.

The bond energy, which is defined in Eq. (6.1) in terms of the *band* density of states $n_d(E)$, may be broken down into the contributions from individual pairs of bonds.[83,84] The local density of states $n_i(E)$ which is associated[85] with the localized orbital ψ_i on the atom at site $\mathbf{R}_i$ is given[86] by

$$n_i(E) = -(1/\pi) \lim_{\zeta \to 0+} \operatorname{Im} G_{ii}(E + i\zeta) \tag{6.3}$$

where

$$G_{ii}(E + i\zeta) = \langle \psi_i | (E + i\zeta - \mathscr{H})^{-1} | \psi_i \rangle \tag{6.4}$$

is the diagonal element of the Green function corresponding to the TB Hamiltonian

$$\mathscr{H} = -\nabla^2 + \sum_i v(\mathbf{r} - \mathbf{R}_i) \tag{6.5}$$

The factor $E \operatorname{Im} G$ in Eq. (6.1) may, therefore, be replaced by $\mathscr{H} \operatorname{Im} G$, since from Eq. (6.4) the Green function satisfies the matrix equation

$$(E - \mathscr{H})G = I \tag{6.6}$$

where I is the unit matrix.

The bond energy for the total lattice then takes the form

$$U_{\text{bond}} = (2\mathscr{N})^{-1} \sum_{\substack{i,j \\ i \neq j}} \mathscr{H}_{ij} \left(-\frac{2}{\pi} \operatorname{Im} \int^{E_F} G_{ji}(E)\, dE \right) \tag{6.7}$$

where $\mathscr{N}$ is the number of atoms and the limit as $\zeta \to 0_+$ is implicitly assumed. For d bands the matrix elements $\mathscr{H}_{ij}$ may be written in terms of the *bond integrals* $dd\sigma$, $dd\pi$, and $dd\delta$ between the neighboring atoms i and j provided

[80] J. Harris, *Phys. Rev. B: Condens. Matter* [3] **31,** 1770 (1985).

[81] W. M. C. Foulkes, to be published.

[82] A. P. Sutton, M. W. Finnis, D. G. Pettifor, and Y. Ohta, to be published.

[83] M. W. Finnis and D. G. Pettifor, *Springer Ser. Solid-State Sci.* **58,** 120 (1985).

[84] A. H. MacDonald and R. Taylor, *Can. J. Phys.* **62,** 796 (1984).

[85] Assuming one orbital per site for simplicity of notation.

[86] V. Heine, in Ref. 79.

nonorthogonality and three-center contributions can be neglected.[78] The expression inside the large parentheses is the *bond order*,[87] because it gives the difference in the number of electrons associated with the bonding state $\psi_b = (\psi_i + \psi_j)/\sqrt{2}$ compared to the antibonding state $\psi_a = (\psi_i - \psi_j)/\sqrt{2}$ as $2G_{ij} = G_{bb} - G_{aa}$. Equation (6.7) is the natural starting point from which to evaluate the forces between atoms in tightly bound molecules[87] and solids.[82,83,88] We will see in the next section that transition metal heats of formation comprise two terms, one reflecting the change in the bond integral, the other the change in the bond order.

The width W of the d band in Fig. 14 may be obtained[89] by fitting the second moment μ_2 of the model rectangular density of states to that of the correct density of states. In general, the pth moment $\mu_p^{(i)}$ of the local density of states $n_i(E)$ is given by

$$\mu_p^{(i)} = \int_{-\infty}^{\infty} E^p n_i(E)\,dE = \langle\psi_i|\mathscr{H}^p|\psi_i\rangle \tag{6.8}$$

Using the completeness of the localized basis ψ_j, the second moment $\mu_2^{(i)}$ may, therefore, be written

$$\mu_2^{(i)} = \sum_j \langle\psi_i|\mathscr{H}|\psi_j\rangle\langle\psi_j|\mathscr{H}|\psi_i\rangle = \sum_j \mathscr{H}_{ij}^2 \tag{6.9}$$

so that the second moment $\mu_2^{(i)}$ is the sum over all the hopping paths of length two which start and finish at the ith atom. If the hopping or bond integrals extend only to the nearest-neighbor shell of atoms, and if $E_i = \mathscr{H}_{ii}$ is taken as the zero of energy, then $\mu_2^{(i)} = zt^2(R_0)$, where z is the shell coordination number, $t(R_{ij})$ is the bond integral[90] $\mathscr{H}_{ij}$, and R_0 is the nearest-neighbor distance. Since the second moment of the rectangular density of states per orbital is $W^2/12$, the band width W is given by

$$W = \sqrt{12z}\,t(R_0) \tag{6.10}$$

The bond energy, therefore, varies as the square root of the number of neighbors rather than linearly, thereby reflecting the unsaturated nature of the covalent bond in metals.[86]

[87] C. A. Coulson. *Proc. R. Soc. London, Ser. A* **169,** 413 (1939).

[88] D. J. Chadi, *Phys. Rev. B: Condens. Matter* [3] **29,** 785 (1984).

[89] F. Cyrot-Lackmann, *J. Phys. Chem. Solids* **29,** 1235 (1968).

[90] Expression (6.9) is trivially extended to the case of non-degenerate orbitals. The second moment corresponding to the density of states of all *five d* orbitals on a given site is $z(dd\sigma^2 + 2dd\pi^2 + 2dd\delta^2)$. This is equivalent to $5zt^2$ where t is the appropriate hopping integral for use in eq. (6.10). Within this second moment approximation to the density of states the hopping integral $t(\underset{\sim}{R}_{ij})$ shows no dependence on the direction $\hat{\underset{\sim}{R}}_{ij}$. The angular character of the orbitals is lost.

Consider now the simplest possible model of the binding in transition metals where the repulsive interatomic potential varies exponentially as

$$\varphi(R) = aN_d^2 e^{-2\kappa R} \tag{6.11}$$

and the hopping or bond integral behaves as

$$t(R) = bN_d e^{-\kappa R} \tag{6.12}$$

where a and b are constants for a given series. The prefactors N_d^2 and N_d are suggested by the respective dependence of $\varphi(R)$ and $t(R)$ on the atomic charge density. The explicit influence[16,37] of the single valence s electron per atom is neglected. Assuming only nearest-neighbor hopping and substituting Eqs. (6.10)–(6.12) into Eq. (6.2), we find the following equilibrium expressions for the band width, nearest-neighbor distance, cohesive energy, and bulk modulus

$$W = [3b^2/5a]N_d(10 - N_d) \tag{6.13}$$

$$R_0 = \kappa^{-1} \ln\{10a\sqrt{z}/[\sqrt{3}b(10 - N_d)]\} \tag{6.14}$$

$$U_{\text{coh}} = (3b^2/200a)[N_d(10 - N_d)]^2 \tag{6.15}$$

and

$$B = (2\sqrt{2}\kappa^2/9R_0)U_{\text{coh}} \tag{6.16}$$

Figure 16 compares the theoretical predictions with the experimental values across the $4d$ series, assuming one valence s electron per atom.[91] The "experimental" values of the band width[90] are taken from the first-principles

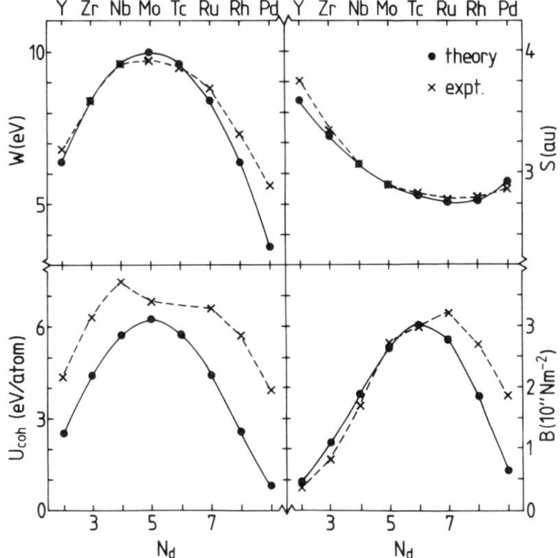

FIG. 16. The theoretical (●) and experimental (×) values of the equilibrium band width, Wigner–Seitz radius, cohesive energy, and bulk modulus of the $4d$ transition metals.

calculations in Table 2 of Pettifor.[91] We have found that the observed equilibrium atomic volumes are fitted by values of the inverse decay length κ, which vary almost linearly across the series.[37] Here we have taken

$$\kappa = 0.435 + 0.075\, N_d \tag{6.17}$$

The ratio $b^2/a = \frac{2}{3}$ eV is obtained from Eq. (6.13) by choosing a band width $W = 10$ eV for Mo with $N_d = 5$. Fitting the observed Mo Wigner–Seitz radius of 2.92 a.u. gives from Eq. (6.14) $a/b = 18.0$. Thus, $a = 216$ eV and $b = 12$ eV for the 4*d* series.

The predicted cohesive energy is about 2 eV/atom too small. This is seen from Fig. 15 to be due primarily to the neglect of the hybridization between the NFE *sp* and TB *d* bands. However, since this *sd* hybridization contribution is practically constant across the series, it may be neglected when predicting the trends in the heats of formation of transition metal alloys.[92] The simple model predictions become increasingly poorer towards the noble metal end of the series, where the *sp* electrons contribute significantly to the cohesion as the *d* bond weakens.[16–18]

An important feature of this rectangular bond model is that the cohesive energy is structure independent,[93] since the coordination number z does not enter Eq. (6.15). This follows directly from the repulsive energy varying as $z\exp(-2\kappa R)$, whereas the attractive energy varies as $\sqrt{z}\exp(-\kappa R)$. The model, therefore, predicts zero values for the structural energy differences and volume-preserving shear elastic constants.[94] This is consistent with the fact that the small energy differences between structures are determined by the higher moments, primarily μ_5 and μ_6.[95,96] The fcc, bcc, and hcp structural energies[97] for a TB *d* band of width 10 eV are shown in Fig. 17.[99] The curves predict the structural trend[18,37] from hcp → bcc → hcp → fcc → bcc as N_d increases, which is in agreement with experiment for the nonmagnetic[100]

[91] D. G. Pettifor, *J. Phys. F* **7**, 613 (1977).

[92] M. O. Robbins and L. M. Falicov, *Phys. Rev. B: Condens. Matter* [3] **29**, 1333 (1984).

[93] Provided only nearest neighbour bonds are important and the repulsive and bond parameters are transferable from one system to another.

[94] F. Ducastelle, *J. Phys. (Orsay, Fr.)* **31**, 1055 (1970).

[95] P. Turchi and F. Ducastelle, *Springer Ser. Solid-State Sci.* **58**, 104 (1985).

[96] D. G. Pettifor and R. Podloucky, *Phys. Rev. Lett.* **53**, 1080 (1984); *J. Phys. C* **19**, 315 (1986).

[97] The structural energy has been defined as the difference between the bond energy resulting from the correct density of states $n_d(E)$ and that resulting from a skew-rectangular density of states (see fig. 2 of Ref. 98) with the same second and third moments as that of the fcc lattice.

[98] D. G. Pettifor, *Solid State Commun.* **28**, 621 (1978).

[99] N. R. Beer, Ph.D. Thesis, Imperial College, London (1985).

[100] The magnetic contribution to the binding energy can be of the same order of magnitude as the structural energy. For example, bcc iron has a magnetic energy of −0.3 eV/atom [J. F. Janak and A. R. Williams, *Phys. Rev. B: Solid State* [3] **14**, 4199 (1976)]. It, therefore, plays a crucial role in determining the most stable structures of magnetic 3d systems [see, for example, H. Hasegawa and D. G. Pettifor, *Phys. Rev. Lett.* **50**, 130 (1983)].

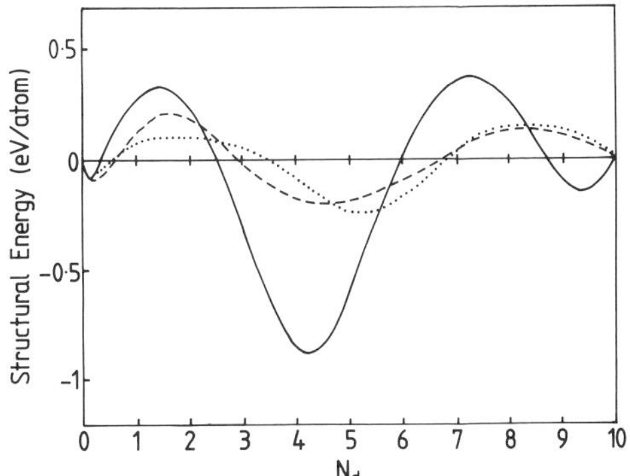

FIG. 17. The bcc (———), fcc (– – –), and hcp (· · ·) structural energies for a tight-binding *d* band of width 10 eV (after Beer[99]).

transition metals (apart from the predicted bcc stability at the noble metal end of the series). The shear elastic constants corresponding to a given lattice are also known to oscillate about zero as a function of band filling, being negative in part of the band-filling region, where the lattice is itself energetically not favored.[101]

The equilibrium nearest-neighbor distance for an element with valence N_d does, of course, depend on the coordination number z through Eq. (6.14). This may be written in terms of the bond number $n = N_d/z$, which gives the number of electrons per bond contributed by a given atom.[102] It takes the identical form to that proposed by Pauling,[102] namely

$$R_n = R_1 - (1.15/\kappa)\log_{10} n \tag{6.18}$$

where R_1 is the bond length associated with a single pair of electrons. For Mo $\kappa = 0.81$ a.u.$^{-1}$, so that the prefactor $(1.15/\kappa)$ takes the value 0.75 Å, which compares with Pauling's empirical value of 0.6 Å for the metallic bond. The values of κ given by Eq. (6.17) lead to a volume dependence of the band width W which is in good agreement with the first-principles calculations.[91] For example, the model and first-principles logarithmic derivatives for Y, Mo, and Rh are (1.3, 1.3), (1.4, 1.4), and (1.7, 1.6), respectively, which are close to the value of $\frac{5}{3}$ predicted theoretically by Heine.[103]

The concept of the size of an atom has been useful in understanding the way atoms bind and pack together in alloys.[104] The equilibrium atomic volume or

[101] M. W. Finnis, K. L. Kear, and D. G. Pettifor, *Phys. Rev. Lett.* **52,** 291 (1984).

[102] L. Pauling, "The Nature of the Chemical Bond," 3rd. ed. Cornell Univ. Press, Ithaca, New York, 1960.

[103] V. Heine, *Phys. Rev.* **153,** 673 (1967).

[104] W. B. Pearson, "The Crystal Chemistry and Physics of Metals and Alloys." Wiley, New York, 1972.

internuclear separation of the elemental metal is normally taken as the measure of atomic size. However, the skewed-parabolic variation shown by S in Fig. 16 is the direct result of the competition between the repulsive interatomic Coulomb term and the attractive parabolic bonding term. No such behavior would be observed for isolated $4d$ transition atoms, since their valence clouds contract monotonically across the series as the nuclear charge increases.[46]

A definition of the relative size of two atoms can be made[96] by generalizing the classical concepts used in ionic systems, where the ions are ascribed definite hard-core radii and packed together until they touch. For the case of metallic atoms where the interatomic repulsion varies smoothly rather than discontinuously, a relative size factor $\mathscr{R}$ may be defined by

$$\mathscr{R} = (r_A/r_B)^3 \tag{6.19}$$

where $2r_{A(B)}$ is the internuclear separation at which a pair of $A(B)$ atoms attains some given reference repulsive energy, i.e.

$$\varphi_{\text{rep}}(2r_A) = \varphi_{\text{rep}}(2r_B) = \varphi_{\text{ref}} \tag{6.20}$$

Taking the reference state as the repulsive energy of Mo at its equilibrium internuclear separation of 5.28 a.u., it follows from Eqs. (6.11) and (6.17) that

$$r_A = (2.67 + \ln N_a)/(0.87 + 0.15N_A) \tag{6.21}$$

in atomic units. The corresponding volumes $\frac{4}{3}\pi r_A^3$ decrease, therefore, monotonically across the series, with Pd having a relative size about one-half of Y. This difference in size of the atoms implies that the hopping or bond integrals in a transition metal alloy will be dependent on the particular pair of atoms forming the bond.

7. Transition Metal Heats of Formation

The difference between the rigid-band and common-band descriptions of bonding in binary AB alloys is illustrated in Fig. 18. The rigid-band model assumes that the alloy density of states is a superposition of the elemental densities of states. The binding energy is lowered as charge flows from the one band to the other in order to equilibrate the chemical potentials. Harrison[105] has shown that this leads to values for the cohesive energy of the alkali halides that are consistent with the predictions of the classical ionic model of Born.[106] In the latter model the energy of the ionic solid compared with that of the free

[105] W. A. Harrison, *Phys. Rev. B: Condens. Matter* [3] **31,** 2121 (1985).
[106] M. Born, *Ergeb. Exakten. Naturwiss.* **10,** 387 (1931).

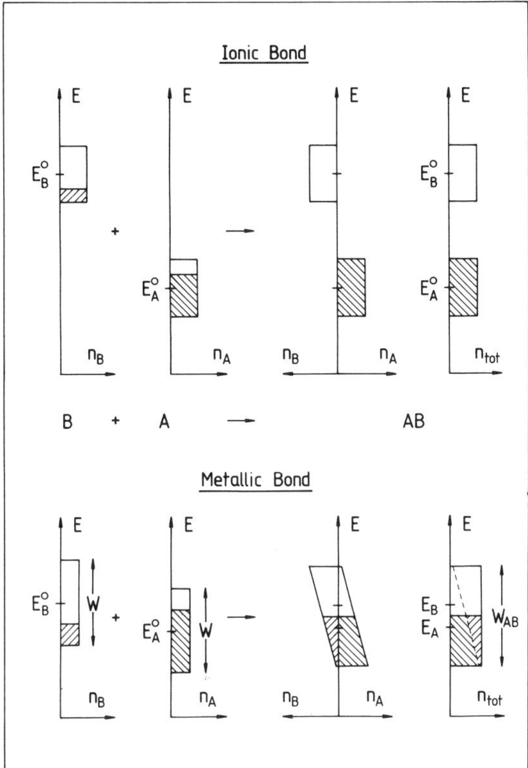

FIG. 18. Schematic representation of the ionic rigid-band and metallic common-band models of bonding in binary AB systems. E_A^0 and E_B^0 give the free-atom energy levels, whereas the positions of E_A and E_B in the metallic bond reflect the small shift which takes place on alloy formation in order to maintain local charge neutrality.

atoms is given per AB atom pair by

$$U_{\text{ionic}} = \mathscr{I} + \mathscr{A} + U_{\text{Mad}} \tag{7.1}$$

where $\mathscr{I}$ is the ionization energy of the alkali atom, $\mathscr{A}$ is the electron affinity of the halogen atom, and U_{Mad} is the attractive electrostatic Madelung energy of the ionic lattice.

Harrison[105] has identified $\mathscr{I}$ with the free atomic energy level E_B^0 in Fig. 18 (that is, $\mathscr{I} = -E_B^0$) and $\mathscr{A}$ with $E_A^0 + \Phi$, where Φ is the intra-atomic Coulomb integral. Therefore,

$$U_{\text{ionic}} = -(E_B^0 - E_A^0) + (U_{\text{Mad}} + \Phi) \tag{7.2}$$

For the alkali halides the second term in parentheses is generally small compared to the first because the *inter*atomic Madelung and *intra*-atomic

Coulomb energies tend to cancel. For example, NaCl has $U_{\text{Mad}} + \Phi = -8.9 + 10.3 = -1.4$ eV per atom pair, which is only 16% that of the first term, -8.8 eV. Consequently, the ionic binding energy of the alkali halides may be approximated by

$$U_{\text{ionic}} \approx -(E_B^0 - E_A^0) \tag{7.3}$$

This is just the prediction of the rigid-band model in Fig. 18 when all the atoms in the elemental reference systems A and B are taken infinitely far apart.

The common-band model, on the other hand, assumes that the local density of states on a given atom changes when the atom is taken from its elemental environment to that of the alloy. The band in the alloy is formed by the quantum-mechanical overlap of neighboring atomic wave functions so that the local densities of states have a common band width. This is illustrated in Fig. 18 within the rectangular band approximation, in which the total AB alloy density of states is also assumed to be constant.[98,107–109] The band width W_{AB} is determined by the second moment μ_2^{AB} of the alloy's average density of states per atom. That is, $\mu_2^{AB} = \frac{1}{2}(\mu_2^A + \mu_2^B)$, where the partial moments μ_2^A and μ_2^B are obtained from Eq. (6.9) by summing over all the hopping paths of length two.[110] It has the simple form

$$W_{AB} = [1 + 3(\Delta E/W)^2]^{1/2} W \tag{7.4}$$

where $\Delta E = E_B - E_A$. The hopping or bond integrals have been taken in Fig. 18 to be site independent so that $W_A = W_B = W$.

The local density of states $n_A(E)$ and $n_B(E)$ are constrained to have the same band width W_{AB}. Moreover, from Eq. (6.9) their centers of gravity μ_1^A and μ_1^B must coincide with the atomic energy levels E_A and E_B, respectively. This may be achieved[107] by simply skewing the partial density of states as shown in Fig. 18. If the skew-rectangular density of states $n_A(E)$ takes the values $10(1 + \alpha)/W_{AB}$ and $10(1 - \alpha)/W_{AB}$ at the bottom and top of the band, respectively, then

$$\alpha = 3\,\Delta E/W_{AB} \tag{7.5}$$

The d charge transfer $q_d e$ which accompanies an atomic energy level separation ΔE may, therefore, be calculated by filling up the bands to the Fermi energy. It is given by

$$q_d = \tfrac{1}{2}\Delta N_d + \tfrac{3}{10}\bar{N}_d(10 - \bar{N}_d)\,\Delta E/W_{AB} \tag{7.6}$$

where $\bar{N}_d = \frac{1}{2}(N_d^A + N_d^B)$ and $\Delta N_d = N_d^B - N_d^A$. The positions and labeling of

[107] D. G. Pettifor, *Phys. Rev. Lett.* **42,** 846 (1979); *J. Magn. Magn. Mater.* **15–18,** 847 (1980).

[108] C. M. Varma, *Solid State Commun.* **31,** 295 (1979).

[109] R. E. Watson and L. H. Bennett, *Phys. Rev. Lett.* **43,** 1130 (1979); CALPHAD: *Comput. Coupling Phase Diagrams Thermochem.* **5,** 25 (1981).

[110] M. Cyrot and F. Cyrot-Lackmann, *J. Phys. F* **6,** 2257 (1976).

the bands in Fig. 18 correspond to $\Delta E > 0$ and $\Delta N_d < 0$. Thus, the first term drives charge from the high-valence A atom to the low-valence B atom, whereas the second term takes charge in the opposite direction from the higher to lower atomic energy level.

Unlike the ionic solids, which are insulators, transition metal alloys are metallic. It is natural, therefore, to choose a set of localized basis orbitals ψ_i such that local charge neutrality is maintained.[96] This may be achieved since the "tails" of basis functions, which are based on resonant orbitals,[111] muffin-tin orbitals,[112] or Slater orbitals[113] are not unique. The same *total* alloy charge density may be obtained by different localized-orbital basis sets corresponding to slightly different *individual* atomic charge densities and charges. Neutral atoms may be achieved within the simple TB model, in which the atomic orbitals are assumed environment independent, by taking the intra-atomic Coulomb integral $\Phi = \infty$.[114] It follows from Eqs. (7.6) and (7.4) that local charge neutrality is obtained within the rectangular band model for an atomic energy-level separation $\Delta E_{AB} = E_B - E_A$ given by

$$\Delta E_{AB} = -W\Delta N/\{\tfrac{9}{25}[\bar{N}_d(10 - \bar{N}_d)]^2 - 3(\Delta N_d)^2\}^{1/2} \tag{7.7}$$

The above expressions, Eqs. (7.5)–(7.7), are valid provided $\alpha \leq 1$. This corresponds to the constraint

$$|\Delta N_d| \leq \tfrac{1}{5}\bar{N}_d(10 - \bar{N}_d) \tag{7.8}$$

so that the above simple model is applicable to all transition metal AB alloys with $\Delta N_d \leq 4$.

Figure 19 demonstrates unambiguously that the common-band model is an accurate description of the bonding in transition metal alloys. Williams *et al.*[28] considered the three ordered compounds MoTc, NbRu, and ZrRh, which have the same number of average valence electrons per atom but increasing valence difference $\Delta N = 1,3$, and 5, respectively. The total densities of states corresponding to the CsCl structure are remarkably similar for all three compounds, being close to that expected for an elemental bcc lattice.[30] Moreover, the partial d densities of states (which are defined[28] with respect to atomic spheres rather than atomic basis functions) show the skewing predicted by Eq. (7.5). The ionic rigid-band model is, therefore, not a valid description of the bonding in metallic systems.

The simple TB model for the transition metal binding energy presented in Section II,6 can be extended to the case of binary alloys. The binding energy

[111] D. G. Pettifor, *J. Phys. C* **5**, 97 (1972).
[112] O. K. Andersen, *Phys. Rev. B: Solid State* [3] **12**, 3060 (1975).
[113] J. W. Davenport, *Phys. Rev. B: Condens. Matter* [3] **29**, 2896 (1984).
[114] G. Allan and M. Lannoo, *J. Phys. Chem. Solids* **37**, 699 (1976); *Phys. Status Solidi B* **74**, 409 (1976).

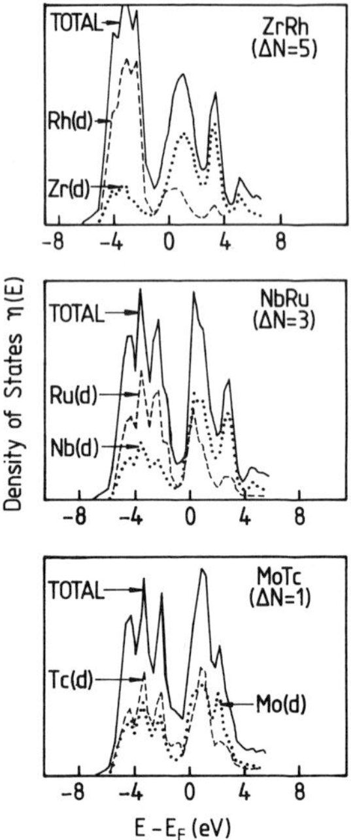

FIG. 19. The densities of states of three ordered AB compounds MoTc, NbRu, and ZrRh, which have the same number of average valence electrons per atom but increasing valence difference $\Delta N = 1$, 3, and 5, respectively. The total (———) and partial density of states with respect to the A (···) and B (– – –) sites are given. (after Williams *et al.*[28]).

per atom is written

$$U_{AB} = U_{\text{rep}}^{AB} + U_{\text{bond}}^{AB} \tag{7.9}$$

where

$$U_{\text{bond}}^{AB} = \tfrac{1}{2}\left(\int^{E_F} (E - E_A) n_A \, dE + \int^{E_F} (E - E_B) n_B \, dE\right) \tag{7.10}$$

is the bond energy between neutral atoms. The repulsive energy is given as the sum over the interatomic pair potentials $\varphi_{AA}(R)$, $\varphi_{BB}(R)$, and $\varphi_{AB}(R)$, where the first two are given by Eq. (6.11) and the latter is approximated by their geometric mean. The repulsive energy per atom for a close-packed *disordered* AB alloy with $z = 12$ is, therefore,

$$U_{\text{rep}}^{AB} = 1.5a(N_A e^{-\kappa_A R_0} + N_B e^{-\kappa_B R_0})^2 \tag{7.11}$$

where R_0 is the nearest-neighbor distance in the alloy.

The bond energy corresponding to the TB densities of states in Fig. 18 can be evaluated explicitly. It follows from Eq. (7.10) that the bond energy can be written in terms of the total density of states $n_{AB}(E)$ as

$$U_{\text{bond}}^{AB} = \int^{E_F} E n_{AB}(E)\, dE - \tfrac{1}{2}(N_A E_A + N_B E_B) \tag{7.12}$$

Assuming a rectangular *total* density of states $n_{AB}(E)$ with a band width W_{AB} corresponding to an energy-level separation ΔE [c.f. Eq. (7.4)], the bond energy becomes

$$U_{\text{bond}}^{AB}(\Delta E) = -W_{AB}\bar{N}_d(10 - \bar{N}_d)/20 - \tfrac{1}{4}\Delta N_d \Delta E \tag{7.13}$$

This expression is stationary with respect to small variations in ΔE for just that value of $\Delta E = \Delta E_{AB}$ which results from filling up the skew-rectangular *partial* densities of states and requiring local charge neutrality. Thus, this simple model is internally consistent.

Substituting Eqs. (7.4) and (7.7) into Eq. (7.9), the bond energy per atom is given by

$$U_{\text{bond}}^{AB} = f_{\text{bond}}(\bar{N}_d, \Delta\bar{N}_d)U_{\text{bond}}(\bar{N}_d) \tag{7.14}$$

where

$$f_{\text{bond}}(\bar{N}_d, \Delta N_d) = \{1 - \tfrac{25}{3}(\Delta N_d)^2/[\bar{N}_d(10 - \bar{N}_d)]^2\}^{1/2} \tag{7.15}$$

and

$$U_{\text{bond}}(\bar{N}_d) = -W\bar{N}_d(10 - \bar{N}_d)/20 \tag{7.16}$$

$U_{\text{bond}}(\bar{N}_d)$ is just the bond energy within the virtual crystal approximation (VCA), in which all atoms are assumed to be identical and described by the average properties of the A and B constituents. The prefactor $f_{\text{bond}}(\bar{N}_d, \Delta N_d)$ represents the loss of bond order with respect to this average VCA state due to the actual mismatch in the atomic energy levels on the A and B sites ΔE_{AB}.

The band width W_{AB} of a *disordered* AB alloy can be found directly from Eq. (6.9) for the second moment of the densities of states.[110] The hopping or bond integrals $t_{AA}(R)$, $t_{BB}(R)$, and $t_{AB}(R)$ are site dependent, the first two being given by Eq. (6.12) and the latter by their geometric mean, which follows from resonant[111] or muffin-tin orbital[112] TB theories. The width W_{AB} is given by Eq. (7.4) with[98]

$$W = \tfrac{1}{2}[W_A(V_{AB}) + W_B(V_{AB})] \tag{7.17}$$

where $W_A(V_{AB})$ and $W_B(V_{AB})$ are the widths of the elemental metals at the volume V_{AB} of the alloy. For a close-packed alloy with $z = 12$

$$W = 6b(N_A e^{-\kappa_A R_0} + N_B e^{-\kappa_B R_0}) \tag{7.18}$$

The equilibrium interatomic distance R_{AB} of the disordered alloy is found from Eqs. (7.9), (7.11), (7.14), and (7.18) to satisfy

$$\tfrac{1}{2}(N_A e^{-\kappa_A R_{AB}} + N_B e^{-\kappa_B R_{AB}}) = (b/a)[\bar{N}_d(10 - \bar{N}_d)/20] f_{\text{bond}}(\bar{N}_d, \Delta N_d) \tag{7.19}$$

Therefore, the total binding energy at equilibrium is given by

$$U_{\text{coh}}^{AB} = f_{\text{bond}}^{2}(\bar{N}_d, \Delta N_d)U_{\text{coh}}(\bar{N}_d) \tag{7.20}$$

where

$$U_{\text{coh}}(\bar{N}_d) = (3b^2/200a)[\bar{N}_d(10 - \bar{N}_d)]^2 \tag{7.21}$$

$U_{\text{coh}}(\bar{N}_d)$ is the energy of the alloy specified by the average VCA state, which corresponds from Eq. (6.15) to an elemental transition metal with valence $\bar{N}_d$. The prefactor f_{bond}^2 reflects the loss of binding energy due to the decrease in bond order which results from the energy-level mismatch ΔE_{AB} in the binary alloy.

The heat of formation of the disordered AB alloy follows directly by comparing the equilibrium binding energy of the alloy, Eq. (7.20), with that of the elemental metals, Eq. (6.15). It is given by

$$\Delta H = \Delta H_{\text{integral}}^{\text{bond}} + \Delta H_{\text{order}}^{\text{bond}} \tag{7.22}$$

where

$$\Delta H_{\text{integral}}^{\text{bond}} = (3b^2/200a)(5 - \bar{N}_d)^2(\Delta N_d)^2 \tag{7.23}$$

and

$$\Delta H_{\text{order}}^{\text{bond}} = -(b^2/400a)[3\bar{N}_d(10 - \bar{N}_d) - 50](\Delta N_d)^2 \tag{7.24}$$

where a fourth-order contribution $(3b^2/3200a)(\Delta N_d)^4$ may be neglected for small ΔN_d.

The origin of these two terms in Eq. (7.22) may be found directly by using the equilibrium energy difference theorem.[115] This states that the total energy difference ΔU between two systems in equilibrium under a binding-energy law of the type in Eq. (7.9) is given to first order in $\Delta U/U_{\text{coh}}$ by the difference in their *bond* energies, provided that the bond lengths have been adjusted so that the two lattices have the same repulsive energy. Thus, the change in binding energy when two metals are brought together to form a binary alloy may be written

$$\Delta H = U_{\text{bond}}^{AB}(\hat{V}_{AB}) - \tfrac{1}{2}[U_{\text{bond}}^{A}(V_A) + U_{\text{bond}}^{B}(V_B)] \tag{7.25}$$

where the prepared volume $\hat{V}_{AB}$ of the AB alloy is such that

$$U_{\text{rep}}^{AB}(\hat{V}_{AB}) = \tfrac{1}{2}[U_{\text{rep}}^{A}(V_A) + U_{\text{rep}}^{B}(V_B)] \tag{7.26}$$

We must stress that the prepared volume $\hat{V}_{AB}$ is not the true equilibrium volume V_{AB} of the alloy, whose equilibrium lattice spacing is given by Eq. (7.19). It follows from Eqs. (7.11) and (7.17) that the repulsive energy

[115] D. G. Pettifor, *J. Phys. C* **19**, 285 (1986).

constraint Eq. (7.26) implies

$$W(\hat{V}_{AB}) = \{[W_A^2(V_A) + W_B^2(V_B)]/2\}^{1/2} \tag{7.27}$$

for the disordered AB alloy.

The heat of formation may, therefore, be considered within a two-step process. The first involves the change in the bond energy due to the change in the *bond integrals* which arises from the different bond lengths in the elemental metals and the prepared binary alloy. It has the form

$$\Delta H^{\text{bond}}_{\text{integral}} = \tfrac{1}{2}\{[W(\hat{V}_{AB}) - W_A(V_A)]N_A(10 - N_A)/20 + [W(\hat{V}_{AB}) - W_B(V_B)]N_B(10 - N_B)/20\} \tag{7.28}$$

where, from Eq. (7.17), $W(\hat{V}_{AB})$ is the elemental band width associated with the average of the t_{AA} and t_{BB} hopping parameters in the disordered alloy. Using the values for the band widths W_A and W_B given by Eq. (6.13), this bond-integral contribution is identical to Eq. (7.23) to second order in $(\Delta N_d)^2$. For this rectangular band model it is always repulsive.

The second step allows these elemental bands of width W to mix and form a common band as shown in the lower panel of Fig. 18, thereby changing the *bond order*. The change in energy is

$$\Delta H^{\text{bond}}_{\text{order}} = U^{AB}_{\text{bond}} - \tfrac{1}{2}[U_{\text{bond}}(N_A) + U_{\text{bond}}(N_B)] \tag{7.29}$$

where the bond energies are given by Eqs. (6.1) and (7.14) with W from Eq. (7.27). To second order in $(\Delta N_d)^2$

$$\Delta H^{\text{bond}}_{\text{order}} = \{-\tfrac{1}{80} + \tfrac{5}{24}[\bar{N}_d(10 - \bar{N}_d)]^{-1}\}W(\Delta N_d)^2 \tag{7.30}$$

The first contribution inside the curly brackets represents the change in the bond energy within the virtual crystal approximation, in which each atom in the alloy is characterized by $\bar{N}_d$ electrons. $\Delta H^{\text{bond}}_{\text{VCA}}$ is attractive since the bond energy is concave upwards as a function of N_d so that $U_{\text{bond}}(\bar{N}_d) < \tfrac{1}{2}[U_{\text{bond}}(N_A) + U_{\text{bond}}(N_B)]$. The second contribution in Eq. (7.20) represents the loss of bonding due to the atomic energy-level mismatch in the alloy. Since it follows from Eqs. (7.27) and (6.13) that

$$W = (3b^2/5a)\bar{N}_d(10 - \bar{N}_d) = W(\bar{N}_d) \tag{7.31}$$

to lowest order, Eqs. (7.30) and (7.24) are identical to second order in $(\Delta N_d)^2$.

Figure 20 shows the different contributions[116] to the normalized heats of formation $\Delta H/(\Delta N_d)^2$ for the case of the $4d$ transition metal alloys with

[116] The ΔH predictions of the rectangular bond model are symmetric about $\bar{N}_d = 5$. In practice, due to the size difference of the atoms the total alloy density of states will be skewed as shown in fig. 2 of Ref. 98. This leads to an additional contribution to ΔH, namely $-\tfrac{1}{48}(5 - \bar{N}_d)\Delta W \cdot \Delta N_d$ where $\Delta W = W_B(V_{AB}) - W_A(V_{AB})$.

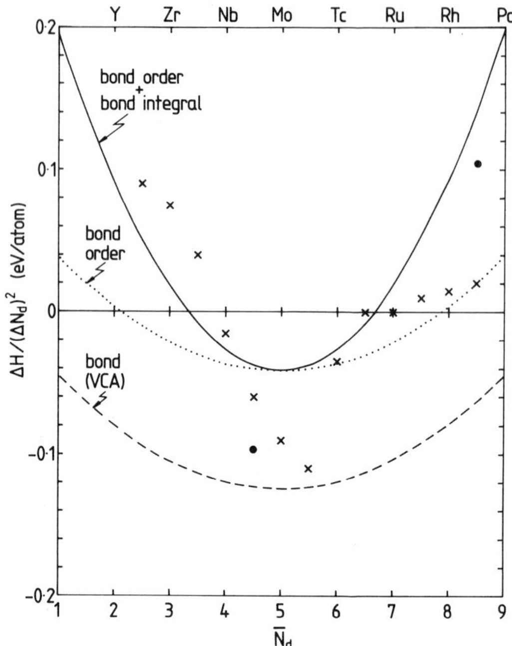

FIG. 20. The contributions to the normalized heats of formation $\Delta H/(\Delta N_d)^2$ for the case of 4*d* transition metal alloys with either $\Delta N_d = 1$ or 2. (×) and (●) represent the Miedema and experimental[92] values, respectively, for disordered alloys.

$\Delta N_d = 1$ or 2, taking $b^2/a = \frac{2}{3}$ eV. We see that the simple TB bond model reproduces the behavior[117] predicted by the Miedema scheme for *disordered* alloys. The theory supports the earlier suggestion by Brewer[118] that the most stable transition metal alloys would comprise elements from groups at opposite ends of the transition metal series, such as Y and Pd. These groups have very few bonding electrons, since they have nearly empty or full *d* shells. Mixing these elements together would result in a dramatic increase in the bond order, since the electrons would be shared in the bonding states of the alloy corresponding to a half-full band. The behavior of the bond-order contribution cannot be reproduced by the ionic rigid-band model. The former contribution changes sign and is repulsive for nearly empty or full *d* bands. The latter is always attractive.[10,11,110] The difference accounts for the oscillation

[117] The rectangular band density of states has also been shown in Ref. 98 to reproduce the behaviour of the TB-CPA disordered binary alloy calculations of J. Van der Rest, F. Gautier, and F. Brouers, *J. Phys. F* **5**, 2283 (1975).

[118] L. Brewer, *Science* **161**, 115 (1968); L. Brewer and P. R. Wengert, *Metall. Trans.* **4**, 83 (1973).

in the error of the ionic model which Williams *et al.*[28] observed across the $4d$ transition series in Fig. 5.

The ordering energy cannot be discussed within the rectangular bond model since it fits only the lowest three moments μ_0^{AB}, μ_1^{AB}, and μ_2^{AB} of the total alloy density of states, whereas the ordered and disordered states differ at the fourth moment μ_4^{AB} when the bond integrals are site independent.[92,95] Ducastelle and Gautier[119] have developed a perturbation theory with respect to the disordered state described by the coherent potential approximation (CPA).[120] They have shown that the *second*-order theory accurately describes the change in the electronic energy on ordering, so that the ordering energy may be written explicitly in terms of effective pair potentials φ_{ij} between atoms i and j. For the case where $\Delta E/W$ is small

$$\varphi_{ij} = \chi_{ij}(\Delta E)^2 \tag{7.32}$$

where the response function χ_{ij} is defined by

$$\chi_{ij} = \frac{1}{\pi} \lim_{n \to 0+} \operatorname{Im} \int^{E_F} G_{ij}^2(E + i\eta)\, dE \tag{7.33}$$

G_{ij} are the off-diagonal matrix elements of the Green function for the virtual crystal in which all atoms are the same and characterized by the average valence $\bar{N}_d$.

The nature of these response functions χ_{ij} may be illustrated by considering a simple cubic lattice of s orbitals with nearest-neighbor hopping integrals $ss\sigma = h$. The Green function linking nearest neighbors G_{100} is given explicitly in terms of the site-diagonal Green function G_{000} by

$$6hG_{100} = EG_{000} - 1 \tag{7.34}$$

where the atomic energy level E_s has been taken as energy zero. Equation (7.34) follows because

$$\begin{aligned} EG_{000} - 1 &= \langle \psi_0 | E(E - \mathscr{H})^{-1} - 1 | \psi_0 \rangle \\ &= \sum_{\mathbf{R}} \langle \psi_0 | \mathscr{H} | \psi_{\mathbf{R}} \rangle \langle \psi_{\mathbf{R}} | (E - \mathscr{H})^{-1} | \psi_0 \rangle \\ &= 6hG_{100} \end{aligned} \tag{7.35}$$

Thus, the imaginary part of $-G_{100}(E)$, which enters the bond order in Eq. (6.7), is negative for bonding states $E < 0$ but positive for antibonding states with $E > 0$.

[119] F. Ducastelle and F. Gautier, *J. Phys. F* **6**, 2039 (1976); G. Treglia, F. Ducastelle, and F. Gautier, *ibid.* **8**, 1437 (1978).

[120] B. Velicky, S. Kirkpatrick, and H. Ehrenreich, *Phys. Rev.* **175**, 747 (1968).

Explicit expressions for more distant Green functions cannot be obtained by the above procedure, since the different Green functions cannot be uncoupled from the equations generated. Instead the Lanczos recursion algorithm[121] may be used[122] to expand the real-space Green functions G_{ij} in terms of the Green functions $\hat{G}_{0n}$ which link the nth recursive orbital $\hat{\psi}_n$ to the orbital at the origin $\hat{\psi}_0 = \psi_0$. In particular,

$$G_{100} = \sqrt{1/6}\,\hat{G}_{01} \tag{7.36a}$$

$$G_{110} = \sqrt{2/27}\,\hat{G}_{02} - \sqrt{32/176715}\,\hat{G}_{04} + \cdots \tag{7.36b}$$

$$G_{111} = \sqrt{6/85}\,\hat{G}_{03} - \sqrt{54/6761}\,\hat{G}_{05} + \cdots \tag{7.36c}$$

where the imaginary parts of the first three recursive Green functions are given by

$$\begin{aligned} \operatorname{Im}\hat{G}_{0n} &= 6^{-1/2}\varepsilon \operatorname{Im} G_{000}, && n = 1 \\ &= 54^{-1/2}(\varepsilon^2 - 6)\operatorname{Im} G_{000}, && n = 2 \\ &= 510^{-1/2}\varepsilon(\varepsilon^2 - 15)\operatorname{Im} G_{000}, && n = 3 \end{aligned} \tag{7.37}$$

with $\varepsilon = E/h$. The first two terms in Eqs. (7.36b) and (7.36c) represent the exact[123] Green functions G_{110} and G_{111} very well except near the Van Hove singularities at $\varepsilon = \pm 2$, which are smeared out. We see from Eqs. (7.36) and (7.37) that, whereas Im G_{100} changes sign only once through the band, Im G_{110} changes twice and Im G_{111} changes three times. These nodes are expected from the behavior of the moments.[124,125]

Figure 21 shows the first-, second-, and third-nearest-neighbor response functions χ_1, χ_2, χ_3 for the simple cubic lattice.[126] They were evaluated analytically[122] by keeping the first term for the Green function in Eq. (7.36) and using Tonegawa's[127] analytic expression for the site-diagonal Green function G_{000}. The effective pair potentials with respect to the disordered lattice $\varphi_m = \chi_m(\Delta E)^2$ are seen to oscillate with band filling and with distance. The dominant nearest-neighbor interaction φ_1 behaves as expected from the variation of ΔH in Fig. 20. The ordering energy, therefore, will be negative for

[121] C. Lanczos, *J. Res. Natl. Bur. Stand.* **45,** 255 (1950); R. Haydock, V. Heine, and M. J. Kelly, *J. Phys. C* **8,** 2591 (1975); R. Haydock, in Ref. 79.

[122] D. G. Pettifor, *in* "Atomistics of Fracture" (R. Latanision, ed.), p. 293. Plenum, New York, 1983; also unpublished.

[123] T. Moriya and H. Hasegawa, *J. Phys. Soc. Jpn.* **48,** 1490 (1980).

[124] F. Ducastelle and F. Cyrot-Lackmann, *J. Phys. Chem. Solids* **32,** 285 (1971).

[125] V. Heine and J. H. Samson, *J. Phys. F* **13,** 2155 (1983).

[126] The Green functions and response functions for fcc and bcc lattices are discussed by G. Treglia and A. Bieber, *J. Phys. (Orsay, Fr.)* **45,** 283 (1984) and by Ref. 83 respectively.

[127] T. Tonegawa, *Prog. Theor. Phys.* **51,** 1293 (1974).

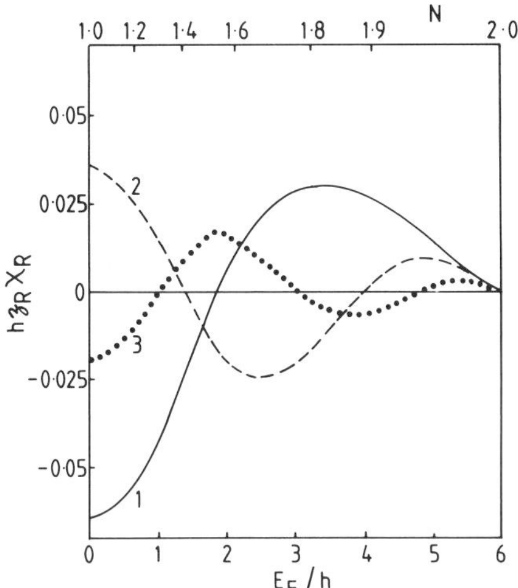

FIG. 21. The first-, second-, and third-nearest-neighbor response functions χ_1, χ_2, and χ_3 for the simple cubic lattice[122] as a function of the band filling N or the normalized Fermi energy E_F/h, where h is the nearest-neighbor hopping integral between s orbitals. z_R is the coordination number of the Rth-nearest neighbor.

a nearly half-full band, corresponding to unlike nearest-neighbor pairs being more stable than like, whereas it will be positive for nearly empty or full bands, corresponding to the tendency for unlike atoms to segregate in order to have only like nearest neighbors.[92] We should note that the weaker, more distant interactions φ_2 and φ_3 play a crucial role in distinguishing between different ordered structures which have the same type of first-nearest-neighbor atoms with respect to a given lattice.[128] For example, φ_2 is positive in Fig. 21 for a nearly half-full band. This is consistent with the $L1_2$ (Cu_3Au) structure with all *like* second-nearest neighbors being preferred over the DO_{22} (Al_3Ti) structure with some *unlike* second-nearest neighbors for transition metal fcc alloys with an average band filling $\bar{N}_d \simeq 5$.[95,129,130]

Finally, Fig. 22 demonstrates the importance of the structural energy contribution to the heats of formation of *ordered* transition metal alloys.[12] Although the positive to negative to positive trend in ΔH with average band filling $\bar{N}_d$ is shown by both the fcc and bcc lattices, the heats of formation corresponding to the bcc lattice are much more stable than those of the fcc lattice around $\bar{N}_d = 4.5$, whereas the fcc lattice has the more stable values around $\bar{N}_d = 6.5$. The origin of this structural dependence can be seen from the

[128] M. Kaburagi and J. Kanamori, *Prog. Theor. Phys.* **54,** 30 (1975).

[129] A. Bieber and F. Gautier, *Solid State Commun.* **38,** 1219 (1981).

[130] A. Bieber, F. Ducastelle, F. Gautier, G. Treglia, and P. Turchi, *Solid State Commun.* **45,** 585 (1983).

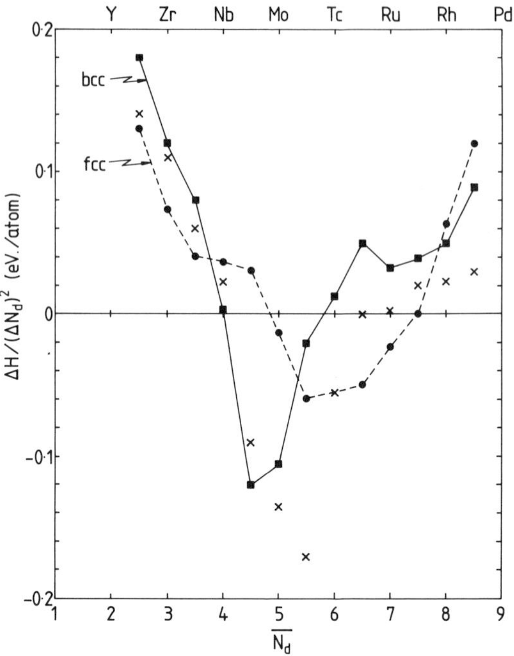

FIG. 22. The normalized LDF heats of formation $\Delta H/(\Delta N_d)^2$ with respect to the bcc (■) and fcc (●) lattices for the case of ordered 4*d* transition metal compounds with either $\Delta N_d = 1$ or 2. (×) give the Miedema values for ordered alloys (after Williams *et al.*[12]).

behavior of the structural energy of the elemental *d* bands in Fig. 17. The contribution to $\Delta H^{\text{bond}}_{\text{order}}$ due to crystal structure can be estimated by writing

$$\begin{aligned}\Delta H_{\text{struct}} &= U_{\text{struct}}(\bar{N}_d) - \tfrac{1}{2}[U_{\text{struct}}(N_A) + U_{\text{struct}}(N_B)] \\ &= -\tfrac{1}{8}\left(\frac{d^2 U_{\text{struct}}}{dN_d^2}\right)_{\bar{N}_d} (\Delta N_d)^2 \end{aligned} \tag{7.38}$$

Assuming that the curvature in the bcc and fcc structural energies around $\bar{N}_d = 5$ can be modeled by $U_{\text{struct}} = A\sin(3\pi N_d/10)$, we have

$$(\Delta H_{\text{bcc}} - \Delta H_{\text{fcc}})/(\Delta N_d)^2 \simeq \tfrac{1}{10}(U_{\text{bcc}} - U_{\text{fcc}})_{\bar{N}_d} \tag{7.39}$$

The oscillatory behavior of the bcc and fcc heats of formation with respect to each other in Fig. 22, therefore, simply reflects the fcc → bcc → fcc → bcc trend in stability shown by the structural energies in Fig. 17. Moreover, Eq. (7.39) predicts[131] a value of approximately -0.1 eV/atom for ΔH_{bcc} −

[131] The skewing of the fcc curve in fig. 22 with respect to $\bar{N}_d = 5$ is not a structural effect (c.f. fig. 17) but is due to atomic size differences which skew the *d* bond energy (see Ref. 116).

ΔH_{fcc} for $\bar{N}_d = 4.5$, which is consistent with the energy difference observed in Fig. 22. The Miedema values in Fig. 22 implicitly include these structural effects because the scheme has been fitted to reproduce the observed signs of the heats of formation. The predicted LDF values for a given lattice[12] should not be compared directly with the empirical values of ΔH unless both elemental metals and the ordered alloy have this lattice as their ground state. Otherwise, an extra contribution must be included which reflects the change in lattice.[37,132,133]

III. Significance of the Miedema Parameters

8. Significance of the Miedema Parameters and Conclusion

The predictions of the Miedema scheme[7,8] for the signs of the heat of formation of binary AB alloys are shown in Fig. 23. This two-dimensional plot $(\mathscr{M}_A, \mathscr{M}_B)$ is obtained by assigning a unique relative ordering number $\mathscr{M}$ to every element in the periodic table. This so-called Mendeleev number $\mathscr{M}$ was defined by running a one-dimensional string through the two-dimensional periodic table, thereby arranging the elements in sequential order.[134] It was introduced to display the structural information of ordered binary compounds with a given stoichiometry A_mB_n within a single two-dimensional structure map.[37,115,135]

The alloys with positive and negative heats of formation tend to be well segregated in Fig. 23. For ease of reference the alkali metals, transition metals, and Groups IIIB–VB elements in the figure have been labeled *s*, *d*, and *p*, respectively, even though electrons with other angular momentum character may be important in their bonding (e.g., elemental Si forms sp^3 hybrids). The Miedema predictions are least successful for the *sp*–*sp* alloys, where incorrect assignments are occasionally made.[136] For example, the alloys of Si with the alkali elements in the top right-hand corner have the wrong sign. The islands of positive *pd* heats of formation contain the 5*p* and 6*p* elements In and Tl, Sn and Pb, or Sb and Bi. The small islands of positive *dd* heats of formation contain either the magnetic constituents Fe and Mn[100] or the actinide elements U and Pu.

The *dd* region in Fig. 23 is seen to comprise a wide domain of negative heats of formation which is centered on an average *d* band filling of $\bar{N}_d = 5$. This empirical behavior is predicted by the theoretical TB bond model since the

[132] J. F. Janak and A. R. Williams, unpublished (1978); see Fig. 2 of Ref. 7.

[133] H. L. Skriver, *Phys. Rev. B: Condens. Matter* [3] **31,** 1909 (1985).

[134] D. G. Pettifor, *New Sci.* **110,** No. 1510, 48 (1986).

[135] D. G. Pettifor, *Solid State Commun.* **51,** 31 (1984).

[136] Table IV of Ref. 6.

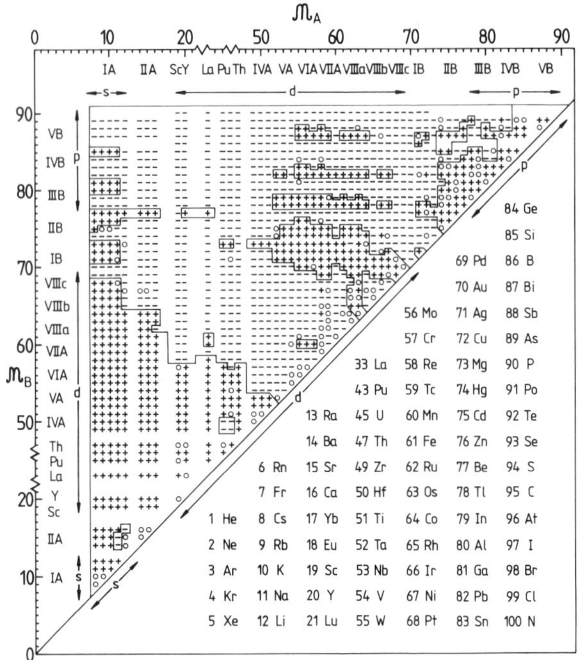

FIG. 23. The sign of the heats of formation predicted by the Miedema scheme[7,8] for binary AB alloys. Each alloy is characterized by some point $(\mathscr{m}_A, \mathscr{m}_B)$, where the values of the Mendeleev number $\mathscr{m}$ for the constituent elements[134] are given explicitly in the figure. (Oxygen, fluorine, and hydrogen take the values 101, 102, and 103, respectively.) (○) correspond to alloys with predicted heats of formation such that $|\Delta H| \leq 0.01$ eV/atom. The lines have been drawn to segregate the plusses and minuses.

bond-order contribution is negative in this region (see Fig. 20).[137] In fact, the bond-order contribution Eq. (7.24) can be made to look like an attractive electronegativity term[102]—$P(\Delta X_d)^2$ in the range $|\bar{N}_d - 5| < 5/\sqrt{3}$, where it is negative. If X_d is represented by a cubic function of N_d, then the coefficients may be found[37] by constraining $-P(\Delta X_d)^2$ to vanish for $\bar{N}_d = 5 \pm 5\sqrt{3}$ and to take the value $-\frac{1}{24}(\Delta N_d)^2$ to second order for a half-full band. (This latter value follows from Eq. (7.24) with $b^2/a = \frac{2}{3}$eV and $\bar{N}_d = 5$.) We find in eV/atom that

$$\Delta H^{\text{bond}}_{\text{order}} \approx -\tfrac{2}{3}(\Delta X_d)^2, \qquad \text{for } |\bar{N}_d - 5| \leq 5/\sqrt{3} \tag{8.1}$$

where

$$X_d = 1.8 - \tfrac{1}{2}N_d[1 - \tfrac{1}{50}N_d(15 - N_d)] \tag{8.2}$$

[137] The small cluster of alloys with negative heats of formation centred on CoPt and NiPt can be accounted for within the TB band model by fitting band widths for the 3d, 4d, and 5d series (see Ref. 109).

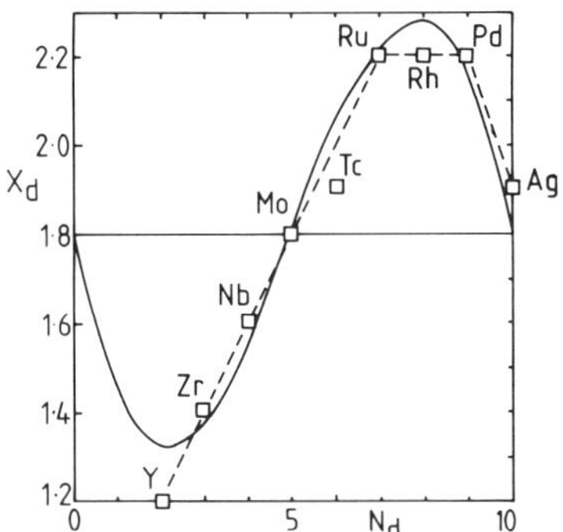

FIG. 24. The d-bond electronegativity[37] X_d compared to Pauling's[102] values (□) for the 4d series.

The constant 1.8 was chosen to reproduce Pauling's electronegativity value for Mo.[102] We see from Fig. 24 that this *theoretical* d-bond electronegativity scale X_d reproduces the Pauling[102] values across the 4d series. This is perhaps not totally unexpected, as Pauling's scale was chosen to reproduce changes in *empirical* homonuclear and heteronuclear bond energies.

Thus, the bond-order contribution can be *approximated* in the region where ΔH is negative by an electronegativity-type term $-P(\Delta X_d)^2$. Moreover, the electronegativity difference ΔX_d is a measure of the flow of charge from one *unrelaxed* atom to another, since it follows from Eq. (8.2) that

$$q_d = -\Delta X_d = \tfrac{1}{2}\Delta N_d - \tfrac{3}{100}\bar{N}_d(10 - \bar{N}_d)\Delta N_d \tag{8.3}$$

This is just the charge transfer predicted by the common-band model before the free-atom charge clouds are allowed to relax, namely Eq. (7.6), because for the 4d series[107] $\Delta E = E_B^0 - E_A^0 = -\Delta N_d$ eV and $W_{AB} \approx 10$ eV, so that $\Delta E^0/W_{AB} \approx -\frac{1}{10}\Delta N_d$. However, we must stress that the bond-order contribution in transition metals is not ionic in origin but reflects directly the change in the quantum-mechanical bond order between *neutral* atoms. Equation (8.1) is only an approximate fit to $\Delta H^{\text{bond}}_{\text{order}}$ in the region where it is negative. It clearly cannot reproduce the positive behavior in Fig. 20 for $|\bar{N}_d - 5| > 5/\sqrt{3}$.

The identification of the Miedema parameter φ^* with the work function or chemical potential of the elemental metals finds no theoretical support within quantum mechanics. Both the bond-order contribution, Eq. (7.24), and the common-band charge flow between unrelaxed atoms, Eq. (7.6), exist even for the case when the elemental chemical potentials are the same.[98,107] The flow of electrons and the resultant change in bonding is not driven by the difference in the elemental work functions because *all* the electrons throughout the band respond on alloying and not just those in the vicinity of the Fermi level. This

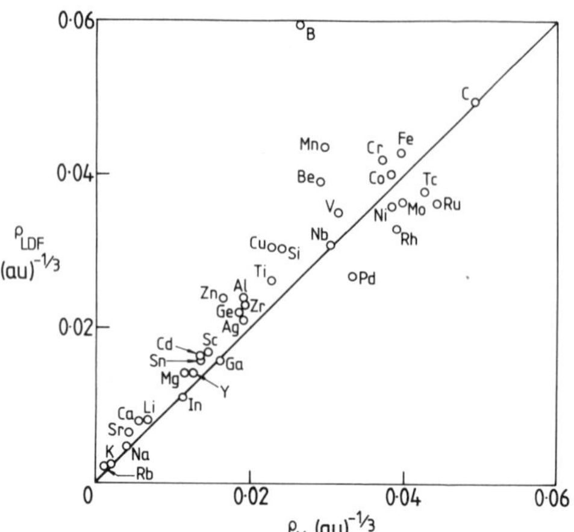

FIG. 25. The similarity between the theoretical LDF values of the Wigner–Seitz electron density and the Miedema coordinate ρ_M in atomic units (after Williams et al.[28]).

can be seen by comparing, in Fig. 18, the skewed partial density of states $n_A(E)$ in the AB alloy with the rectangular density of states $n_A(E)$ in the pure metal.

The Miedema parameters φ^* and $\rho_M^{1/3}$ have been identified[1–8] with the experimental work functions and the cube root of the theoretical charge densities, respectively, because they display similar trends across the periodic table. For example, Fig. 25 shows the approximately linear variation between the LDF values of the charge density at the Wigner–Seitz radius[28] and the Miedema values.

However, this identification of the Miedema parameters with physical coordinates must be treated with caution, since the heats of formation are sensitive even to very small changes in the values of φ^* and $\rho_M^{1/3}$. Consider, for example, the alloys of iron with s and d elements, for which the simplest form of Miedema's scheme, namely Eq. (2.1), is applicable. ΔH will then be positive or negative depending on whether $\Delta\varphi^*$ is less than or greater than $\sqrt{Q/P}\,\Delta\rho_M^{1/3}$. This is illustrated in Fig. 26, where the two lines with slopes $\pm\sqrt{Q/P}$ separate the regions of positive and negative heats of formation.[6] As a first approximation φ^* and $\rho_M^{1/3}$ vary linearly across the figure with the slope $\sqrt{Q/P}$. Therefore, as Hodges has stressed,[138] it is the *deviation* from this linearity which is responsible for the observed heats of formation. Since the deviations in Fig. 26 are only about 10% of the absolute values, the identification of the Miedema parameters φ^* and $\rho_M^{1/3}$ with the work function and charge density, respectively, can be misleading. For example, the sign of the FePd, FeRh, and FeCa heats of formation would be changed by taking

[138] C. H. Hodges, *J. Phys. F* **7**, 1687 (1977).

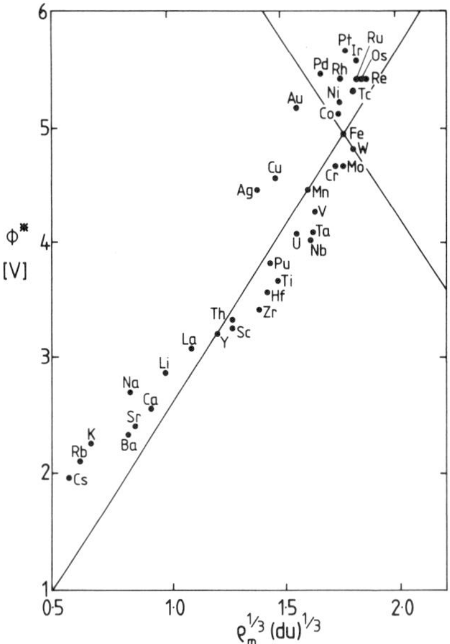

FIG. 26. The graphical construction which determines whether the heats of formation of iron alloys with *s* and *d* elements are either positive or negative. The straight lines have slopes $\pm\sqrt{Q/P}$, respectively, and separate the regions of positive and negative heats of formation (after Miedema *et al.*[6]).

the theoretical values of the charge density $\rho_{\mathrm{LDF}}^{1/3}$.[28] Similarly, the sign of the FeNb, FeTa, and FeZr heats of formation would be changed by taking the experimental values of the work function φ.[139]

The Miedema parameters φ^* and $\rho_{\mathrm{M}}^{1/3}$ are phenomenological coordinates[140] which have been adjusted to reproduce the empirical behavior observed in Fig. 23. It was found by trial and error[141] that the charge density rather than the atomic volume, for example, gave better separation at the first iteration, so that the required modifications of the initial input charge-density parameters would be the smaller. However, we have seen for the case of *sd* and *dd* alloys that the modifications are still too large to identify $\Delta\varphi^*$ and $\Delta\rho_{\mathrm{M}}^{1/3}$ with differences in the work function and charge density, respectively.

In addition, the *pd* domain in Fig. 23 could be reproduced only by introducing a second attractive contribution $-R_pR_d$ and assigning a metallization energy to B, C, N, Si, Ge, N, and P. This latter energy represents the energy required to transform the observed elemental ground state to the

[139] H. B. Michaelson, *J. Appl. Phys.* **48,** 4729 (1977).

[140] D. G. Pettifor and R. Podloucky, *Phys. Rev. Lett.* **55,** 261 (1985).

[141] See §10 of Ref. 5.

metallic state. The Group IVB elements C, Si, and Ge were given[8] the metallization energies of 1.00, 0.34, and 0.25 eV/atom, which are smaller than the theoretically predicted tetrahedral to close-packed values of 4.08, 0.53, and 0.44 eV/atom, respectively.[142] The metallic reference state, therefore, does not correspond to the close-packed fcc or bcc lattices that were used by Gelatt *et al.*[14] in their LDF study of the heats of formation of 4*d* transition metals with 2*p* and 3*p* elements. On the other hand, the TB bond model of Eq. (7.9) can explain the observed trends in structural behavior across the *pd* domain in Fig. 23, predicting the relative stabilities of the NaCl, CsCl, NiAs, MnP, and boride lattices.[96]

In conclusion, therefore, the Miedema parameters φ^*, $\rho_M^{1/3}$, R_p, and R_d should be regarded as phenomenological parameters because quantum mechanics provides no justification for the physical basis of the "macroscopic atom" picture.[1–8] The negative contribution to the heats of formation in metallic alloys is not driven by the difference in the elemental work functions or chemical potentials as the rigid-band ionic model assumes. Although there is a redistribution of charge on the formation of the alloy, the constituent atoms may be taken to be perfectly screened by the metallic environment and, therefore, to remain charge neutral. The heat of formation arises from the creation of a common band within the alloy. For transition metal alloys this leads to an attractive bond-order contribution for nearly half-full bands, which is consistent with the earlier arguments of Brewer[118] and which behaves approximately as Pauling's[102] empirical electronegativity term $-P(\Delta X)^2$. The repulsive contribution results from the change in the bond integrals due to changes in bond lengths on alloy formation.

However, even though the physical concepts underpinning the Miedema model are not supported by quantum mechanics, the importance of the scheme rests in its compatibility with known experimental data and its ease of applicability to a multitude of problems in alloy cohesion.[1–8,21–23,143] The very simple analytic quantum-mechanical models discussed in this article can elucidate the origin of bonding in metals and alloys, but they will never replace phenomenological schemes for interpolating accurate heats of formation.

Acknowledgments

I should like to thank Mike Finnis and Adrian Sutton for many stimulating conversations about the tight-binding method.

[142] M. T. Yin and M. L. Cohen, *Phys. Rev. Lett.* **45,** 1004 (1980); *Solid State Commun.* **38,** 625 (1981); *Phys. Rev. Lett.* **50,** 2006 (1983); M. T. Yin, Ph.D. Thesis, University of California, Berkeley (1982).

[143] F. R. de Boer, R. Boom, W. C. M. Mattens, A. R. Miedema, and A. K. Niessen, "Cohesion and Structure" (F. R. de Boer and D. G. Pettifor, eds.), Vol. 1. North-Holland, Amsterdam, 1987.

Electronic Shell Structure and Metal Clusters

WALT A. DE HEER AND W. D. KNIGHT

Department of Physics, University of California, Berkeley, California

M. Y. CHOU AND MARVIN L. COHEN

Department of Physics, University of California, and Materials and Molecular Research Division, Lawrence Berkeley Laboratory, Berkeley, California

I. Development of Cluster Physics

... these atoms exist in the unbounded void, being entirely separate from each other... they differ in shape, size, position, and arrangement... they move through the void, overtaking each other and colliding... because they fit together in shape, size, position, or arrangement, they become interlocked and so remain in association... this is the origin of composite bodies... Leukippos

1. Historical Perspective

In the ancient world, atoms were widely discussed by Leukippos and Lucretius long before atomic structure was understood. Understanding followed a long period of study and speculation, some of which was surprisingly close to what we now believe to be the truth. The Greek studies concentrated primarily on the shapes and arrangements of associated atoms, and it was only after clues to the internal electronic structure were elaborated that modern atomic structure developed. The idea of association of atoms in molecules also developed over a long period, culminating with the electronic band structure of solids. Now, the atom by atom building of sophisticated microstructures is creating objects of both fundamental interest and technical importance.

Much is to be learned about how a crystal grows by the aggregation of atoms to form the ultimate extended solid structure. The study of the physical properties of states intermediate between the atom and the solid is called cluster physics. Lacking a precise definition, we shall say that a cluster is a stable group of a few or a few hundred identical atoms or molecules. The present review article mainly treats metal clusters.

Successful views of the structure of matter have been based on the idea of stability. Among atoms there are the nonreactive noble gases and noble metals. Closely related to the reactivities of atoms are the ionization potentials, which show a regular periodic progression with high values for the noble gases. Actually, we can distinguish among bulk metals, semiconductors, and insulators according to the respective ionization potentials of the constituent.

Atomic theory depends on the application of angular momentum conditions in the Coulomb field. The independent-particle shell model of the nucleus was based on the application of angular momentum conditions to a system bound in a spherical harmonic oscillator or square-well potential. Application of the angular momentum conditions produced shell closings which agreed nicely with the observed high-stability "magic number" nuclei. As we shall see in this and later parts, the ideas of shell structure that have worked well in atomic and nuclear physics are applicable to metal clusters.

2. Recent Developments

The first three meetings of the International Conference on Small Particles and Inorganic Clusters at Lyons (1976), Lausanne (1980), and Berlin (1984) marked the extremely rapid development of common interest among a wide variety of physical scientists. The proceedings of these conferences[1–3] collectively contain a record of the rapid growth of the field during these years.

[1] *J. Phys. (Orsay, Fr.)* **38,** C2 (1977).
[2] *Surf. Sci.* **106** (1981).
[3] *Surf. Sci.* **156** (1985).

A compelling stimulus in these studies has been the desire to understand how the extended solid develops from the growing cluster aggregates. Much of the early work involved experiments on the optical properties of metal clusters grown in a dielectric matrix.[4] Although the results contained much valuable information, it proved difficult to produce controlled size distributions. The application of molecular beam techniques[5] was an important advance because the production and analysis of free clusters avoids problems of matrix interactions and permits the selection of clusters of identical mass. It was found[5] that the trend in ionization potentials of small sodium clusters was toward the bulk work function. As for other trends, extended x-ray absorption fine structure (EXAFS) results showed contraction of the lattice constant[6] for copper clusters on a carbon support. Similar results have been obtained for gold[7] and also for silver.[8] The growth processes were studied in preliminary ways in terms of elementary kinetics, but available data were limited. As for structure, work has been reported[9,10] in the 2–4 nm diameter range for gold crystals, suggesting that when such clusters grow to contain a few hundred atoms, crystallinity has developed. Electron diffraction experiments on metal clusters in beams[11] reveal crystalline structures in Pb and In in the size range of $N \sim 2000$ atoms with variation in lattice constant and axial ratio as a function of size. This evidence for the early development of crystals in cluster growth represents a small fraction of what we would like to know about the "cluster state," which lies between the atom and the extended solid.

One of the earliest attempts to study clusters theoretically[12] visualized the growth of a crystal by addition of successive atoms to a bcc lattice and gave a perturbation theory of electronic energies and equilibrium lattice spacings for up to eight sodium atoms and up to five copper atoms. A more recent calculation[13] carries the procedures to 79 copper atoms. In recent years, increasingly elaborate theoretical methods have been employed to explain the

[4] U. Kreibig, *Z. Phys.* **234,** 307 (1970).

[5] E. J. Robbins, R. E. Leckenby, and P. Willis, *Adv. Phys.* **16,** 739 (1967).

[6] G. Apai, J. F. Hamilton, J. Stohr, and A. Thompson, *Phys. Rev. Lett.* **43,** 165 (1979).

[7] A. Balerna, E. Bernieri, P. Picozzi, A. Reale, S. Santucci, E. Burattini, and S. Mobilio, *Surf. Sci.* **156,** 206 (1985).

[8] P. A. Montano, H. Purdum, G. K. Shenoy, T. I. Morrison, and W. Schulze, *Surf. Sci.* **156,** 230 (1985).

[9] L. D. Marks and D. J. Smith, *J. Cryst. Growth* **54,** 425 (1981).

[10] L. R. Wallenberg, J.-O. Bovin, and G. Schmid, *Surf. Sci.* **156,** 256 (1985).

[11] A. Yokozeki and G. Stein, *J. Appl. Phys.* **49,** 2224 (1978).

[12] H. S. Taylor, H. Eyring, and A. Sherman, *J. Chem. Phys.* **1,** 68 (1933).

[13] B. Delley, D. E. Ellis, A. J. Freeman, E. J. Barends, and D. Post, *Phys. Rev. B: Condens. Matter* [3] **27,** 2132 (1983).

energies and properties of clusters. The statistical model of Kubo[14] stimulated a number of further investigations[15] of small particles, and among other things suggested the idea that a small particle or cluster may be considered a large molecule with only a small number of degeneracies.

3. Types of Clusters

The packing of spheres under the influence of pair interactions has for some time been an attractive approach to the analysis of rare gases and molecular crystals. Models for this sort of geometrical packing have been used with some success,[16] and experiments on both neutral[17] and ionized[18] clusters give evidence for systematic features in the abundance spectra of xenon and argon corresponding to especially stable icosahedral or cubooctahedral structures. It has been proposed[19] that clusters of silicon may be based on a primitive form of the crystal structure of the extended solid governed by the covalent bonding. Beyond the well-known solvated complexes which have long been called clusters, aggregations of molecules such as NaCl, H_2O, SF_6, GaAs, C_2H_6, and many others are known.

The observation of prominent features[20] in the abundance spectra of sodium clusters prompted the examination of the data as a possible example of shell structure. The resulting analysis of electronic structure of the monovalent alkali metals agreed in most respects with the experimental results. In view of the prevailing tendency to analyze clusters in terms of sphere packing models and Lennard–Jones potentials, the experimental observation and explanation of electronic shell structure for the alkalis stimulated considerable experimental and theoretical activity along these lines, and, at present, the idea is being tested as a general organizing principle for the study of metallic clusters.

In the following Section II, we shall set forth the basic electronic shell theory for metal clusters along with some expectations for the limits of its applicability.

[14] R. Kubo, *J. Phys. Soc. Jpn.* **17,** 975 (1962).
[15] R. Denton, B. Möhlschlegel, and D. J. Scalapino, *Phys. Rev. B: Solid State* [3] **7,** 3589 (1973).
[16] M. R. Hoare and P. Pal, *J. Cryst. Growth* **17,** 77 (1972).
[17] O. Echt, K. Sattler, and E. Recknagel, *Phys. Rev. Lett.* **47,** 1121 (1981).
[18] J. A. Harris, R. S. Kidwell, and J. A. Northby, *Phys. Rev. Lett.* **53,** 2390 (1985).
[19] L. Bloomfield, R. Freeman, and W. L. Brown, *Phys. Rev. Lett.* **54,** 2246 (1985).
[20] W. D. Knight, K. Clemenger, W. A. de Heer, W. A. Saunders, M. Y. Chou, and M. L. Cohen, *Phys. Rev. Lett.* **52,** 2141 (1984).

II. Basic Electronic Shell Theory of Metal Clusters

4. Introduction

In subatomic physics, the concept of shell structure provides explanations for various properties of atoms and nuclei. The shells are a direct consequence of quantum mechanics for these systems, and, in principle, shell structures could be present in other physical systems having similar physical structure. Recent experimental and theoretical considerations, which will be discussed later, have shown that an electronic shell theory is appropriate for small clusters of atoms of certain metals. The underlying structure of the theory for these clusters resembles the older well-established shell theories for atoms and nuclei. In all cases, the system consists of interacting fermions confined to lengths ranging from $\sim 10^{-12}$ cm in nuclei, $\sim 10^{-8}$ cm in atoms, to $\sim 10^{-7}$–10^{-6} cm in the clusters under consideration. Although the fundamental interaction mechanisms have different physical origins, the major common feature is that these quantum systems consist of fermions which can be described reasonably well by an effective one-particle spherical potential, and this results in shell structure associated with the fermion states.

In the atomic case, shell structure results from spherical symmetry and from the properties of the electromagnetic interactions between the electrons and the nucleus. Since the nucleus acts almost like a point charge as far as electronic properties are concerned, the Coulomb potential generated by the nuclear charge is taken to be spherically symmetric. In the absence of external fields, the spherical symmetry remains even after the mutual interactions between electrons are included in the effective net potential. As a result of this symmetry, eigenstates of the Hamiltonian can be characterized by the orbital angular momentum l, and the electronic system can then be viewed as a series of shells with $2l + 1$ degeneracy (neglecting spins). These angular momentum shells determine the structure of the periodic table.[21] In this mode, electrons may be described as moving in an effective, spherical one-particle potential.

The shell structure in nuclei is similar.[22] Here the fermions are strongly interacting nucleons with complex interactions that are not fully understood. Despite the complexity, it is known that nuclei with certain "magic numbers" of nucleons are unusually stable.[23] To explain this behavior, a Fermi gas model of the nucleus was proposed. In this model, it is assumed that each

[21] See, for example, R. M. Eisberg, "Fundamentals of Modern Physics," p. 407. Wiley, New York, 1967.

[22] See, for example, H. Frauenfelder and E. M. Henley, "Subatomic Physics," Chapter 15. Prentice-Hall, Englewood Cliffs, New Jersey, 1974.

[23] J. H. Bartlett, *Phys. Rev.* **41,** 370 (1932); W. M. Elsasser, *J. Phys. Radium* [7] **4,** 549 (1933); **5,** 625 (1934).

nucleon moves in a net effective nuclear potential and that the Pauli principle strongly suppresses collisions between nucleons.[24] The exact form of the potential has not been derived from first principles; however, a successful model consists of an attractive spherically symmetric potential function in the form of a rounded-edge square well with strong inverted spin-orbit coupling.[25] When the Schrödinger equation is solved, the eigenstates (shells) are occupied in ascending order.

The nuclear shell model predicts successfully the "magic numbers," nuclear spins, and parities of nuclei.[26] Further modification took into account the deformation associated with the correlated or collective motion of the nucleons (Nilsson model[27]) and predicted the magnetic dipole and electric quadrupole moments.[28] The success of the nuclear shell model confirms the appropriateness of the smooth one-particle potential approximations for a system of nuclei. Although this approach involves a simplified model for a complex problem, much of the basic physics of the system is reflected in this model.

It is impressive that a shell-model theory applies for both atoms and nuclei despite the differences in the forces involved. A similar situation also exists for certain metal clusters where it is found experimentally that the cluster size dependence on stability, ionization potentials, static polarizability, and other properties exhibits distinct shell structure. These were first observed for sodium and potassium clusters.[20,29–31]

Recently, shell structure has also been observed in copper, silver, and gold[32] and in cadmium and zinc.[33] These observations provide insight into determining the important factors in describing these systems. Unlike nuclei, the fundamental interactions in clusters are known. The electron–electron

[24] V. F. Weisskopf, *Helv. Phys. Acta* **23,** 187 (1950).

[25] M. G. Mayer, *Phys. Rev.* **74,** 235 (1948); **75,** 1969 (1949); **78,** 16 (1950).

[26] A thorough discussion can be found in M. G. Mayer and J. H. D. Jensen, "Elementary Theory of Nuclear Shell Structure." Wiley, New York, 1955.

[27] S. G. Nilsson, *Mat.-Fys. Medd. K. Dan. Vidensk. Selsk.* **29,** No. 16 (1955).

[28] For a review, see M. E. Bunker and C. W. Reich, *Rev. Mod. Phys.* **43,** 348 (1971).

[29] W. D. Knight, W. A. de Heer, K. Clemenger, and W. A. Saunders, *Solid State Commun.* **53,** 445 (1985).

[30] W. D. Knight, K. Clemenger, W. A. de Heer, and W. A. Saunders, *Phys. Rev. B: Condens. Matter* [3] **31,** 2539 (1985).

[31] W. A. Saunders, K. Clemenger, W. A. de Heer, and W. D. Knight, *Phys. Rev. B: Condens. Matter* [3] **32,** 1366 (1985).

[32] I. Katakuse, I. Ichihara, Y. Fujita, T. Matsuo, T. Sakurai, and H. Matsuda, *Int. J. Mass Spectrom. Ion Proc.* **67,** 229 (1985).

[33] I. Katakuse, T. Ichihara, Y. Fuiita, T. Matsuo, H. Matsuda, *Int. J. Mass Spectrom. Ion Proc.* **69,** 109 (1986). We are grateful to the authors for communicating their results to us before publication.

mutual interactions have been studied extensively, but computational complexity makes the difficulty of a theoretical investigation increase with cluster size. However, if electrons are delocalized and the detailed atomic positions do not play an important role, one can depict these confined electrons as subject to an effective smooth potential in the cluster. These conditions are expected to hold in some metal clusters, and a shell theory can then be developed to study and predict properties of these clusters. In this section, a shell model is described for metal clusters, and some of the theoretical results will be compared with experimental data in the following sections.

5. Quantum Chemistry and Solid-State Physics

Since clusters are systems with characteristics intermediate between atoms (or molecules) and solids, theoretical studies of their properties often employ techniques borrowed from applications in both atomic and solid-state science. Currently, the study of the electronic structure of atoms and molecules is in the realm of quantum chemistry, where considerable progress has been made in the past two decades. At present, there exist rigorous *ab initio* approaches for solving the many-electron Schrödinger equation in the Hartree–Fock self-consistent scheme supplemented by configuration interaction to take account of correlation effects.[34] Successful applications have been achieved for various properties in many different systems.[35] However, studies of large molecules usually require large numbers of basis functions (which means large numbers of integrals to be evaluated and large matrices to be diagonalized) to obtain a moderately accurate description of these systems. Hence, the calculation is ultimately limited by the capacity of modern computers. There are other popular theoretical methods which are semiempirical and less sophisticated, including the Hückel molecular-orbital (HMO) method, the neglect of differential overlap (NDO) method, the perturbation configuration-interaction method using localized orbitals (PCILO), the $X\alpha$ method, the consistent force field (CFF) method, etc.[36] These methods bypass some of the previous difficulties by utilizing empirical parameters, but they still require considerable computer time when applied to the study of clusters.

[34] A detailed discussion is given by H. F. Schaefer, III, "The Electronic structure of Atoms and Molecules." Addison-Wesley, Reading, Massachusetts, 1972; "Quantum Chemistry: The Development of ab initio Methods in Molecular Electronic Structure." Oxford Univ. Press, London and New York, 1984.

[35] For a review, see "Applications of Electronic Structure Theory, Modern Theoretical Chemistry" (H. F. Schaefer, III, ed.), Vol. 4, Plenum, New York, 1977.

[36] For a review of these methods and their applications, see "Semiemprical Methods of Electronic Structure Calculation, Modern Theoretical Chemistry (G. A. Siegal, ed.)," Vols. 7 and 8. Plenum, New York, 1977.

In contrast with the quantum chemistry methods, the theoretical study of solids utilizes translational symmetry and Bloch's theorem to handle infinite systems. To apply this approach directly to clusters, a supercell is often constructed which consists of a single cluster surrounded by a region of empty space. The supercell is then repeated periodically, and standard band-structure methods used for extended solids are then applicable. Although there are many state-of-the-art techniques which can be applied to do accurate calculations, the scale of the calculations, unfortunately, often demands the use of supercomputers. With increasing cluster size, the complexity of the calculation and the number of degrees of freedom in arranging the atomic positions dramatically increase. Hence, this area is also dependent on computer capacity. Because of the difficulties discussed above, detailed theoretical studies of the structural and electronic properties of metal clusters have been performed only for clusters either with small numbers of atoms or with a limited number of atomic configurations.[37–39]

The observed shell structure in alkali metal clusters appears to be of electronic origin; therefore, a main objective in this case is to deal with the properties of electrons in these clusters without constraints on the size of the cluster. Hence, to examine clusters over a large range of sizes and to study the trends in various properties, it is necessary to simplify the problem. A model is needed which preserves the important physical characteristics and is applicable to large and small clusters.

For example, one may use a simple three-dimensional harmonic-oscillator potential or a square well or one of intermediate shape such as a Woods–Saxon[40] potential. The order of level filling depends on the shape of the potential. Figure 1 is an example of the construction of a potential which represents a transition between the harmonic oscillator and the square well.[26]

Because of the delocalized nature of valence electrons found in simple metal crystals, a common approximation in the study of these metal crystals is to leave out the structure of the lattice and to replace it with a uniform positively charged background. This is the jellium approximation (in which the electron–electron interactions can be studied more easily).[41,42] Many electronic properties of metals have been calculated in this scheme. Most applications including the dielectric functions, specific heat, spin suscepti-

[37] J. L. Martins, J. Buttet, and R. Car, *Phys. Rev. B: Condens. Matter* [3] **31,** 1804 (1985).

[38] Although it is not expected that the shell model will give accurate detailed results for clusters smaller than N = 8, it is interesting to notice that the distortion parameters for 3,4 and 5,6 are not inconsistent with the results of Ref. 37.

[39] A. Cleland and M. L. Cohen, *Solid State Commun.* **55,** 35 (1985).

[40] R. D. Woods and D. S. Saxon, *Phys. Rev.* **95,** 577 (1954).

[41] D. Pines, "Elementary Excitations in Solids," Chapter 3. Benjamin, New York, 1981.

[42] G. D. Mahan, "Many-Particle Physics," Chapter 5. Plenum, New York, 1981.

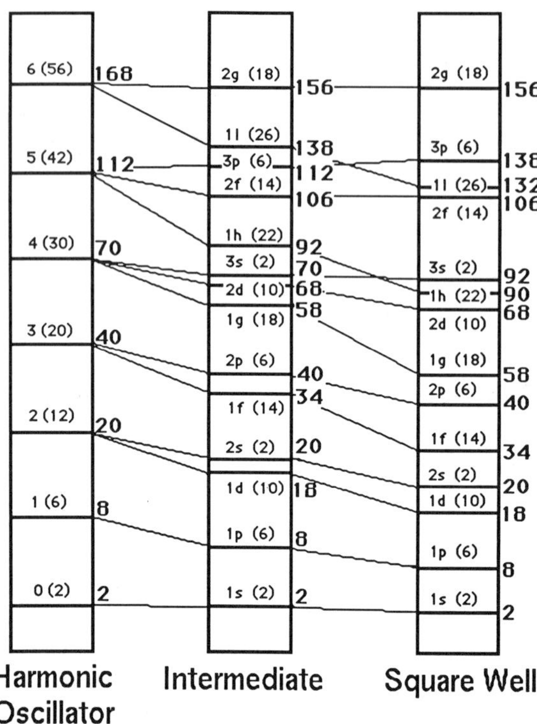

FIG. 1. Energy-level occupations for three-dimensional harmonic oscillator, intermediate, and square potential wells. (After M. G. Mayer and J. H. D. Jensen, "Elementary Theory of Nuclear Shell Structure." Wiley, New York, 1955.)

bility, cohesive energy, etc., give results in fair agreement with studies which consider the detailed lattice structure.[41,42] This approximation was also successfully applied to the study of the electronic structure of metal surfaces.[43]

The justification of the jellium approximation in these simple metal systems comes from the fact that the effective electron–ion potential or pseudopotential[44,45] is weak and that the crystal structures are relatively close packed.[46] An orthogonal criterion[47] between the wave functions of the core electrons and those of the valence electrons causes the pseudopotential generated by the ionic core (nucleus plus core electrons) to be weak since the net effect is a repulsion which pushes the valence electrons away from the

[43] N. D. Lang, *Solid State Phys.* **28,** 225 (1970).

[44] W. A. Harrison, "Pseudopotentials in the Theory of Metals." Benjamin, New York, 1966.

[45] M. L. Cohen and V. Heine, *Solid State Phys.* **24,** 37 (1970).

[46] Section 6 of Ref. 43.

[47] J. C. Phillips and L. Kleinman, *Phys. Rev.* **116,** 287 (1959).

vicinity of the nucleus and weakens the Coulomb attraction. This weak pseudopotential can be considered as a perturbation on the free-electron states for metals with mainly *s* and *p* conduction electrons.[44,45] It is expected that, in general, this feature persists when atoms aggregate to form clusters. These considerations motivate the use of the jellium model to study clusters over a large size range. The computational requirements are small and the model contains most of the required major electronic features; hence it should be capable of reproducing the observed shell structure. Since the jellium model does not contain the effects of discrete lattice structure, major discrepancies with experiment should emerge if these contributions are important.

6. Spherical Jellium Model and Shell Structure

As discussed earlier, in the jellium model the effect of ionic cores is simulated by a uniformly positively charged background. We now add a further restriction and assume spherical clusters; extensions to nonspherical cases will be discussed in Subsection 8. The positive density of the charged sphere, n^+, is assumed to be the same as the electron density, n^-, in the bulk, that is,

$$n^+ = (\tfrac{4}{3}\pi r_s^3)^{-1} \tag{6.1}$$

within the sphere. This defines r_s, which is the bulk Wigner–Seitz radius for the electron system. The radius of a jellium sphere with Z total valence electrons is

$$R = Z^{1/3} r_s \tag{6.2}$$

To evaluate the total energy of the system, the electron–electron interactions need to be added. To do this, the local-density-functional scheme is employed.[48,49] This approximation is described below.

In the density-functional formalism,[48] the total energy of an interacting electron gas is expressed as a functional of density only

$$E_{el} = E_{kin} + E_{xc} + E_{Coul} + E_{ext} \tag{6.3}$$

that is, the kinetic, exchange-correlation, Coulomb, and external-field energy terms. The external-field term is used to account for the electrostatic potential generated by the positively charged sphere.

By using the variational method, one obtains a set of Schrödinger-like equations from Eq. (6.3).[49] The many-body effects in the system are replaced by an effective one-electron potential which significantly simplifies the problem. Here an electron moves in the average potential created by the other

[48] P. Hohenberg and W. Kohn, *Phys. Rev. B: Condens. Matter* [3] **136,** 864 (1964).

electrons. The exact form of the exchange-correlation contribution to this effective potential is not determined, but different functional approximations have been made. A commonly used local approximation[49] treats a small neighborhood of the electron system as behaving like jellium at the local density. Thus, one can utilize the exchange-correlation results obtained from studies on a homogeneous electron gas. There are several suggested expressions of the exchange-correlation energy.[50] For calculations of the type considered here, they usually give similar results. Once the form of the exchange-correlation potential is chosen, the one-particle equations are solved self-consistently to obtain the energy levels and the position-dependent charge density of the electrons. The one-electron potential depends on the electronic charge density. A self-consistency condition is imposed, which requires that the input potential not deviate appreciably from the potential evaluated using the output wave functions to calculate the electron density. This criterion ensures that the calculation is internally consistent and likely to be correct within the approximation scheme used. This local-density approximation does not yield accurate excited-state properties for solids or clusters; however, it is accurate for the ground state.

Once the wave functions and eigenvalues are computed, the total energy of an N-atom cluster, $E(N)$, can be calculated for each N. The self-electrostatic energy of the uniformly charged sphere is included. This contribution to the energy varies smoothly as the cluster gets larger. In contrast, the electronic energy shows discontinuities at certain cluster sizes of the existence of well-separated discrete energy levels. When the one-particle equations are solved self-consistently in spherical coordinates, the eigenstates are labeled by the principal quantum number n, where $n - 1$ is equal to the number of nodes of the radial wave function and by the angular momentum quantum numbers l and m_l. Therefore, n is taken to be 1 for the lowest energy state of a given l. The same convention is used for nuclear levels, but this differs from the atomic physics convention. Conventional symbols for the angular momentum are used; namely, s, p, d, f, g, $h, \ldots$, respectively, correspond for $l = 0, 1, 2, 3, 4, 5, \ldots$. The high degeneracy, $2(2l + 1)$, of an energy level results from the spherical symmetry and is a property of this model. In a real cluster, however, it is expected that crystal-field effects will usually resolve these degeneracies but in most cases should not destroy the shell structure. With the above considerations, it is expected that the spherical jellium model should be capable of providing a rudimentary description of the physical features of the metal clusters.

[49] W. Kohn and L. J. Sham, *Phys. Rev. A* **140,** 1133 (1965).

[50] For example, L. Hedin and B. I. Lundqvist *J. Phys. C* **4,** 2064 (1971); J. Perdew and A. Zunger, *Phys. Rev. B: Condens. Matter* [3] **23,** 5048 (1981), Appendix C.

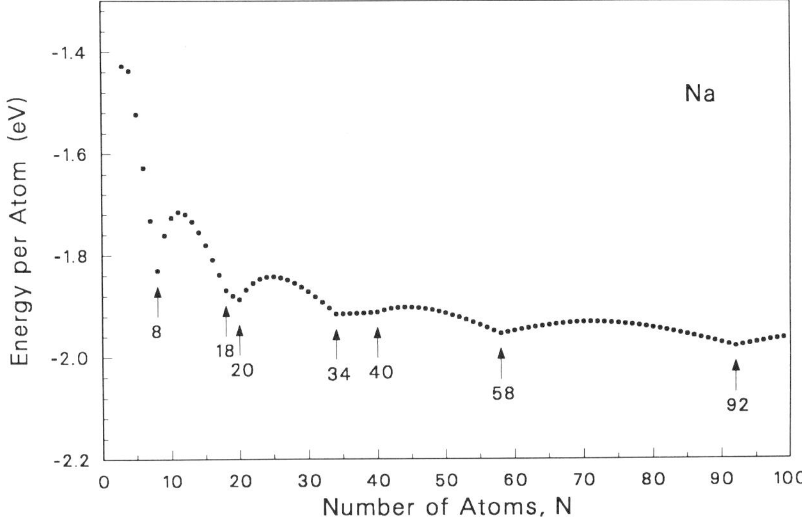

FIG. 2. The calculated total energy per atom in the spherical jellium model for sodium ($r_s = 3.93$ a.u.) as a function of cluster size.

If a perfectly degenerate shell is partially filled, the spins tend to line up so that the total energy can be lowered in a manner consistent with the Pauli exclusion principle and the Coulomb interactions between electrons. In the presence of a crystal field, it is difficult to determine the spin configuration of the ground state without a knowledge of the actual energy splittings. In addition, it has been shown that the exchange stabilization is less important in the presence of a nonspherical perturbation.[51] Therefore, in the present description, the calculation is performed without consideration of spin polarization.

7. Results for Some Simple Metal Clusters

In this section, we will discuss the calculated results within the spherical jellium model for clusters of monovalent alkali metals[52] and multivalent simple metals.[53] The total energy per atom, $E(N)/N$, is plotted in Fig. 2 as a function of cluster size for sodium. It is a smooth curve except for small kinks at $N = 8, 18, 20, 34, 40, 58, 92, \ldots$, where the total energy changes abruptly. The bulk limit is expected to be around -2.2 eV, which is the sum of kinetic,

[51] Y. Ishii, S. Ohnishi, and S. Sugano, *Phys. Rev. B: Condens. Matter* [3] **33,** 5271 (1986).

[52] M. Y. Chou, A. Cleland, and M. L. Cohen, *Solid State Commun.* **52,** 645 (1984).

[53] M. Y. Chou and M. L. Cohen, *Phys. Lett. A* **113,** 420 (1986).

exchange, and correlation energies of the homogeneous electron gas with the background of jellium. To examine the discontinuities in the total energy for specific cluster sizes, we define a quantity

$$\Delta_2(N) = E(N+1) + E(N-1) - 2E(N) \tag{7.1}$$

which is related to the second derivative of the energy with respect to N and represents the relative binding energy change for clusters with N atoms compared to clusters with $N+1$ and $N-1$ atoms. The quantity $\Delta_2(N)$ is one of the measures for relative stability and is independent of the reference energy of free atoms. If an energy level is just filled by the electrons in a cluster of N atoms and the next available level is separated from this filled level by a perceivable energy gap, the total cluster energy will have a jump from $E(N)$ to $E(N+1)$. This gives rise to a peak in $\Delta_2(N)$. A peak in $\Delta_2(N)$ indicates that a cluster with N atoms is relatively stable. It is similar to the atomic case described earlier where the noble gases are represented by atoms with filled electronic shells. The stability suggests that this cluster should have a larger abundance in the mass spectrum than a cluster with $N+1$ or $N-1$ atoms. It also turns out that the magnitude of the peak in $\Delta_2(N)$ is close to the immediate energy gap in eigenvalues for the closed shell.

In Fig. 3, the calculated $\Delta_2(N)$ for lithium, sodium, and potassium (with $r_s = 3.25$, 3.93, and 4.86 a.u., respectively[54] is shown for N up to 95. The peaks in $\Delta_2(N)$ for sodium appear at $N = 8$, 18, 20, 34, 40, 58, and 92 with the filled orbitals indicated in Fig. 3b. This is consistent with the experimental data which will be discussed in detail in Section IV. For lithium and potassium (Fig. 3a and c), the peaks in $\Delta_2(N)$ are at the same positions as for sodium. However, the general trend found in Fig. 3 is that the magnitudes of the peaks decrease from lithium to potassium. In particular, only a small step appears at $N = 40$ for potassium. This can be understood by examining the one-electron effective potentials for $N = 40$ lithium, sodium, and potassium clusters shown in Fig. 4. The potentials are smooth and flat within the sphere and behave like r^{-1} for large r. It is clear from electrostatic theory that the smaller the radius of a uniformly charged sphere (with the charge fixed), the deeper the electrostatic potential for the electrons inside the sphere. For these jellium spheres, the radii are scaled with r_s. This results in the differences in the depth of the effective potentials. Because the effective potential is shallower in potassium, the spacing between the energy levels is smaller, and hence, the gaps and $\Delta_2(N)$ are smaller.

This can be further illustrated by the eigenvalue plot as a function of r_s for $N = 40$ in Fig. 5. As r_s gets larger, the gap between the $2p$ and $1g$ levels decreases, and almost vanishes at the r_s value of potassium. One can, therefore,

[54] C. Kittel, "Introduction to Solid State Physics." Wiley, New York, 1976.

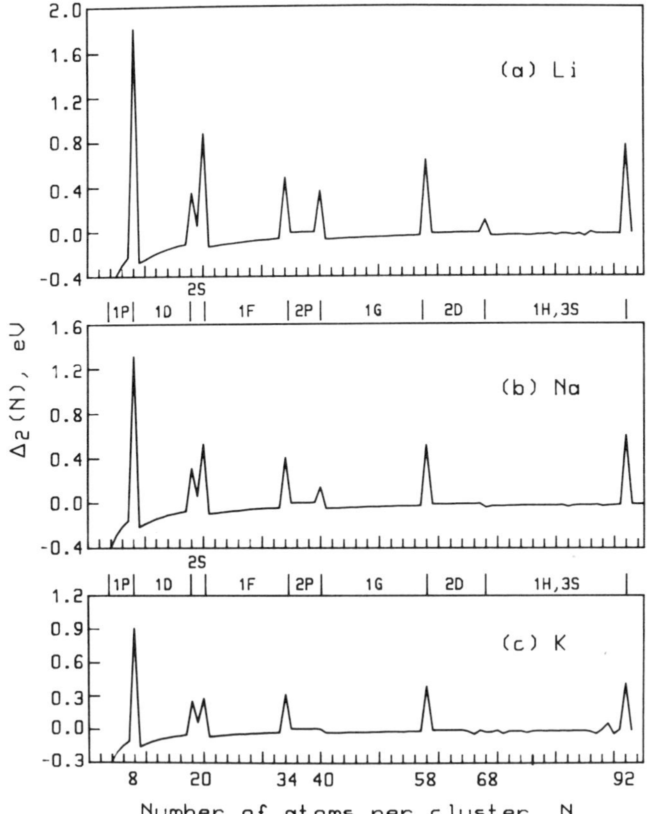

FIG. 3. The relative binding energy change $\Delta_2(N)$ versus N, the number of atoms per cluster for (a) lithium; (b) sodium; and (c) potassium. The labels correspond to filled shell orbitals. [After Chou *et al.*[52]]

expect that the gross features in the mass spectrum are less distinct in potassium than in sodium. It is also interesting to find that the average depths of the wells in Fig. 4 are close to the sum of the bulk Fermi energy, 4.72 eV (3.23, 2.12 eV)[54] and the experimental work function, 2.32 eV (2.7, 2.39 eV)[55] for lithium (sodium, potassium). This illustrates how some bulk properties relate to the properties of the clusters.

Two other similar calculations for small jellium spheres were performed independently[56,57] and yielded similar results. Shown in Fig. 6 are the charge

[55] N. D. Lang and W. Kohn, *Phys. Rev. B: Solid State* [3] **3,** 1215 (1971).

[56] W. Ekardt, *Phys. Rev. B: Condens. Matter* [3] **29,** 1558 (1984).

[57] D. E. Beck, *Solid State Commun.* **49,** 381 (1984).

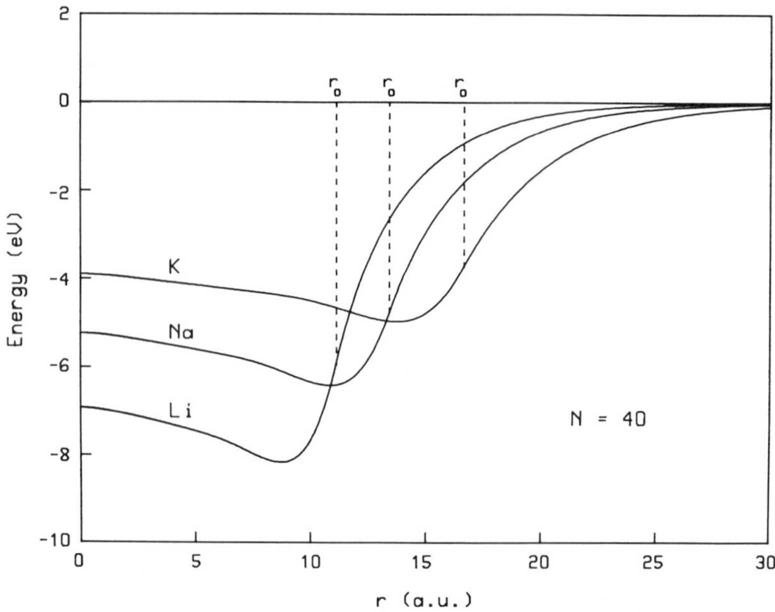

FIG. 4. The effective radial electronic potential of a jellium sphere with 40 atoms for lithium, sodium, and potassium. The radii of the jellium spheres are indicated. [After Chou *et al.*[52]]

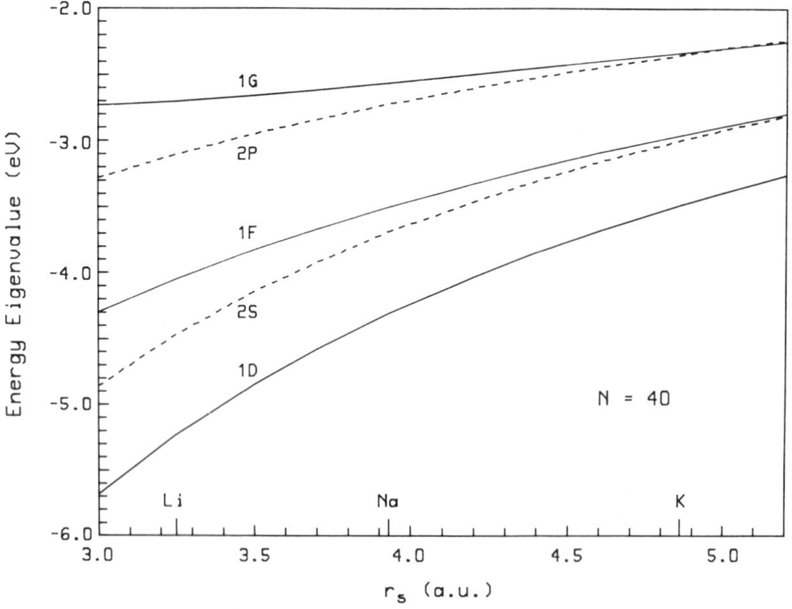

FIG. 5. The electron eigenvalues of a jellium sphere with 40 atoms are plotted as a function of r_s. The levels with the quantum number $n = 1$ and 2 are plotted with (———) and (– – –) lines, respectively. The $2p$ level is just filled for $N = 40$. [After Chou *et al.*[52]]

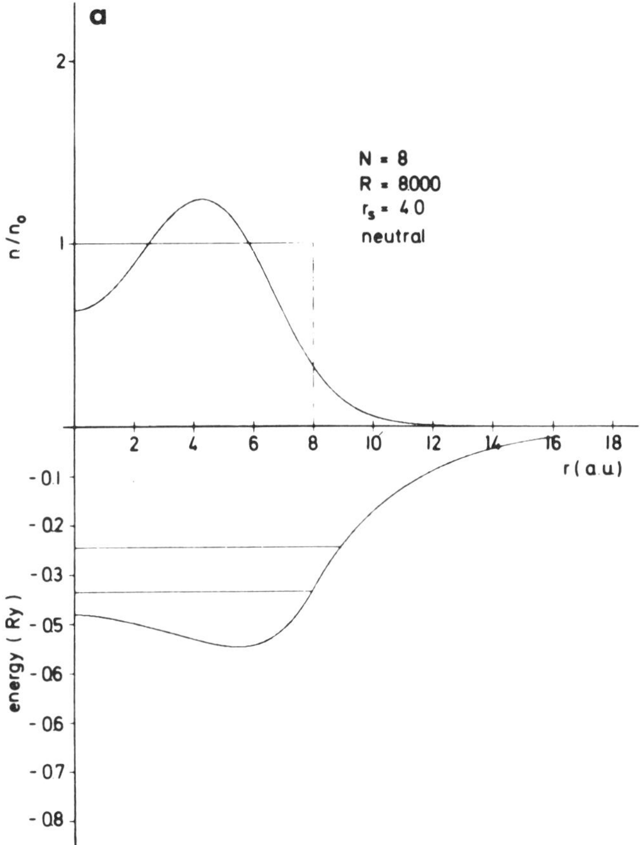

FIG. 6. The electron charge density, effective potential, and eigenvalues for several closed-shell clusters of sodium: (a) $N = 8$; (b) $N = 58$; and (c) $N = 198$. [After Ekardt.[56]]

density, the effective potential, and the energy-level structure for some cluster sizes with filled shells ($r_s = 4.0$ a.u.) obtained by Ekardt.[56] It can be clearly seen that the gaps between energy levels become smaller when N increases. Eventually these energy levels will evolve into energy bands in solids when N is sufficiently large.

The electronic charge densities in Fig. 6 have small oscillations within the spheres and decay to zero outside the spheres. For the highest particle number studied, $N = 198$, the electron density near the jellium sphere boundary appears to have variations which are very similar to the Friedel oscillations in the charge density found near the surface of an infinite plane.[56]

Further justification of the electronic shell structure and the spherical jellium model can be found in the results of self-consistent pseudopotential

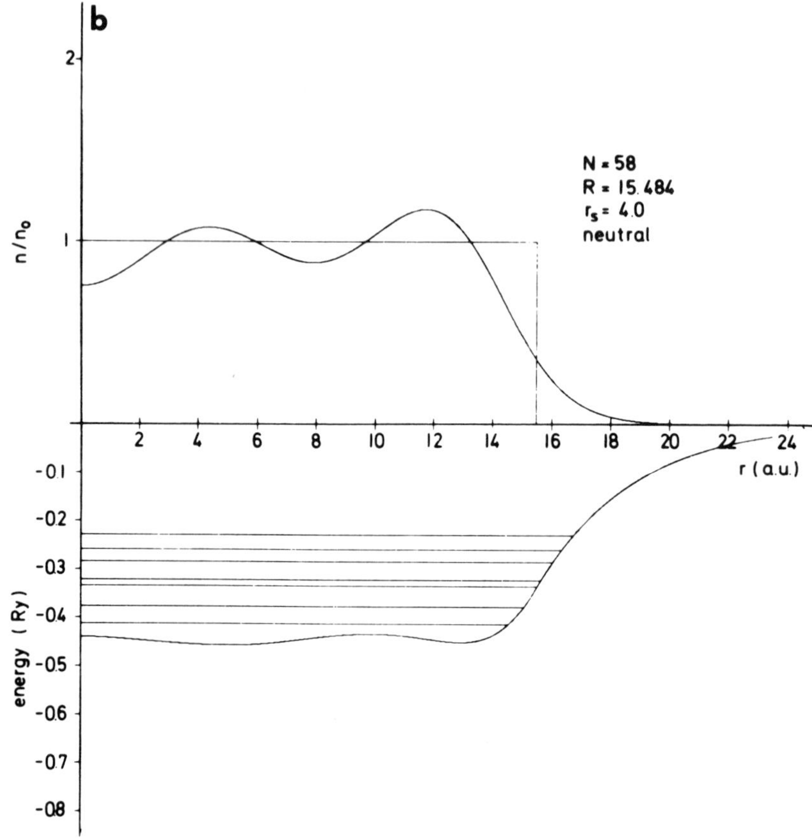

FIG. 6 (*Continued*)

local-density-functional calculations for some small sodium clusters.[37,39] In a study (using a linear combination of Gaussian atomic orbitals for Na_N with $N \leq 8$ and $= 13$) by Martins *et al.*,[37] the bonding in the clusters is found to be weak, nonlocalized, and nondirectional, and there is no sign of preference for dimerization, namely, no development of covalent bonding characteristics. Moreover, the calculated molecular orbitals can be classified as "*s*-like," "*p*-like," "*d*-like," etc., according to their global shape. These orbitals are occupied by electrons in a well-defined order as the number of electrons in the cluster increases. When the Jahn–Teller effect is taken into account,[37] the electronic structure is the principal factor in the determination of the equilibrium geometries of these clusters. For example, in the cases of Na_3, Na_4, $Na_4{}^+$, and $Na_5{}^+$ (where positive clusters lack one electron), one *s*-like and one *p*-like orbital are occupied; hence, the stable clusters have a planar, elongated

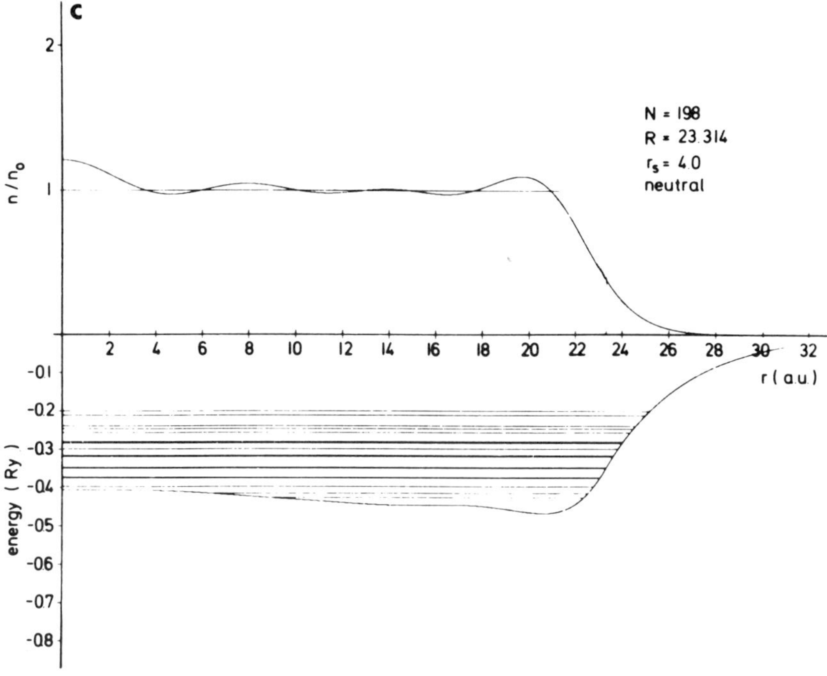

FIG. 6 (*Continued*)

structure. While the Na_7, Na_8, and $Na_8{}^+$ clusters, with three *p*-like orbitals (p_x, p_y, and p_z) all occupied, have very regular three-dimensional configurations with almost identical nearest-neighbor distances. Although these small clusters are probably beyond the lower limit in N where the jellium approximation is expected to be valid, the delocalized nature of electrons is, nevertheless, still a common characteristic feature for these small alkali metal clusters.[38]

Similar results are also found[39] in a pseudopotential study using the supercell method the plane-wave basis sets for Na_{13} and Na_{15} assuming fcc- and bcc-type structures, respectively, for these clusters, while constraining the nearest-neighbor distance to the bulk value. The self-consistent charge density is found to be delocalized. The ordering in energy of the angular momentum eigenstates is almost identical to the spherical jellium results, and the energy eigenvalues of these two different structures agree with the jellium results to within 10% for nearly all states, as shown in Table I. These findings suggest that the main electronic features are fairly insensitive to the actual geometrical structure and support the use of the spherical jellium model as a reasonable and feasible method to study larger clusters.

TABLE I. ELECTRON EIGENVALUES (IN eV) FOR 13- AND 15-ATOM CLUSTERS[a]

	Na_{13}		Na_{15}	
State	Jellium	Cluster	Jellium	Cluster
1s	−5.037 (2)	−4.805 (2)	−5.097 (2)	−4.933 (2)
1p	−3.985 (6)	−3.675 (6)	−4.104 (6)	−3.930 (6)
1d	−2.718 (10)	−2.498 (6)	−2.916 (10)	−2.930 (4)
		−2.255 (4)		−2.636 (6)
2s	−2.265 (2)	−2.195 (2)	−2.505 (2)	−2.433 (2)

[a] Degeneracies are noted in parentheses. The electrons occupy the lowest energy levels. [Taken from Table I in Cleland and Cohen.[39]]

Simple metal clusters of part of the third-row elements in the periodic table (sodium, magnesium, and aluminum) have also been studied in the spherical jellium model.[53] For clusters with the same number of valence electrons, Z, the numbers of atoms per cluster, N, are different because of the different valence of the constituent atoms. The calculated effective electronic potentials for spheres of sodium, magnesium, and aluminum with 66 electrons and r_s values of 3.93, 2.65, and 2.07 a.u.[54] respectively, are plotted in Fig. 7. As discussed

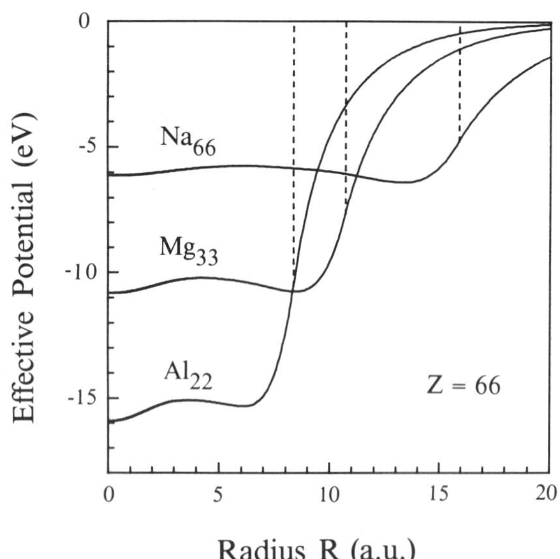

FIG. 7. The effective radial electronic potential of a jellium sphere with 66 electrons for sodium, magnesium, and aluminum. The radii of the jellium spheres are indicated by (---): 15.88 a.u. (Na), 10.71 a.u. (Mg), and 8.25 a.u. (Al). [After Chou and Cohen.[53]]

above, the smaller the sphere radius, the deeper the potential. The depth is again consistent with the sum of the Fermi energy and the work function in the jellium calculation for metal surfaces.[55] (The sums are 6.2, 11.7, and 16.4 eV for $r_s = 4$, 2.5, and 2 a.u., respectively.)

Figure 8 shows the calculated function $\Delta_2(N)$ for sodium, magnesium, and aluminum. Magnesium is a divalent metal with N equal to $Z/2$, and the positions of the peaks in Fig. 8b are shifted to one-half of their previous N numbers in sodium. There is an additional peak at $N = 34$ ($Z = 68$) since the effective potential of magnesium is deeper than that of sodium (Fig. 7). This causes the gaps between levels and the resulting peaks in $\Delta_2(N)$ to become larger. It also suggests that the abundance differences near the peaks or steps in the mass spectrum of magnesium should be more significant than those of sodium when the experimental conditions are kept the same. For the trivalent metal aluminum with N equal to $Z/3$, peaks in $\Delta_2(N)$ occur at clusters with numbers of total valence electrons close to the shell-closing numbers in Fig. 8c. For example, the peaks at $N = 13$ ($Z = 39$) and $N = 19$ ($Z = 57$) are responsible for the shell-closing number 40 and 58, respectively. The peaks of

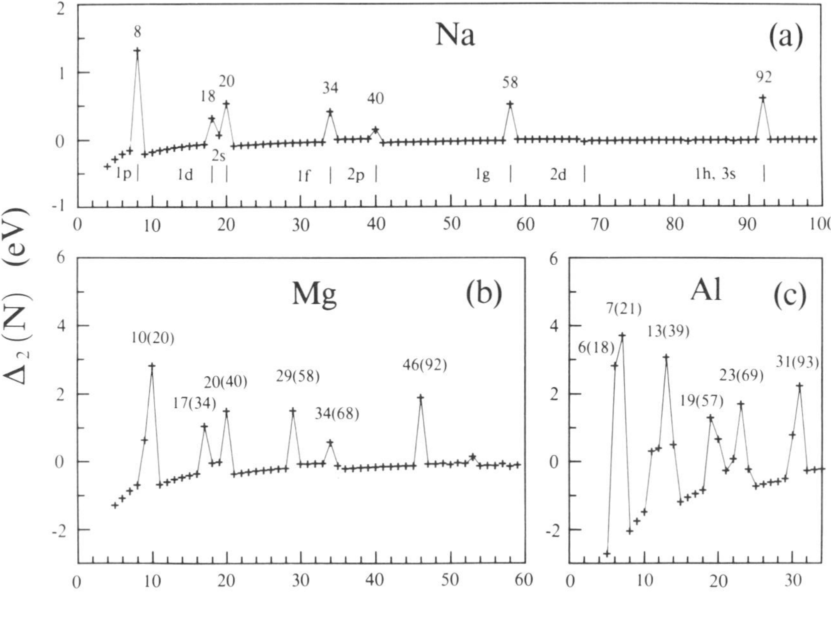

FIG. 8. The second differences of the total energy versus N for (a) sodium; (b) magnesium; and (c) aluminum. The labels in (a) correspond to filled shell orbitals. Numbers in parenthesis are total numbers of valence electrons. [After Chou and Cohen.[53]]

$\Delta_2(N)$ in Fig. 8c are found not to be as distinct as those in Fig. 8a and b, because the shell-closing numbers are seldom integer multiples of 3. In addition, the smearing of the ionic cores is expected to be a better approximation for singly charged cores than for multiply charged cores. However, certain discontinuities in the mass spectrum might still be found for aluminum, since the effective potential is the deepest of the three. (See also Section IV for related experimental results.)

8. The Spheroidal Model

Although the detailed configuration of the atomic arrangement in simple metal clusters does not seem to play an important role in the study of their physical properties, the distortion from a spherical shape does change the symmetry and thus the energy-level splittings of the system. The spherical assumption results in a great computational advantage for the calculation and is, as will be discussed in Section IV, very successful in correlating the prominent features of the mass spectra. However, there is evidence of many small features in the mass spectra which correspond to the existence of subshells. This indicates that an aspherical correction is needed to fully describe these systems, and the simplest appropriate deformation is to change the geometry from a sphere to a spheroid with axial symmetry. Even this first-step distortion can give rise to substantial complexities in solving the one-particle equation within the local-density-functional scheme, since self-consistency has to be obtained for differential equations with two variables instead of one. We are unaware of any calculations of this kind which have been done. Here we discuss briefly a simplified spheroidal model by Clemenger[58] to examine the general behavior of this axially symmetric deformation and its consequences.

Clemenger's model[58] for clusters resembles the spheroidal model for nuclei by Nilsson.[27] Instead of solving the complicated self-consistent equations, a simple harmonic-oscillator single-particle Hamiltonian is used

$$H = \frac{p^2}{2m} + \tfrac{1}{2} m\omega_0^2(\Omega_\perp^2 \rho^2 + \Omega_z^2 z^2) - U\hbar\omega_0(L^2 - \langle L^2 \rangle_n) \tag{8.1}$$

which can be solved more easily. The first and second terms represent the kinetic and potential energies. The z axis is taken as the axis of symmetry and $\rho^2 = x^2 + y^2$. When the oscillating frequencies along and perpendicular to the z axis (Ω_z and $\Omega_\perp$) are equal, we have a spherical oscillator with eigenvalues $(n + \frac{3}{2})\hbar\omega_0$, when n is a positive integer or zero. The third term contains an

[58] K. Clemenger, *Phys. Rev. B: Condens. Matter* [3] **32,** 1359 (1985).

anharmonic term depending on L^2, where L is the angular momentum operator.[59] This term has been shown to be equivalent to an r^4 perturbation in the spherical case.[60] Its purpose is to flatten the bottom of the potential well and to make it resemble a rounded square-well potential similar to those shown in Figs. 3 and 6. In the case of the spherical oscillator, this term separates the states with different angular momentum quantum number l associated with each n, yielding eigenvalues closer to those of a rounded square-well potential. An extra term $\langle L^2 \rangle_n$ [which is an average of L^2 over states with the same n and is equal to $n(n+3)/2$] is added to keep the average energy spacing between oscillator shells of different n's constant. The coefficient U is an adjustable parameter which was determined by comparing the resulting energy spectrum with the spherical jellium results. This gives values in the range of 0.04 and 0.08 for most sodium clusters under consideration.

For an aspherical system, the oscillating frequencies $\Omega_\perp$ and Ω_z are different. With cylindrical symmetry, the energy levels are either two- or fourfold degenerate. The shape of the cluster at each N is described by the equipotential surface, which is a spheroid. If the same volume is assumed as in the spherical case, $\Omega_\perp$ and Ω_z are not independent. Therefore, for clusters with each N, there is one independent geometric parameter η [defined as $2(z_0 - \rho_0)/(z_0 + \rho_0)$ in Ref. 58] which describes how prolate or oblate the cluster is. This "distortion parameter" is determined by minimizing the total energy calculated by summing the electronic eigenvalues of occupied states. The optimal values of η for sodium clusters of different sizes with $U = 0.04$ are discussed in Appendix A. The function $\Delta_2(N)$ obtained under this circumstance has, in addition to the peaks that appear in Fig. 2b, many smaller subshell-filling peaks. Fine-structure features are found at $N = 10$, 14, 18, 26, 30, 34, 36, 38, 44, 46, 50, 54, 60, 62, 66, 68,.... The comparison with the experimental mass spectrum will be discussed in Section IV and in Appendix B. In Appendix A, the discussion of spheroidal distortions will be extended.

9. Limits of Extended Shell Theory

In the previous sections, we used the spherical jellium model to examine clusters of simple metals and discussed the resulting shell structure. The whole approach neglects the ionic structure, assuming that the atomic potentials are small perturbations and emphasizing the behavior and mutual interactions of valence electrons. In real clusters, the crystal-field effect will split the high

[59] An analog of the orbital angular momentum, defined in appendix A of Ref. 27 as $\vec{l}_t$.

[60] A. Bohr and B. Mottelson, "Nuclear Structure," Vol. 2, pp. 593–595. Benjamin, Reading, Massachusetts, 1975.

degeneracy of the eigenvalues associated with spherical symmetry. The magnitude of the effect depends on the strength of the pseudopotentials and the locations of the ionic cores. If crystal-field effects are not large enough to destroy the delocalized property of electrons, this picture of electronic shell structure is still expected to be valid. The spheroidal model with axial symmetry discussed in the last section gives several minor shells without loss of the main shell features. For sodium, which is characterized by a weak pseudopotential, the introduction of cubic symmetry sets apart certain energy levels within a shell but maintains distinctive energy separations between shells.[39] It is expected that the shell structure associated with delocalized electrons will break down for clusters where covalent bonds are formed or valence electrons are localized (or associated with specific atoms) to a considerable extent. For example, in Si or C stability occurs for clusters in which the lattice energy contributes significantly.

In addition to the simple metals discussed above, the noble metals are expected to be next possible candidates for the application of the jellium model. These monovalent elements have one *s* electron (as the alkalis) outside the closed *d* shells. However, the *d* electrons in the core are relatively less inert compared with the core electrons in the alkalis. Considerable hybridization between the valence *s* electron and the localized *d* electrons is found in these noble-metal crystals. The pseudopotential for the valence *s* electrons can be substantially affected by the *d* electrons. For example, the pseudopotential for the *s* electrons of Au is stronger at the edge of the core than that of *K* because of the influence of *d* electrons, even though both elements have nearly equal core radii.[61] It is this extra binding that causes the noble metals to have much higher ionization potentials than the alkalis. Although we still expect the electronic shell structure to be present in the noble-metal clusters, the stronger pseudopotentials will result in eigenvalues with larger intrashell separations, which can cause structure in the abundance spectra. When the energy-level separation within a shell (for low-symmetry clusters) is appreciable compared with intershell energy gap, clusters with even numbers of electrons are expected to have noticeably higher stability than those with odd numbers of electrons.

The applicability of the shell theory to the transition-metal clusters would not be totally justified. Although there exist valence *s* and *p* electrons which are needed to establish the electronic shell structure, the unfilled, localized *d* orbitals cannot be ignored in the determination of structural and electronic properties. This differentiates the transition-metal from simple metal clusters, and they are excluded in the present discussion.

[61] It is because of the incomplete cancellation in the pseudopotential resulting from the extended orbitals, see the discussion by V. Heine, *Solid State Phys.* **24,** 25 (1970).

III. Production and Detection of Metal Clusters

10. Introduction

In recent years, developments in the production of clusters in the gas phase have resulted in important advances in cluster physics. In particular, the development of laser vaporization sources and secondary ion (SIMS) sources produced cluster beams of even the most refractory metals. Nevertheless, the well-known supersonic nozzle sources and the gas aggregation sources are reliable and easy-to-use alternatives, especially for the low-boiling-point metals. Cluster detectors are usually versions of standard molecular-beam mass spectrometers with modifications to accommodate the larger mass range.

In this part, we will discuss several types of sources and detectors with an emphasis on the apparatus used in the alkali cluster experiments,[62] which consists of a supersonic nozzle source, and a quadrupole mass analyzer–Daly detector system at the downstream end of a two-meter-long beam.

The cluster abundances from supersonic nozzle sources reflect the stabilities of clusters in detail (cf. Section IV). The reason for this involves the production processes in the jet where the clusters are formed and cooled. We will describe these processes and demonstrate that the observed cluster abundances can be understood in terms of frozen thermodynamic equilibrium distributions. For convenience we define rough size ranges as small, medium, and large containing, respectively, <10, <100, and >100 atoms per cluster.

11. Supersonic Nozzle Sources

Supersonic nozzle sources are widely used as a means of producing beams of gas molecules with small velocity spread, which reflects a low translational temperature in the beam. The low temperature results from the adiabatic expansion of the gas into a vacuum. If an inert carrier gas is seeded with a low concentration of nearly saturated metal vapor, clusters will form when cooling results in supersaturation. Although cluster formation in supersonic nozzle sources is not yet completely understood, one can make definite inferences about the production processes on the basis of the observed abundance spectra which show reproducible fine structure under a wide variety of source conditions. Source conditions primarily influence the overall cluster production without seriously altering the fine structure. The conclusion is that the formation process must be sensitive to the binding energies of the clusters. In

[62] W. A. de Heer, Ph.D. Thesis, University of California, Berkeley (1985).

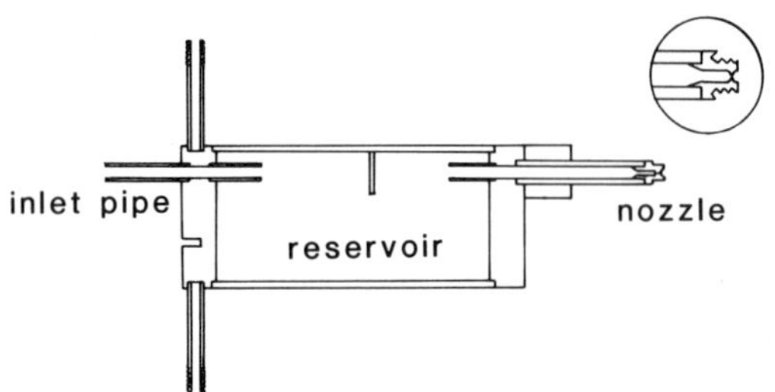

FIG. 9. Supersonic nozzle source chamber. [After de Heer.[62]]

the following, we will examine the properties of the supersonic jet sources. In Appendix B, a thermodynamic model for cluster production is set forth which quantitatively explains the fine structure in the abundance spectra of the alkali clusters.

Figure 9 shows the main chamber of the supersonic nozzle source used[62] in obtaining the alkali cluster mass spectra[20] shown in Figs. 15–17. The nozzle is a 0.0076 cm diameter hole with a 0.015 cm long channel. The nozzle is mounted at the end of a nozzle tube, the other end of which is connected to the reservoir. High-purity metal is melted and removed from sealed glass ampules in an argon-filled glove box and introduced into the reservoir through a heated inlet pipe. After the source is filled, high-purity inert gas at 2–10 atm pressure is introduced, and the source and nozzle are further heated so that the desired metal vapor pressure (50–300 Torr) is obtained. Similar sources have been reported by others.[63–65] The nozzle is kept about 100 K hotter than the reservoir to prevent plugging. In all phases of loading, the metal is handled in a clean inert atmosphere to prevent contamination of the source material, which usually leads to plugging and deterioration of the nozzle performance.

12. Adiabatic Expansion in Free Jets

The adiabatic expansion of noble gases in free jets has been investigated theoretically and experimentally.[66,67] Theoretically, it is found that the translational temperature of the beam is related to the Mach number M (the ratio of the local speed of the beam to the local speed of sound) by

$$T/T_0 = [1 + \tfrac{1}{2}(\gamma - 1)M^2]^{-1} \tag{12.1}$$

[63] R. A. Larsen, S. K. Neoh, and D. R. Herschbach, *Rev. Sci. Instrum.* **45,** 1511 (1974).

[64] M. M. Kappes, R. W. Kunz, and E. Schumacher, *Chem. Phys. Lett.* **91,** 413 (1982).

[65] G. Delecrétaz and L. Wöste, *Surf. Sci.* **156,** 770 (1985).

[66] J. B. Anderson, *in* "Molecular Beams and Low Density Gas Dynamics" (P. P. Wegener, ed.), p. 1. Dekker, New York, 1974.

[67] J. B. Anderson and J. B. Fenn, *Phys. Fluids* **8,** 780 (1985).

where T_0 is the stagnation temperature of the gas just before it enters the nozzle channel, and γ is the specific heat ratio, i.e., $\frac{5}{3}$ in monatomic gases. The Mach number is position dependent: It is zero in the stagnation region; unity inside the nozzle channel; and it increases to a terminal value M_T in the free jet expansion after the transition from hydrodynamic to molecular flow. Experimentally, it is found[67] that the terminal Mach number can be expressed

$$M_T = A(p_0 D/T_0)^{2/5} \tag{12.2}$$

where the constant $A \sim 47$ for He and ~ 90 for Ar, the gas pressure p_0 is in units of Torr, and the nozzle diameter is in cm. The mean speed of the beam u and the rms variation about the mean w are

$$u = \left(\frac{5kT_0}{m}\right)^{1/2} \Bigg/ \left(\frac{3}{M^2} + 1\right)^{1/2}, \qquad w = \left(\frac{2kT}{m}\right)^{1/2} \tag{12.3}$$

The speed ratio $S = u/w$ is more easily determined experimentally than the Mach number, and they are related by $M_T = (5/6)^{1/2}S$ so that the final translation temperature becomes, using Eq. (12.1),

$$\frac{T_{tr}}{T_0} = \frac{3}{M^2} = \frac{2.5}{S^2} \tag{12.4}$$

As an example, if $T_0 = 1000$ K, $p_0 = 2000$ Torr, and $D = 0.0076$ cm, then for Ar the nozzle channel temperature is $0.75T_0 = 750$ K, the terminal Mach number $M_T = 17$, the speed ratio $S = 15.5$, and the final translational beam temperature $T_{tr} = 10.5$ K.

During the free jet expansion into the vacuum, both the density of the gas and the temperature (see Fig. 10) decrease rapidly. The rate of increase of the Mach number[66] is approximately $dM/dx = 2/D$ during the expansion, while the density and the temperature are related by $n/n_0 = (T/T_0)^{3/2}$, and the temperature is given by Eq. (12.1). This gives cooling rates up to $dT/dt \sim 10^{11}$ K/sec[68] for an argon beam downstream from a 0.01 cm diameter nozzle. The temperature and the density are reduced to about 25% and 13%, respectively, within one nozzle diameter from the nozzle exit.

13. Cluster Formation in Pure Vapor Jets

An early study of chemical reactions in supersonic nozzle expansions[69] showed that a chemical equilibrium will be maintained during the expansion as long as the equilibrium concentration does not increase faster than the rate at which the atoms can combine. Since just outside the nozzle the density and

[68] G. Stein, *Surf. Sci.* **156,** 44 (1985).
[69] K. N. C. Bray, *J. Fluid Mech.* **6,** 1 (1959).

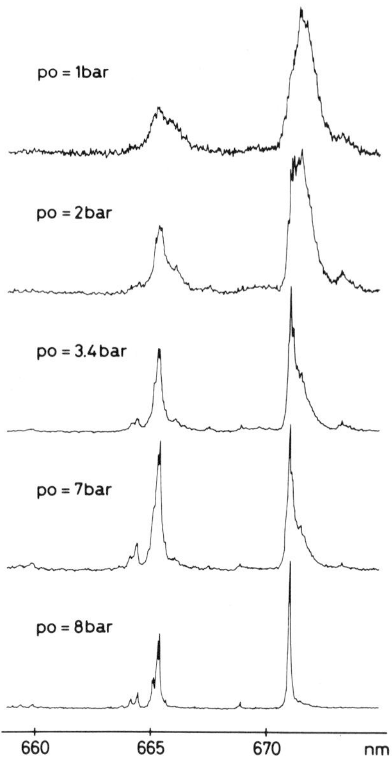

FIG. 10. Two-photon ionization spectrum of Na trimer for different carrier gas pressures, showing cooling. [After Delacrétaz and Wöste.[65]]

temperature drop very rapidly, to a good approximation the transition to a frozen flow occurs there.

The process of cluster formation in the expansion of a pure vapor in a jet is similar to the situation above. The formation of alkali dimers has been studied in some detail. One study[70] confirmed that dimers were in chemical equilibrium in the early stages of the expansion (i.e., in the nozzle channel) and also that the majority of the dimers experimentally observed in the beam were the result of three-body collisions outside the nozzle. In particular, it was found that, of the 15% mole fraction of sodium dimers found in the beam, about 70% are produced outside the nozzle during the free jet expansion.

Further studies[71,72] showed that the dimer formation processes in pure vapor expansion sources are determined by the value of the product p_0D,

[70] R. J. Gordon, Y. T. Lee, and D. R. Herschbach, *J. Chem. Phys.* **54,** 2393 (1971).
[71] K. Bergmann, U. Hefter, and P. Hering, *J. Chem. Phys.* **65,** 488 (1976).
[72] K. Bergmann, U. Hefter, and P. Hering, *Chem. Phys.* **32,** 329 (1978).

where p_0 is the metal vapor pressure. For sodium it was found that for $0.3 < p_0 D < 1$ Torr cm, the dimer concentration in the beam is essentially that of the equilibrium distribution in the throat. For larger values of $p_0 D$, progressively more dimers are produced outside the nozzle. For smaller values, the collision rate in the throat is too low to establish an equilibrium. Experimentally,[72] the dimers produced outside the nozzle are vibrationally much hotter than those which are produced in the throat. The dimer temperatures for $p_0 D \sim 1$ are found to be ~ 150 K with a source temperature of ~ 1000 K,[73] and the rotational temperatures are ~ 55 K.

The pure vapor sources produce some medium-sized clusters but with very low abundances. For example, with 1 atm of sodium vapor pressure in the source, the mole fraction of the dimer is 30% (see Refs. 70 and 134), while only 0.3% Na_3 and 0.1% Na_4 are observed. Clusters up to Na_{20} are observed, but with very small abundances above Na_{16}.[74] The pure vapor expansion source is not ideally suited for the formation of large clusters because the vapor must act both as a heat bath to cool the clusters and as an energy source during the reaction. This leads to contradictory requirements. In seeded sources, the inert carrier gas acts as a cooling agent, and large clusters can be produced.

14. Seeded Nozzle Sources

In a seeded jet source, an inert gas is mixed with a low concentration of a seed (e.g., alkali metal vapor), and the mixture is ejected.[62–65] Studies of polyatomic molecules seeded in an inert carrier show[75,76] that the carrier will cool the internal degrees of freedom unequally. It is found[76] that the rotational temperature of alkali halides in beams of inert gases approaches the translational temperature of the carrier for low (10^{-2}–10^{-4}) seed concentrations. The vibrational temperature, on the other hand, is much higher. For example, the vibrational temperature of CsF carried in Ar is found to be about a factor of 10 greater than the translational temperature over a wide range of source conditions.

If the seed can condense, clusters will be formed in the expansion. This situation differs from the pure vapor expansion in that the carrier gas can serve as a heat bath for the seed. In principle, the cluster formation processes during the expansion should resemble those for pure vapor expansions rather closely. As in the calculations of Refs. 70 and 72, when the vapor pressure is high enough, collisions between seed and seed facilitate the establishment of

[73] M. P. Sinha, A. Schulz, and R. N. Zare, *J. Chem. Phys.* **58,** 549 (1973).

[74] G. Delacrétaz, J. D. Ganière, R. Monot, and L. Wöste, *Appl. Phys.* [*Part*] *B* **B29,** 55 (1982).

[75] O. F. Hagena, *in* "Molecular Beams and Low Density Gas Dynamics" (P. P. Wegener, ed.), p. 93. Dekker, New York, 1974.

[76] H. G. Bennewitz and G. Buess, *Chem. Phys.* **28,** 175 (1978).

equilibrium among the cluster species in the nozzle channel. The carrier gas mediates the equilbration and absorbs most of the heat of formation, which is converted into translation energy of the beam.[77]

Outside the nozzle, the density and temperature rapidly decrease. Since the seed densities are much smaller than the carrier densities (typical mixing ratios should be 1–10%), the seed–seed collisions will cease before the carrier–seed collisions die out so that the cluster cooling continues after the cluster formation processes stop. This continued cooling aids in stabilizing the initially hot clusters.

Significant cooling of the vibrational states of the sodium trimer in seeded molecular beams has been demonstrated[65] for carrier pressures ranging from 1 to 8 atm, where it is observed that spectral features which are broad at 1 atm become narrow at 8 atm (see Fig. 10). Spectra for pure vapor expansion sources under those conditions correspond to a vibrational temperature of about 100 K[72] and appear similar to the 1 atm carried-beam spectra; so it is reasonable that they too represent a vibrational temperature of about 100 K. Since the observed linewidths at 8 atm are about $\frac{1}{5}$ the values at 1 atm, one can estimate that the temperature was reduced by roughly the same fraction.

The conspicuous features in the abundance spectra in Figs. 15–17 suggest a connection with cluster stabilities. The role of the binding energies in governing the structure of the abundance spectra is explored in Appendixes A and B. The model proposed in Appendix B assumes that the abundance spectra essentially reflect a frozen equilibrium distribution of clusters. For this interpretation to be correct, not only must an equilibrium be established inside the nozzle channel, but the distribution must also be frozen without significant alteration in the free expansion. If the cooling is insufficient to stabilize the clusters which were formed in the nozzle channel, then evaporation processes may distort the abundance spectra by shifting the maxima to lower cluster numbers. In Appendix B it is shown that, on the other hand, weak evaporation will tend to enhance the fine structure, since at a given temperature the evaporation rates and stabilities are inversely related. Figure 11a and b show, respectively, abundance spectra of argon-carried potassium clusters with a nozzle 300 and 150 K hotter than the oven. The shift of the abundance maxima to lower values is explained by strong evaporation so that even the most stable clusters lose one or two atoms after leaving the nozzle.

The effect of different carrier gases on the production of sodium clusters was studied[64] for He, Ne, Ar, Kr, and N_2. The heavier noble-gas carriers progressively increased production of the heavier clusters. The vapor and carrier pressures were, respectively, 350 Torr and 1.3 atm, which resulted in a high (35%) mixing ratio. The effects of different gases on the production was

[77] W. A. de Heer, unpublished (1985).

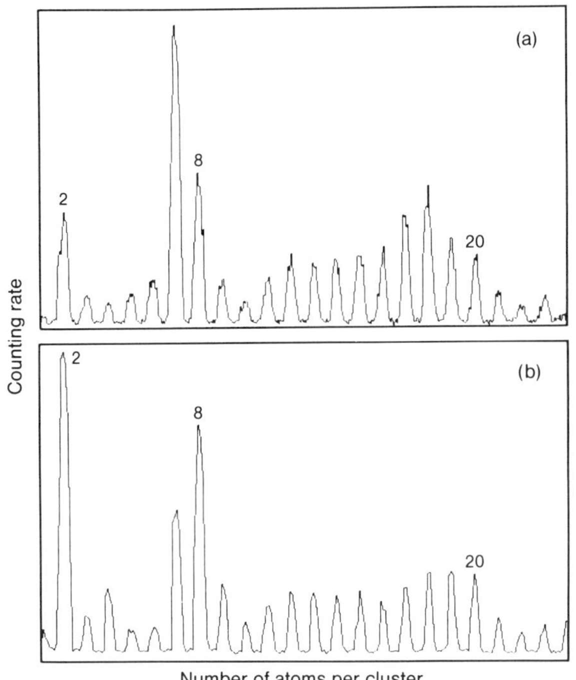

FIG. 11. Abundance spectra versus nozzle temperature, (a) 300 K hotter than oven; (b) 150 K hotter than oven. At the higher temperature the shell edges at 8 and 20 are distorted. [W. D. Knight, K. Clemenger, W. de Heer, and W. Saunders, unpublished (1984).]

explained[64] in terms of varying efficiencies in removing the heat of condensation from the clusters and by the slower velocities for the heavier gases. For nitrogen, very few clusters were formed. This was explained by the lower heat-capacity ratio of diatomic gases.[67,76] In the wide mass range spectra shown in this work, broad features resembling the now-known shell[20] structure appear but were not reported as such. The distortion of the edges in the abundance spectrum was probably caused partly by evaporation from clusters which were too hot, perhaps because the high mixing ratio left too low a concentration of carrier for adequate cooling.

Although temperatures of clusters produced in seeded beams have not been measured, it is possible to find an upper temperature limit based on cluster stabilities. Since the clusters survive the 2 m flight from source to detector,[62] the evaporation rate must be less than 10^3 atoms/sec from a given cluster. An order-of-magnitude estimate of the evaporation rate can be made by assuming that, per unit area, the clusters evaporate at approximately the same rate as the bulk which can be estimated from the vapor pressure to be

$$\frac{dN}{dt} = Av'n_0 \tag{14.1}$$

where A is the area of the cluster, $v' = (kT/2\pi m)^{1/2}$, and n_0 is the saturated vapor density at temperature T. From this expression, using the bulk vapor pressure of sodium, we find that for Na_{20} the vibrational temperature must be less than 500 K. This estimate is refined in Appendix B.

Although the seeded cluster source is a simple and reliable method of producing small and medium-size clusters, its operation requires rather high vapor pressure, and hence it is unsuitable for high-boiling-point metals. It is unlikely that even medium-size copper clusters can easily be produced[78] abundantly by this method. We will next investigate the gas aggregation sources which can produce medium-size clusters of high-boiling-point metals.

15. Gas Aggregation Sources

In the nozzle source, the metal vapor exists in the supersaturated state for a relatively short time, and large clusters are not formed abundantly. The corresponding times in gas aggregation sources are longer, and large clusters are easily formed. One type of gas aggregation source is constructed as in Fig. 12. Details of other designs can be found in the literature.[79–84] In operation, metal vapor from the oven enters the cool condensation chamber, where it mixes with a stream of inert gas at pressure of ~1 Torr more or less, and temperatures of ~77 K. Cluster formation continues until the gas–cluster mixture is discharged through an orifice into a surrounding vacuum chamber. The size distribution is controlled by the temperatures of oven and condensation chamber and by the gas flow rate. High-temperature ovens at ~2300 K[80] have been employed in this method.

Consistent abundance patterns have been observed (cf. Fig. 26) with this type of source. We may therefore infer that the sequence of events similar to those in the nozzle source may be taking place. This implies the operation of thermodynamic processes whose end products show abundances which depend on stabilities, since sharp features in the abundance spectra are not

[78] D. R. Preuss, S. A. Pace, and J. L. Gole, *J. Chem. Phys.* **71,** 3553 (1979).

[79] K. Sattler, J. Möhlbach, and E. Recknagel, *Phys. Rev. Lett.* **45,** 821 (1980).

[80] S. J. Riley, E. K. Parks, C. R. Mao, L. G. Pobo, and S. Wexler, *J. Chem. Phys.* **86,** 3911 (1982).

[81] S. Yatsuya, S. Kosukabe, and R. Uyeda, *Jpn. J. Appl. Phys.* **12,** 1675 (1973).

[82] D. M. Mann and H. P. Broida, *J. Appl. Phys.* **44,** 4950 (1973).

[83] J. Hecht, *J. Appl. Phys.* **50,** 7186 (1979).

[84] F. Frank, W. Schulze, B. Tesche, J. Urban, and B. Winter, *Surf. Sci.* **156,** 90 (1985).

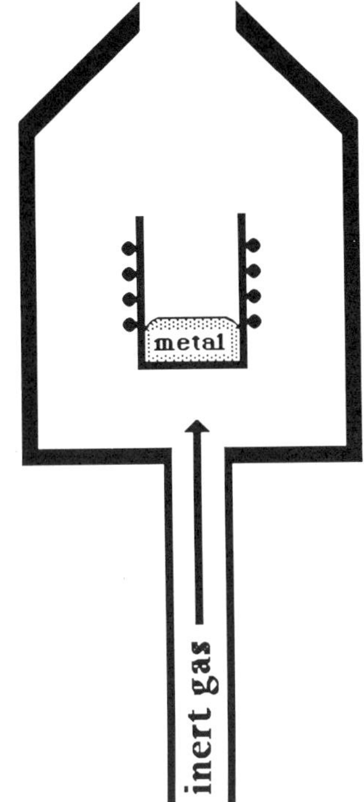

FIG. 12. Gas aggregation source.

accounted for by kinetic arguments. Although a complete analysis has not been given, several aspects of the problem have been treated.[85–88] The gas aggregation source may be adjusted to produce clusters up to sizes containing 10^5 atoms.[84]

16. OVENLESS SOURCES

Several sources of recent development employ photon, ion, electron, or neutral atom beams which locally vaporize small amounts of metal at the bulk surfaces. The local heating is intense, only minute quantities of metal are vaporized, and even the most refractory metals may be used.

[85] R. Rechsteiner and J. D. Ganière, *Surf. Sci.* **106,** 125 (1981).
[86] R. S. Bowles, J. J. Kolstad, J. M. Calo, and R. P. Andres, *Surf. Sci.* **106,** 117 (1981).
[87] F. C. Goodrich, *Proc. R. Soc. London, Ser. A* **277,** 155 (1964).
[88] G. L. Griffin and R. P. Andres, *J. Chem. Phys.* **71,** 2522 (1979).

The production of secondary ions by the bombardment of a surface with energetic (~ 10 keV) rare-gas ions (SIMS) is a well-known and effective technique.[89] In general, positive and negative cluster ions as well as neutrals are formed. The secondary-ion yields are high, e.g., 10^9 $Ag_4{}^+$ ions per cm^2 per sec.[90] The yield depends on the incident energy and the impacting ion, xenon being very effective in producing large clusters, e.g., up to $N \sim 200$ for silver.[32]

It is believed that, for typical impacting ion currents, a group of clusters and all of their constituents atoms result from a single incident ion[91] which vaporizes a volume of $\sim 10^6$ Å^3. Fluorescence experiments[92] indicate temperatures $\sim 10^4$ K for the original dense plasma of ions, atoms, and electrons. The plasma cools during the ensuing rapid expansion when clusters are formed by a process analogous to the nozzle expansion.

The laser vaporization source[93–96] directs intense pulses from, e.g., a Nd:YAG laser onto a ~ 1 mm diameter area of the metal target. Metal from that area vaporizes and is entrained in an inert gas which carries the condensed clusters through a channel to a supersonic nozzle. The method is generally applicable to all metals, including tungsten. It has been highly developed and is widely used as a major cluster source. The abundance spectra produced by the laser vaporization source exhibit features which are found in the secondary-ion sources. Since the cluster ions from the latter are believed to have been born as ions according to their respective stabilities, it is likely that the clusters which are finally detected in the laser vaporization experiments were also born as ions. These features will be discussed further in Section IV and compared with similar features in the neutral abundance spectra produced by nozzle sources.

17. Detectors

The main requirements for a detector are good signal-to-noise ratio, wide mass range and resolution, and single-particle counting capability. Mass selection is usually accomplished with either a quadrupole mass analyzer or a time-of-flight (TOF) mass spectrometer.

[89] F. Honda, Y. Fukuda, and J. W. Rabadais, *Springer Ser. Chem. Phys.* **9,** 18 (1979).

[90] P. Fayet, F. Granzer, G. Hegenbart, B. Pischel, and L. Wöste, *Phys. Rev. Lett.* **55,** 3002 (1985).

[91] R. J. Macdonald and R. F. Garrett, *Surf. Sci.* **78,** 371 (1978).

[92] G. P. Können, A. Tip, and A. E. de Vries, *Radiat. Eff.* **26,** 23 (1975).

[93] T. G. Dietz, M. A. Duncan, D. E. Powers, and R. E. Smalley, *J. Chem. Phys.* **74,** 6511 (1981).

[94] D. E. Powers, S. G. Hansen, M. E. Geusic, D. L. Michalopoulos, and R. E. Smalley, *J. Chem. Phys.* **78,** 2866 (1983).

[95] V. E. Bondybey and J. H. English, *J. Chem. Phys.* **76,** 2165 (1982).

[96] E. A. Rohlfing, D. M. Cox, and A. Kaldor, *Chem. Phys. Lett.* **99,** 161 (1983).

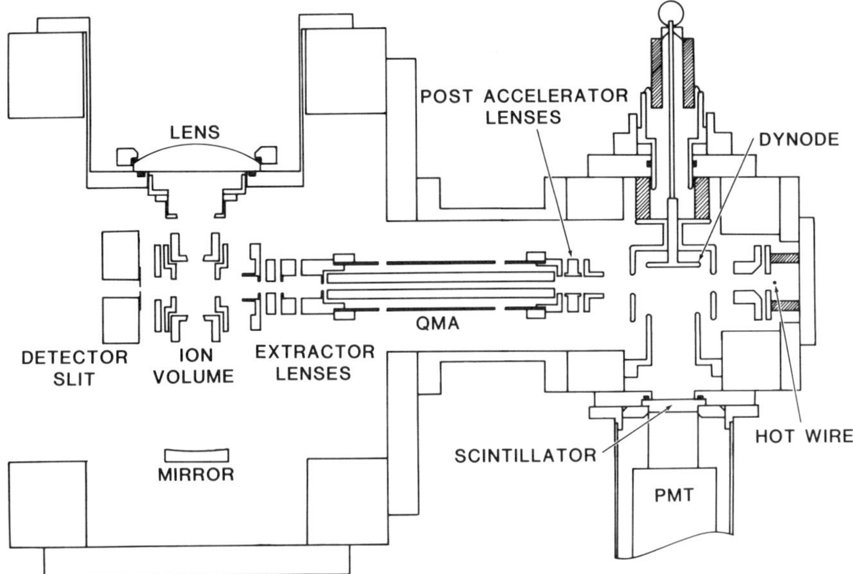

FIG. 13. Cross-section diagram of QMA–Daly detector. [After de Heer.[62]]

Figure 13 shows a cross section of a quadrupole mass analyzer (QMA) based detector.[62] Clusters are photoionized in the ion volume by light from a mercury arc lamp with appropriate filters. Other arrangements include a xenon arc lamp and monochromator or a tunable laser. The cluster ions are extracted from the ion volume and focused on the entrance of the QMA. At the exit of the QMA, a postaccelerator lens system focuses the mass-selected clusters on the 30 kV dynode of a Daly detector.[97] At the dynode, each cluster ion is converted to a pulse of ~ 10 electrons. These are focused on a scintillator, which is viewed by a photomultiplier. Subsequent electronics provides for discrimination against single-electron signals and storage of the ion counts in a multichannel scaler. The QMA is mounted on the beam axis, which accounts for high ion throughput and mass selectivity. The QMA may be set for a single selected mass, or the acceptance mass may be swept. Wide mass range with good resolution is obtained[98] by sweeping the QMA rod potentials with high-resolution computer-controlled digital-to-analog converters.

The TOF mass spectrometer[99] operates in a pulsed mode. Ions are created

[97] N. R. Daly, *Rev. Sci. Instrum.* **31,** 264 (1960).

[98] W. A. Saunders, Ph.D. Thesis, University of California, Berkeley (1986).

[99] K. Sattler, J. Möhlbach, E. Recknagel, and A. Reyes Flotte, *J. Phys. E* **13,** 673 (1980).

by electron or photon impact. A voltage pulse V accelerates all singly charged ions to the same energy. The mass-dependent velocity $v = (2qV/m)^{1/2}$ results in distinguishable flight times for the different clusters. The TOF mass spectrometer need not be swept, since each pulse contains the whole range of cluster masses which are spread out in arrival times at the detector. The resolution is determined primarily by the energy spread of the ions and their spatial distribution at the time the voltage pulse is applied. Resolutions of one percent or better are obtained at 20,000 amu.[100]

18. Overview of a Cluster Beam Apparatus

The beam machine which was used in some alkali cluster experiments[62] incorporates many required capabilities, including particle electric deflection. Figure 14 gives an overview. The supersonic nozzle source is situated at the extreme left, and the particles are mass selected and counted in the detector, which lies approximately 2 m downstream. Following the source, we see the adjustable heated skimmer, a series of collimating slits C 1–4, and, in addition, the mechanical beam chopper, which is used in TOF measurements, and the

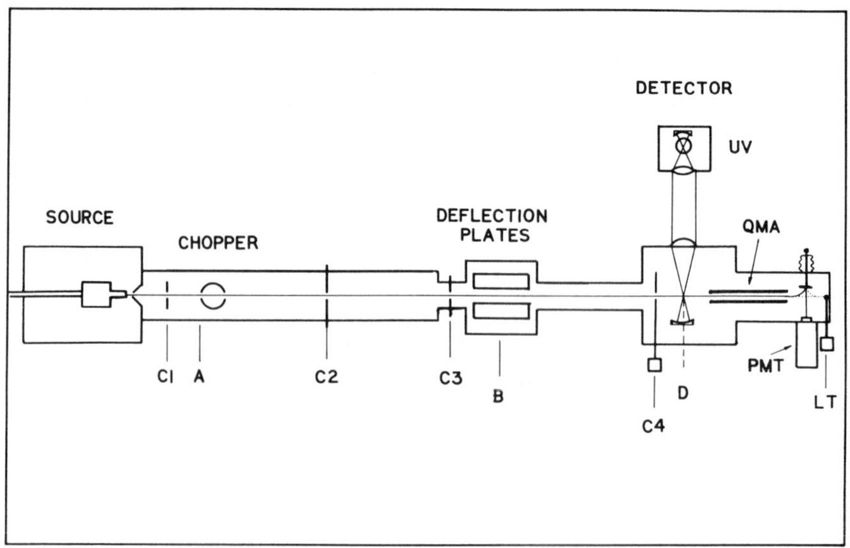

Fig. 14. Apparatus overview, including E-field deflection plates. [After K. Clemenger, Ph.D. Thesis, University of California, Berkeley (1985).]

[100] K. Sattler, *Surf. Sci.* **156,** 292 (1985).

deflection plates. Typical electric deflections (≤ 0.5 mm) are relatively independent of cluster mass because polarizability and mass are both proportional to volume. Ionization and other optical experiments are performed in the ion volume just upstream of the QMA. Typical pressure in the source chamber is 5×10^{-3} Torr with 3 atm pressure of Ar in the source. Pressure in the detector unit is $\sim 10^{7}$ Torr.

Photoionization efficiency profiles, e.g., of K clusters,[31] are accomplished with an appropriately tunable light source whose wavelength is varied synchronously with the channel address in a multichannel scaler. The ionizing light is monitored, and data are normalized by the on-line computer for intensity fluctuations. Beam fluctuations and background are averaged out by an appropriate chopping sequence for both beam and ionizing light.

Polarizability experiments are performed by observing particle deflections resulting from the application of ~ 40 kV to the inhomogeneous field-deflecting plates.[30] The electric dipole moment induced in the clusters interacts with the field gradient and results in deflections at the detector measured by scanning the detector slit across the beam. Deflection is proportional to the squared ratio of applied voltage to velocity. Deflection measurements combined with measurements of velocity distributions permit a determination of cluster polarizabilities to a precision of $\pm 5\%$ or better.

A great variety of apparatus configurations have been used,[101] depending on the particular experiments at hand. It is worth noting that in the machine described above, high-resolution deflection profiles with resulting beam attenuations and with typically small photoionization cross sections give single mass-selected ion counting rates up to 10^3/sec. With wider collimation where resolved deflection profiles are not required, counting rates $> 10^4$/sec are common.

19. Conclusions

The foregoing brief survey of cluster sources provides comparisons among several common types of sources. We have emphasized the supersonic jet source because the basic features of its operation are understood and its practical operation is also simple and straightforward. Final internal cluster temperatures can be predicted approximately. The beam intensity is high, and thus adequate deflection experiments even with long flight paths may be performed. The intensities are also high enough to allow many optical experiments to be attempted.

[101] Cf. Refs. 1, 2, and 3.

The gas aggregation source is also important, and it provides good control of the size distributions and the temperatures as well. Use of the high-temperature oven makes these sources also suitable for use with refractory metals.

The laser vaporization and secondary-ion sources are good alternatives, the latter being particularly good if the study of cluster ions is a primary objective. Although typical beam intensities are lower than for the supersonic jet and gas aggregation sources, mass-selective manipulation is easy, and detection is more efficient. The final temperatures tend to be higher, and in-flight fragmentation is more common than for the other sources. However, the currently known abundance spectra appear to be consistent and range to high masses up to $N \sim 200$.

Returning to the jet source, we understand the basic operation, and it can be used as a prototype for understanding the others. Although it is not immediately obvious that the mechanisms of the jet source have close parallels to the secondary-ion source, the similar structures in the abundance spectra suggest that such parallels must exist. It seems likely that they have a common feature and that the basic stage of cluster production involves a balance of cluster aggregation and division so that clusters of neighboring sizes have relative abundances reflecting an underlying equilibrium.

IV. Cluster Abundance Spectra and Shell Structure

20. Introduction

In this part, we will summarize the observed features in the abundance spectra and correlate them with theoretical models. It will be shown that the abundance spectra are sensitive reflections of the relative stabilities of the clusters. This idea is explored further in Appendix B. The extension of shell theory to mixed and polyvalent clusters is discussed.

21. Pure Alkali Metals

Cluster abundances are shown for sodium[102] in Figs. 15 and 16, and for potassium[103] in Fig. 17. Data were taken with argon-carried seeded beams. Aside from the expected decrease in intensity for the larger clusters, the mass spectra show a number of conspicuous features. Figure 15 displays the

[102] Cf. Ref. 20.

[103] K. Clemenger, Ph.D. Thesis, University of California, Berkeley (1985).

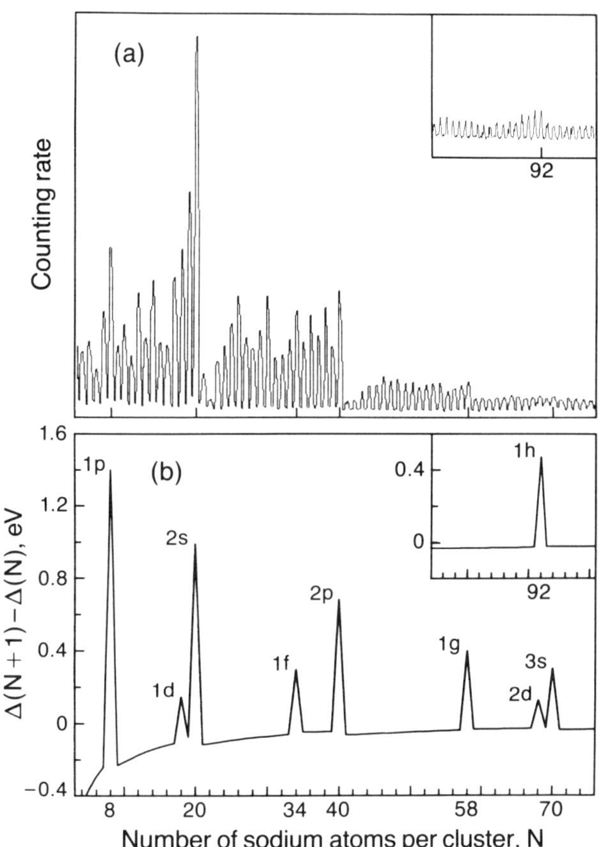

FIG. 15. Na cluster abundance spectrum: (a) experimental, (b) theoretical, using Woods–Saxon-type potential. Peaks at 68–70 are not observed experimentally. [After Knight *et al.*[20]]

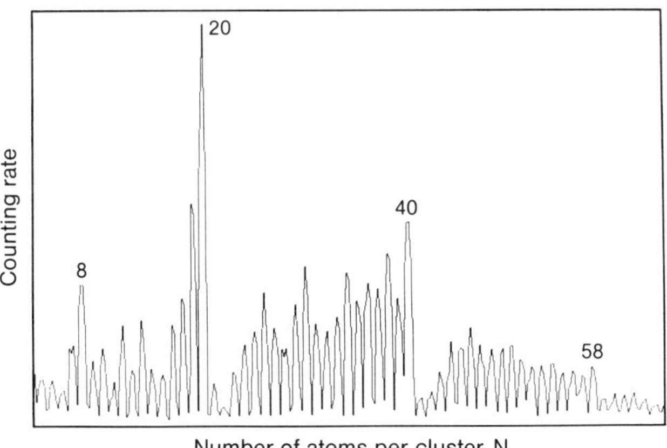

FIG. 16. Na cluster abundance spectrum, emphasizing fine structure. [After Clemenger.[103]]

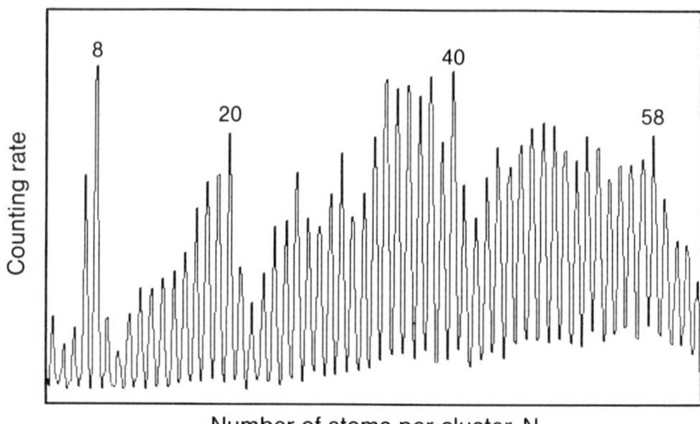

FIG. 17. K cluster abundance spectrum with laser ionization. [After Clemenger.[103]]

intensities of sodium clusters showing major peaks or edges in the overall profile and also a number of minor features. The list of major (*underlined*) and minor features to be found in Figs. 15 and 16 is as follows: number of atoms per cluster, N = *8*, 10, 12, 14, *20*, 26, 30, 34, 36, 38, *40*, 44, 46, 50, 54, *58*.

In addition to the above, a peak at 2 and an edge at 92 are observed. Figure 17 displays potassium mass spectra which show major and minor features. The minor fine structure becomes less distinct for increasing N, and, in general, it is weaker for potassium than it is for sodium. The envelope of the peaks around potassium 20 is broad and rounded, while near sodium 20 it is sharp. All of the significant numbers are even, and the fine structure is characterized by sequences of two or four.

The data for Figs. 15–17 were taken under a variety of conditions, but the features in the mass spectra are not highly sensitive to the details of the source or ionizing conditions. As a further example of the persistence of features under various conditions, mass spectra for pure uncarried beams of sodium,[74] although limited in range and weak above 20, nevertheless show features at 8 and 20. Figure 18 shows a mass spectrum for sodium clusters taken at several carrier gas pressures,[102] other source conditions remaining constant. The prominent features in the profiles are those discussed above.

Overall cluster production is enhanced with emphasis on the larger clusters at increased pressures of both metal vapor and carrier gas. Steady cluster production requires clear nozzle conditions, obtained by operating the nozzle at slightly elevated temperatures relative to the reservoir. However, when the nozzle temperature is raised drastically by $\sim$300 K, the main peaks in the spectrum are distorted; see peaks 7, 8, and 18–20 in Fig. 11.

The shell closings predicted by the jellium model (c.f. Section II) are observed in the abundance spectra. However, 18 and 34 are weak in both Na

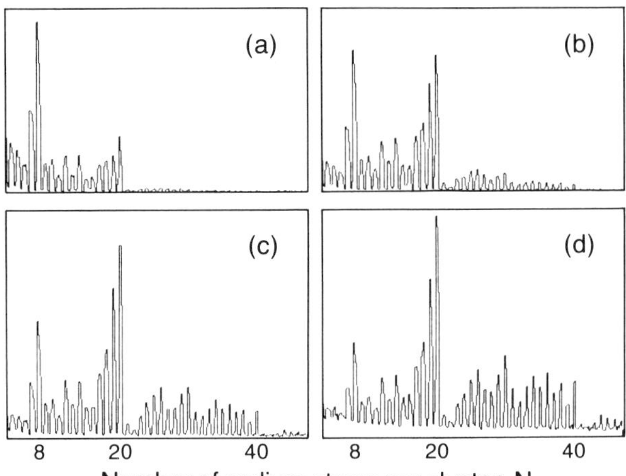

FIG. 18. Na abundance spectrum versus argon carrier gas pressure, at (a) 300 kPa; (b) 400 kPa; (c) 500 kPa; (d) 600 kPa. [After Knight *et al.*[20]]

and K. The peaks at 18 and 34 are less prominent than might have been expected, because of spheroidal distortions (c.f. Appendix A) which modify the gaps in the level structures. As for 19, it is an *s* state superposed on a shell closing and will have high stability like 20 with a large energy gap to the 1*f* shell.

For the 1*d* and 2*s* shells, the Na spectrum rises to a single peak at 20. However, for K the structure is more edgelike than a peak, and the 17–19 peaks are comparable in size. This is in general agreement with the trend of $\Delta_2(N)$, seen in Fig. 3. For Na $\Delta_2(20)$, there is a distinctive peak, while the K $\Delta_2(18-20)$ peaks are of comparable size and smaller than those of Na. Another visible difference between the mass spectra of Na and K is that the edge at 40 is weaker for K than it is for Na. This is consistent with the trend in energy levels, as in Fig. 5, where the energy gap between the 2*s* and 1*f* shells diminishes as r_s increases.

In Fig. 19[58], experimental and theoretical results for spherical and spheroidal clusters are compared. All of the observed fine-structure peaks for $N \geq 10$ (except for 12) correspond with those in the theoretical result derived from Nilsson theory. The extent of the detailed agreement between experiment and theory, which we obtain with a single value for the U parameter, emphasizes the basic importance of the symmetries of the problem. The U parameter was allowed to vary in the Nilsson model for nuclei and gave good correspondence with experiment for properties such as quadrupole moments and neutron activation energies. We can improve agreement with experiment

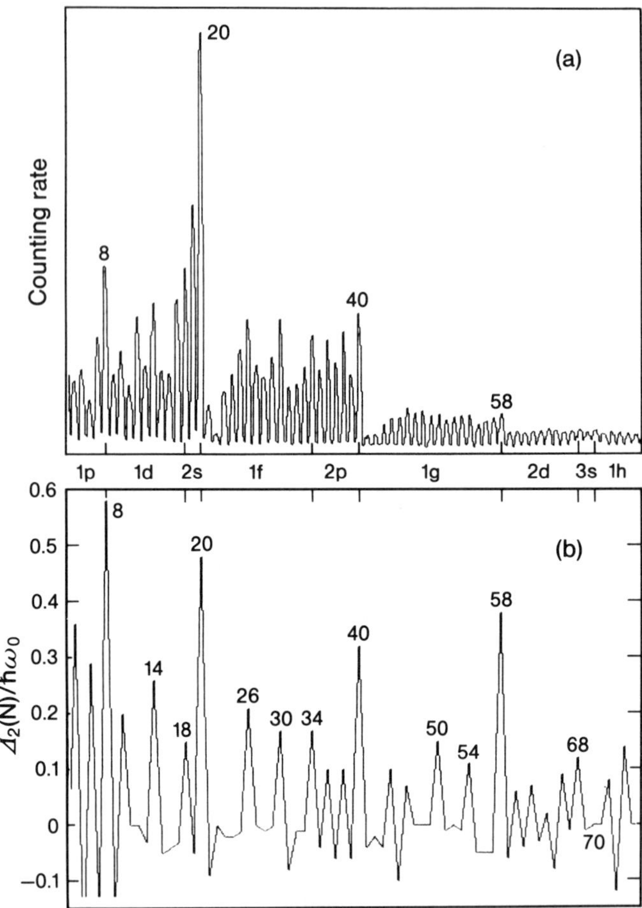

FIG. 19. Spheroidal distortions and Na mass spectrum: (a) experimental; (b) theoretical. [After Clemenger.[58]]

by allowing U to vary from shell to shell, but it is quite satisfactory as presented in Fig. 19 for the single value $U = 0.04$.

Some examples of the agreement in Fig. 19 follow. The fourfold patterns in the $1f$ and $1g$ shells appear correctly, and the twofold patterns in the $2p$ shell correspond to the filling of a prolate subshell at 36 and an oblate shell at 38. The peaks at 18 and 34, which are identified as closings in spherical shell theory, are of a lesser strength because of the distortions and reductions in the corresponding energy gaps. The intensity resolution is inadequate to determine whether 60, 62, etc., appear experimentally. The reduction in energy gaps resulting from spheroidal distortions is of general significance, as will be seen in the following Sections V and VI.

The observation of the cluster abundance peak at 12 and the corresponding result for the ionization potential of 12 (cf. Section VI) suggest that the spheroidal theory is not applicable in this case. Actually, recent estimates[77,98] suggest that, while ellipsoidal distortions for the other clusters do not change the stability predictions appreciably, the added degree of freedom allowed to 12 increases its stability in agreement with experiment.

For small clusters $N < 12$, we might not expect agreement with a model that ignores the effects of the individual cluster structures. The procedure of minimizing the cluster energy by relaxing the shape to conform to the available wave functions appears to work well, not only for the larger clusters discussed above but also for the smaller ones.[37] The latter calculation is for clusters 3–8 and 13. The atomic locations are allowed to relax to conform, e.g., to the p-like orbitals for $N = 3$–7 in the $1p$ shell. Refinement of both experimental and theoretical studies of the structure of the abundance spectra of the alkalis is expected to reveal further essential features of the electronic structure of metal clusters.

22. Noble-Metal Ions

The foregoing discussion relates to neutral cluster systems of pure alkali metals. The major features in the mass spectra identify clusters of special stability associated with the closing of electronic shells at total electron numbers 2, 8, 20, etc. It is reasonable to inquire whether similar relations apply to clusters which are produced as ions. In fact, experimental mass spectra for positive copper cluster ions[94,104,105] show high abundances at 9+ and 21+, representing the number of atoms in the singly ionized clusters and corresponding to total valence electron numbers 8 and 20, which characterize neutral systems of high stability. The pattern for silver ions[105,106] is similar, where abundances of 3+, 9+, and 21+ are conspicuous. In other experiments[107] the observed abundances in the negative-ion spectrum (Fig. 20) of silver are high for 7− and 19−, which are also 8- and 20-electron systems. The mass spectra for copper[104,105] and silver[106] also show a feature at 35+.

We have labeled the monovalent cluster ions according to N, the number of atoms in the cluster. A singly ionized positive cluster will contain one

[104] W. Begemann, K. H. Meiwes-Broer, and H. O. Lutz, *Int. Symp. Mol. Beams, 10th, 1985.*

[105] K. H. Meiwes-Broer, private communication (1986).

[106] F. M. Devienne and J.-C. Roustan, *Org. Mass Spectrom.* **17,** 173 (1982).

[107] G. Hortig and M. Möller, *Z. Phys.* **221,** 119 (1969).

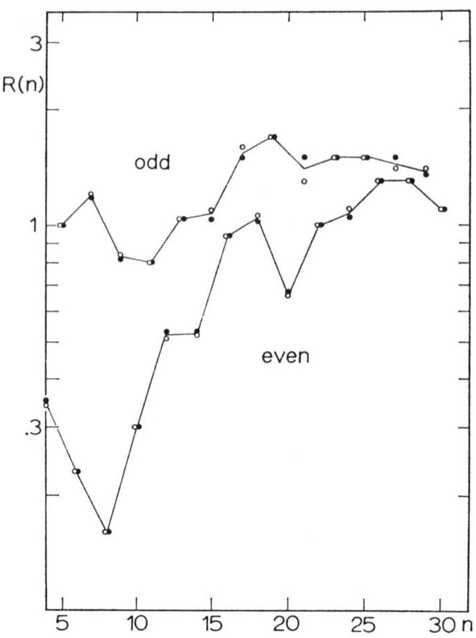

FIG. 20. Ag^- cluster ion abundance spectrum. Note peaks at 7− and 19−. (○) and (●) represent 15 and 10 keV incident energy, respectively. [After Hortig and Müller.[107]]

fewer electron: 9+ contains eight electrons, 21+ contains 20 electrons, etc. For negative ions containing an extra electron, 7− and 19− are, respectively, 8- and 20-electron systems.

Very recently, an extensive study of the secondary-ion (SIMS) spectra[32] (Figs. 21 and 22) for Ag, Cu, and Au positive cluster ions has reported features for Ag at $N = 3+, 9+, 21+, 35+, 41+, 59+, 93+, 139+$, and $199+$. The same sequence is seen for Cu and Au positive cluster ions up to 59+. These latest experiments provide striking confirmation of the shell structure for the noble-metal cluster ions over an extended range. It is interesting to see that the features at 18 and 34 electrons predicted by spherical shell theory appear in the experimental mass spectra of the cluster ions. The relative intensities for 18, 20, 34, and 40 electrons vary somewhat among the three metals:

$$N = 18, \text{Cu} < \text{Ag} < \text{Au}; \qquad N = 20, \text{Au} < \text{Cu} \sim \text{Ag}$$

$$N = 34, \text{Cu} < \text{Ag} < \text{Au}; \qquad N = 40, \text{Au} < \text{Cu} \sim \text{Ag}$$

Since these metals have comparable r_s values (2.67, 3.02, and 3.01 a.u. for Cu, Ag, and Au, respectively), a simple dependence on r_s could not be the only factor in these variations. The occupied d-electron core states are not considered in this model. It is possible that the structure of the pseudo-

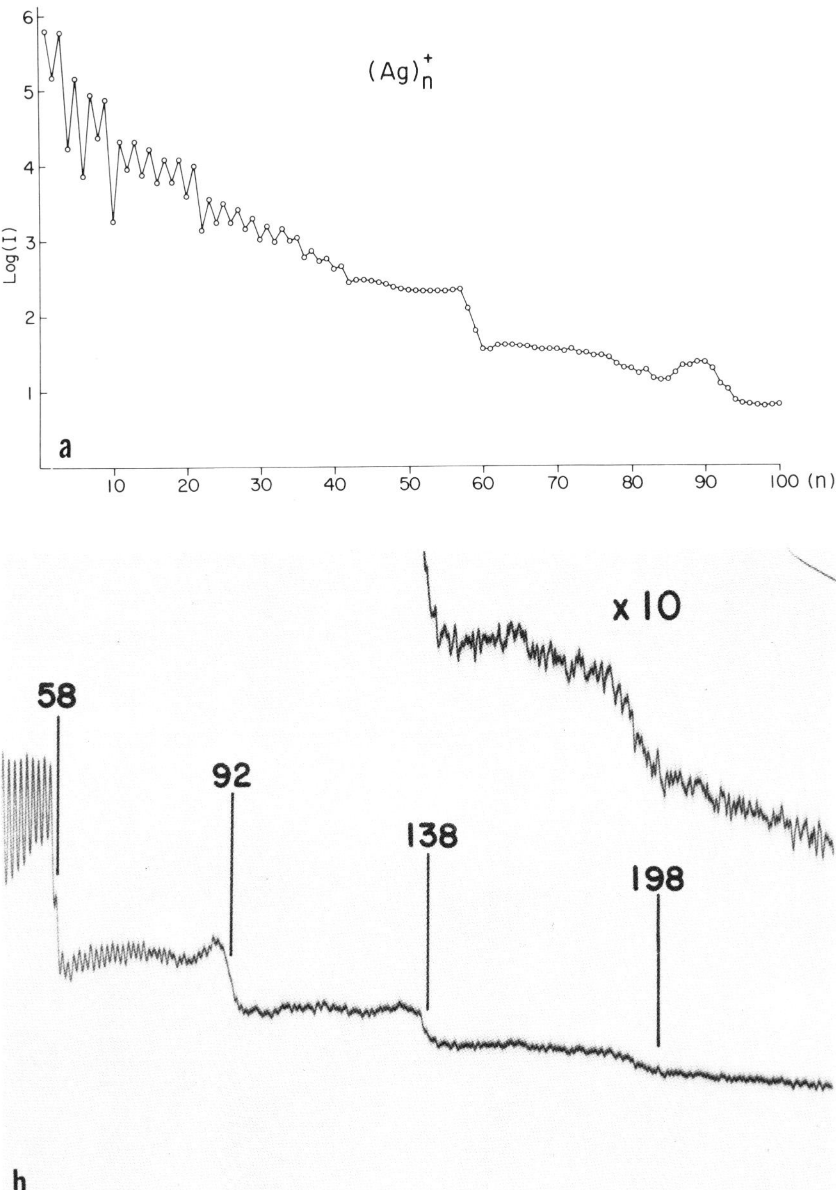

FIG. 21. Abundance spectrum of Ag^+ cluster ions; (a) low mass range; (b) high mass range. [After Katakuse *et al.*[32]]

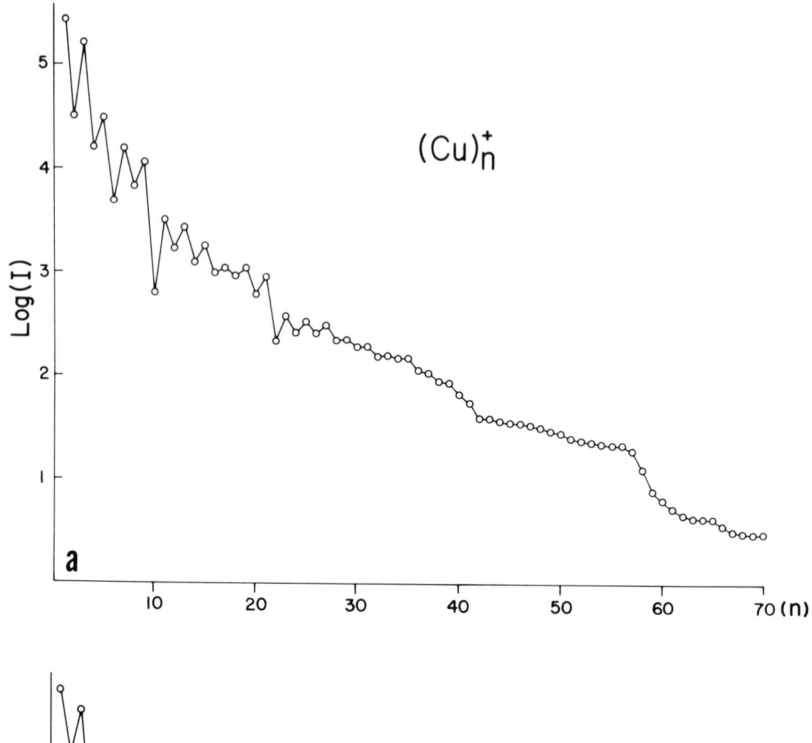

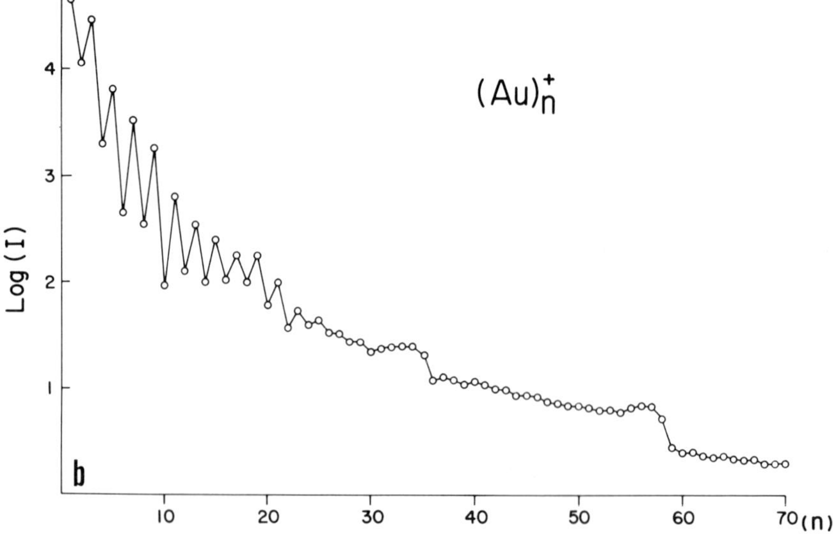

FIG. 22. Abundance spectra of cluster ions: (a) Cu^+; and (b) Au^+. [After Katakuse *et al.*[32]]

potentials for the cores and the *d* states could affect the results. The fact that 18 is observed as a distinctly resolved peak results from the odd–even effect.

No conspicuous fine structure appears in cluster ions of these metals, and the presence of the odd–even alternation and lack of adequate intensity resolution may obscure it. One might infer from this that the ionic structures are not significantly distorted from spheres.

In addition to the shell effects in the cluster ion systems, the odd–even variations in abundances are pronounced, with the odd ions (i.e., those whose parent neutrals contained an odd number of atoms) being the more abundant.[108] The odd–even effect in ions has also been observed in Li cluster ions up through 9+,[109] but it has not been reported in polyvalent metal clusters. In a more extensive recent report,[110] the odd–even alternations have been studied in the noble-metal ions. The singly charged odd positive and negative ions are the more stable, while the even doubly charged positive ions show larger stability. Further analysis has been given by Joyes.[110] The odd–even intensity effect in the noble-metal cluster ions[32] is comparable to the shell effects and extends beyond $N = 40$ for Ag. However, the odd–even effect in the neutral alkali clusters is less intense and less regular. No odd–even alternations are observed for the divalent metals[109] Be and Mg or Zn and Cd.[33] Differential odd–even electron system stabilities are seen in negatively and positively charged cluster ions[110] as well as in some more complex systems, such as[111] Cu_xBe_y, where enhanced intensity is found for x odd, and enhanced intensity of Cu_xAl_y is found for $x + y$ odd.

The odd–even effect in cluster ions is discussed by Joyes[110] in terms of shape-dependent stability. Spin interactions[112] must also be considered. It could be that there exist well-separated energy levels within a shell, resulting from the stronger pseudopotentials for the noble metals compared with the alkalis and from the crystal-field effect, as discussed in Section II,6. Comparing stabilities for neutral and ionized systems, we note the respective conditions of origin. Although both are detected as ions, the spectra of secondary ions and ions from liquid-metal sources represent systems which are born as ions and subject to a set of conditions during the formation process, resulting in high abundances for the more stable species. Although the detailed conditions of formation[91] differ from the those which produce neutrals[113], e.g., in the seeded source, both systems eventuate because of general stability conditions governed by electronic energies and shell structure.

[108] P. Joyes, *J. Phys. Chem. Solids* **32,** 1269 (1971).
[109] M. Leyleter and P. Joyes, *Radiat. Eff.* **18,** 105 (1973).
[110] P. Joyes and P. Sudraud, *Surf. Sci.* **156,** 451 (1985).
[111] M. Leyleter and P. Joyes, *J. Phys. B* **7,** 516 (1974).
[112] D. R. Snider and R. S. Sorbello, *Surf. Sci.* **143,** 204 (1984).
[113] Cf. Part III and Appendix B.

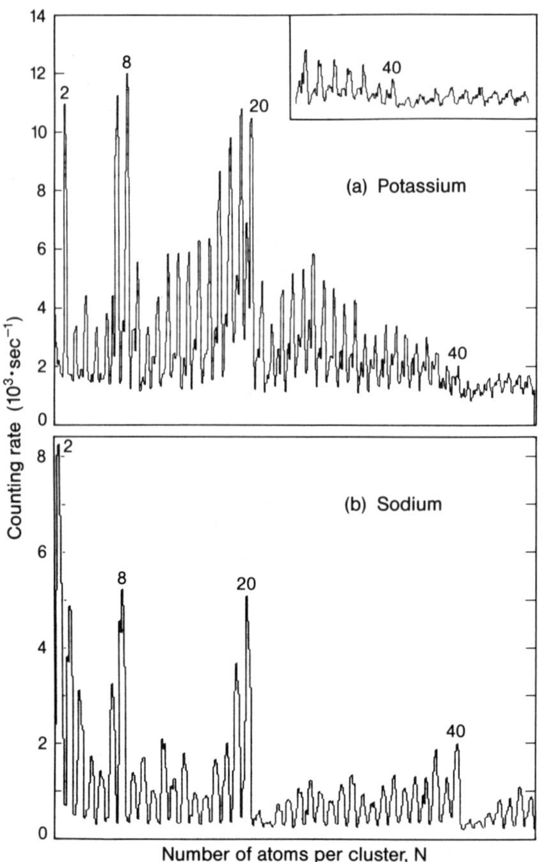

FIG. 23. (a) K–Na mixed cluster spectrum, compared with (b) pure Na spectrum. Mixed cluster peaks are seen between the pure K peaks. [After Knight *et al.*[29]]

23. MIXED CLUSTERS

Mass spectra for K_xNa_y with $y = 1$ show peaks or edges at 8, 20, and 40, as shown in Figs. 23 and 24. Careful examination of the profiles reveals that the mixed cluster feature at 20 is sharper than it is for pure potassium. This might be understood by assuming the Na impurity with smaller r_s to reside near the center of the cluster, causing a local minimum in the potential well[114] near the center. For the spherical wave functions, only *s* electrons have appreciable amplitude near the center. Therefore, this deeper impurity potential would

[114] W. Ekardt, private communication (1985).

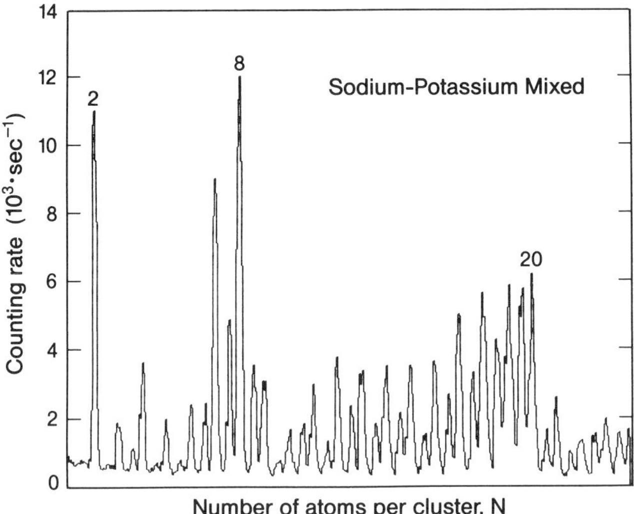

FIG. 24. K–Na high-resolution mixed spectrum. Mixed cluster peaks are seen between the pure K peaks. The overall profile shapes for pure K and mixed clusters are different as the 2*s* shell closing is approached. [After Knight *et al.*[29]]

increase the binding and stability of the extended *s* states and raise the relative intensities of 19 and 20 as observed. The results for mixed clusters K_xCs_y, $y = 1$,[98] indicate that the feature at 20 is reduced corresponding (r_s of Cs is larger) to a local maximum in the potential for *s* states near the center of the cluster.[114] Further study is needed to decide the question as to whether the presence of an impurity affects the spheroidal distortions.

Abundances for some K_xMg_y clusters have been reported[115] showing a maximum intensity at $x = 8$, $y = 1$. Although this is a ten-electron system, it is likely that the Mg electrons are not delocalized. Since the atomic ionization potentials are quite different (7.64 eV for Mg versus 4.34 eV for K), the Mg atom could behave essentially as a neutral inert unit in the cluster with the two Mg valence electrons forming a localized subshell. So the K_8Mg cluster stability may be expected to be like the pure K_8 cluster. The fact that K_6Mg has a higher ionization potential than K_8Mg is explained[115] in terms of a Jahn–Teller distortion (see Note Added in Proof).

In secondary-ion emission experiments,[111] the CuAl and CuBe alloys show the expected alternations, but the series of clusters investigated did not extend beyond a few electrons. The pure Al experiments show an edge at 7, which apparently also shows up for Al_xMg systems. Experiments of this sort will be

[115] M. M. Kappes, P. Radi, M. Schär, and E. Schumacher, *Chem. Phys. Lett.* **119,** 11 (1985).

of further interest when larger clusters are made and studied. Mixed clusters represent a rich area of research and pose a number of new and difficult problems of interpretation.

24. Polyvalent Metals

Aluminum and magnesium are predicted (Fig. 8) to show shell structure based on numbers of delocalized electrons equal to the valence times the number of atoms. Experimental cluster ion abundance spectra are known[105,106] for aluminum (Fig. 25), as well as for the divalent elements zinc and cadmium.[33] The aluminum spectra show features at measured masses 7 and 14, which correspond, respectively, to large stabilities for 20- and 40-electron systems, allowing for the fact that not all shell closing numbers are divisible by the valence number 3. Tentative observation of features in the range 20–23 atoms are also reported.[105] In addition, the evaporative[105] effects are observed to be more severe when flight times are increased in the TOF mass spectrometer. This is evidence for in-flight fragmentation resulting from inadequate cooling in the postproduction process. (See Appendix B for a discussion of evaporation processes).

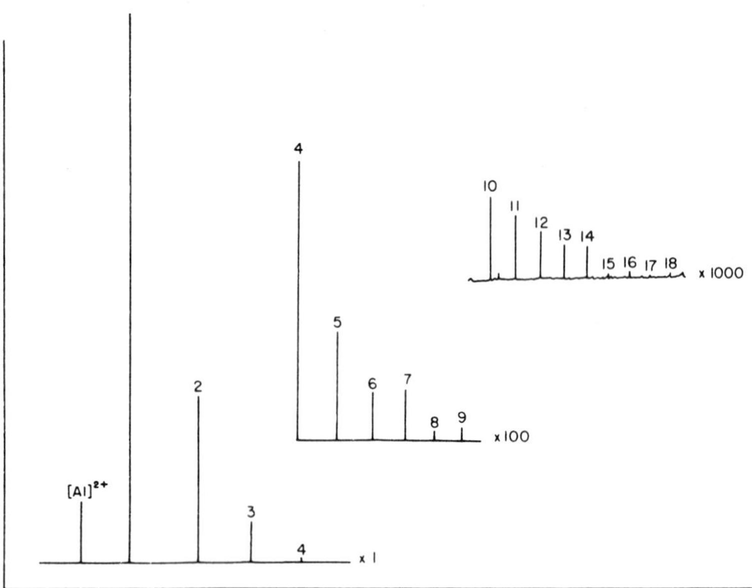

Fig. 25. Abundance spectrum for Al^+ cluster ions. Note edges at 7 and 14. [After Devienne and Roustan.[106]]

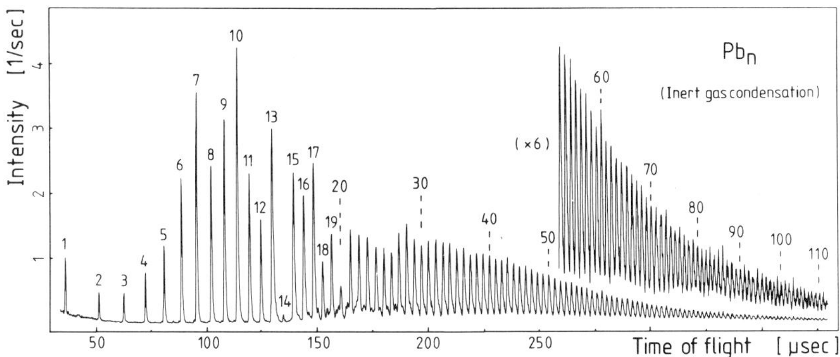

FIG. 26. Mass spectrum of Pb clusters. [After Möhlbach *et al.*[117]]

The results of Katakuse *et al.*[33] for zinc and cadmium show features corresponding to the 2*s*, 1*f*, 2*p*, 1*g*, 3*s*, 1*h*, 2*f*, 1*i* shell closings, according to their interpretation of a model intermediate between a harmonic oscillator and a square well (see Fig. 1). Their earlier[32] results on the noble metals plus these for two divalent elements indicate the possible large extent of the range of materials and size of metallic clusters which may yet be expected to show shell structure.

Additional experimental data for other trivalent atom clusters are needed in order to reconfirm the conclusions drawn for Al. The experimental data[116,117] for Ga and In are insufficient to allow conclusions.

Lead has a unique spectrum (Fig. 26), with features at $N = 7, 10, 13,$ and 19. It has been suggested[100] that these features reflect an icosohedral sphere packing. The features do not correspond to the simple shell structure interpretation. It is likely that the large s–p splitting must be included along with spin-orbit coupling effects. The latter may also produce effects in the Cs mass spectrum, which is yet to be studied. Similar questions apply to transition metals, as discussed in Subsection 9.

25. Conclusions

Experimental abundance spectra correlating with shell models exist for alkali metals, noble metals, divalent zinc and cadmium, and trivalent aluminum. The size range now extends to ~200 atoms per cluster. It seems reasonable to expect that experimental work may be extended to clusters containing hundreds of atoms and to a number of yet-untested elements. The

[116] T. P. Martin, *J. Chem. Phys.* **83,** 78 (1985).

[117] J. Mühlbach, K. Sattler, P. Pfau, and E. Recknagel, *Phys. Lett. A* **87A,** 415 (1982).

shell structure is a general property of clusters and will be used as an organizing principle for their further study. Since shell theory supposes that the electrons in the system are delocalized, we may infer that metallic properties begin early in the sequence from atom to bulk, and we may optimistically hope that further study of the level structure will shed some light on the development of the band structure of bulk crystalline metals.

It might seem remarkable that the features of shell structure appear in such apparently different systems as neutral clusters condensing in a molecular jet source or emerging from an ion-bombarded surface. In fact, it must be clear now that both sources of clusters involve thermodynamic processes which result in unique stability and abundance patterns. We enlarge on this idea in Appendix B and show that relative intensities of neighboring clusters in the Na abundance spectra can be successfully predicted from energies derived from the Nilsson diagram.

V. Static Dipole Polarizability

26. Metallic Surface Properties

The screening of external fields from the interior is a fundamental metallic property which is closely related to the delocalized nature of the conduction electrons. The microscopic configuration of the surface charge in the presence of an applied electric field had been uncertain until recently. The application of the self-consistent uniform jellium background (SCUJB) model[118] gives the electron density in the region of the surface. Oscillations in charge density and a surface charge extending beyond the jellium edge are predicted.[55]

The surface response is important for small particles and, in particular, for clusters with a high surface-to-volume ratio. The nature of the dipole response can be probed by measuring the static polarizability of clusters. This, in turn, can be compared with calculations for clusters. In this Section we discuss the recent theoretical and experimental evaluations of cluster polarizabilities. Relations to shell structure are found experimentally and theoretically.

27. Theories of Polarizability

Lang and Kohn found[55] that, in the presence of an applied electric field, the electron density extends appreciably beyond the jellium edge for a bulk metal surface (Fig. 27). The centroid of the induced electron density locates the

[118] N. D. Lang and W. Kohn, *Phys. Rev. B: Solid State* [3] **12,** 4555 (1970).

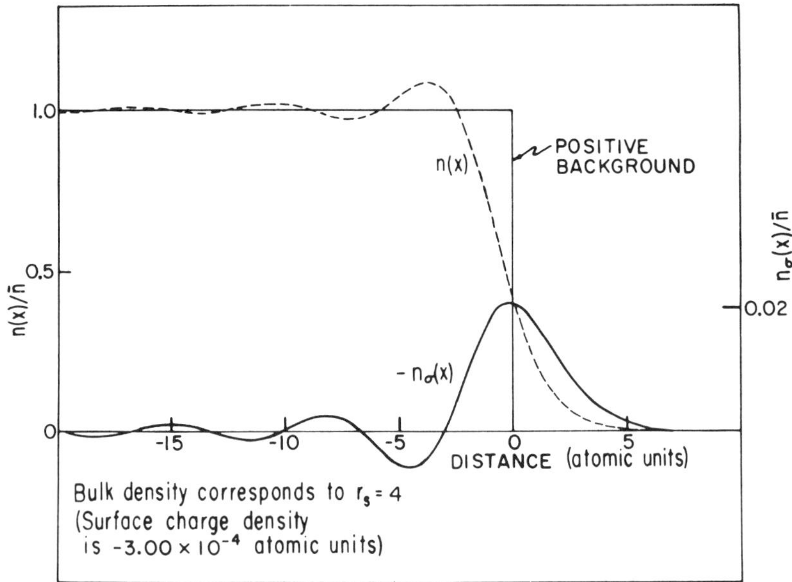

FIG. 27. Electron density and induced charge density at jellium edge of bulk sodium. [After Lang and Kohn.[55]]

position of the image plane, within which the average field is essentially zero, and is found to lie a distance δ outside the jellium boundary. The parameter $\delta_p = 1.3$ a.u. for a Wigner–Seitz radius $r_s = 4$ a.u., which applies to sodium.

Early attempts[119–121] to calculate the polarizabilities of metal clusters assumed them to be spherical and found charge and field distributions in the presence of an external field ε. Various assumptions were made concerning the boundary conditions, potential barriers, electronic response, etc. Different results, both larger and smaller than the classical result, were obtained. The classical solution gives the radial field on axis rising to 3ε at the surface and falling abruptly to zero inside. The induced surface charge corresponds to a polarizability $\delta = R^3$, where R is the radius of the sphere. Some calculations, e.g., Ref. 119, found appreciable field penetration in small clusters with an implied polarizability less than the classical value.

The first application of the jellium model to small spheres[122] was a semiclassical density-functional calculation which evaluated the electron

[119] M. J. Rice, W. R. Schneider, and S. Strässler, *Phys. Rev. B: Solid State* [3] **8,** 474 (1973).

[120] B. B. Dasgupta and R. Fuchs, *Phys. Rev. B: Condens. Matter* [3] **24,** 554 (1981).

[121] W. Ekardt, D. B. Tran Thoai, F. Frank, and W. Schulze, *Solid State Commun.* **46,** 571 (1983).

[122] D. R. Snider and R. S. Sorbello, *Phys. Rev. B: Condens. Matter* [3] **28,** 5702 (1983).

density using a gradient expansion approximation for the kinetic energy functional. The resulting polarizability was expressed in a form based on the classical equation $\alpha = R^3$. The result

$$\alpha = (R + \delta)^3 \tag{27.1}$$

includes the polarizability parameter δ, which is the analog of the Lang–Kohn image plane distance δ_p and indicates that the electron density of the clusters also spills out beyond the jellium edge. The calculated[122] polarizability parameter δ for the cluster is some 30% smaller than its bulk counterpart and is approximately independent of cluster radius. This result implies that a small cluster responds to external fields like a metal sphere whose effective radius is slightly larger than the classical value. For example, 8- and 20-atom clusters of sodium with $R \sim 8$ and 11 a.u., respectively, have predicted polarizabilities of 42% and 30% larger than the classical value assuming $\delta \sim 1$ a.u.

Subsequent SCUBJ calculations were carried out independently by several authors.[123–125] These authors use the SCUJB in the random-phase approximation and arrive at results for the polarizability which are in good agreement with those of Beck,[126] who used the modified Sternheimer relation. The electric field and density profiles analogous to the bulk surface calculation are plotted in Fig. 28,[126] showing the charge spillout, the screening to the field in the interior, and the Friedel oscillations. It is found that the polarizability parameter δ is uniquely size dependent and may be related to shell structure. The wave functions, which tend to be localized near the center of the sphere for the smaller shells, are related to smaller polarizabilities. This follows[126] from the fact that the fractional excess of polarizability over the classical value is linearly proportional to the fraction of the electronic density outside the jellium boundary.

In Fig. 29, we show several theoretical results for the polarizability. The earlier ones indicate low screening and small polarizabilities compared to the classical value. The recent calculations include the physical content of the results of Snider and Sorbello,[122] verifying that the polarizabilities of small clusters should be larger than the classical value, mainly as a result of the charge spillout and the resulting effective increase in the cluster radius. Ekardt cites the importance of the energy-level structure. The elastic deformation of the jellium background has also been studied.[127] It is found that the total

[123] W. Ekardt, *Phys. Rev. Lett.* **52,** 1925 (1984).

[124] W. Ekardt, *Surf. Sci.* **152,** 180 (1985).

[125] M. J. Puska, R. M. Nieminen, and M. Manninen, *Phys. Rev. B: Condens. Matter* [3] **31,** 3487 (1985).

[126] D. E. Beck, *Phys. Rev. B: Condens. Matter* [3] **30,** 6935 (1984).

[127] P. Sheng, M. Y. Chou, and M. L. Cohen, *Bull. Am. Phys. Soc.* [2] **31,** 636 (1986).

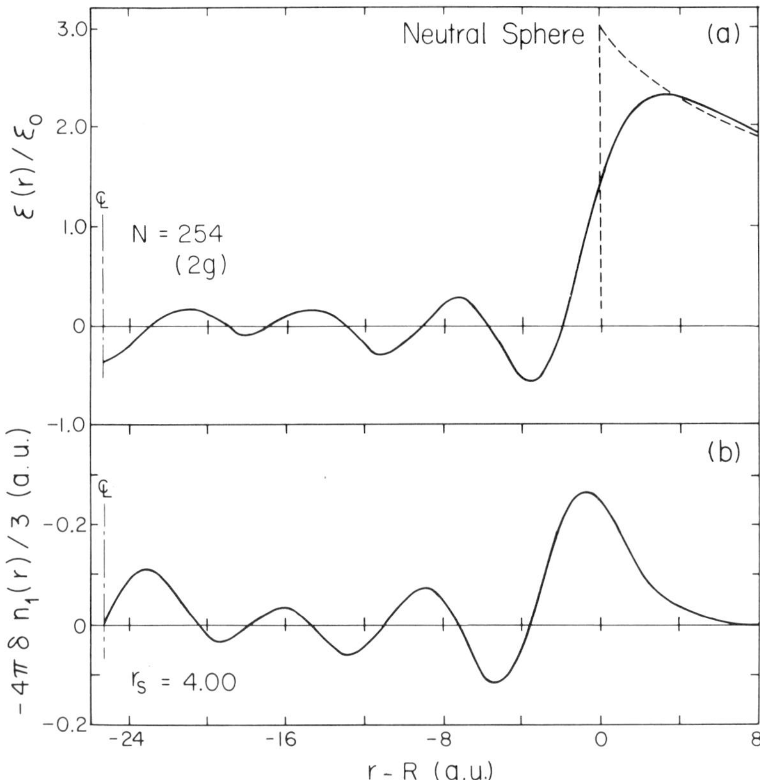

FIG. 28. (a) E field and (b) induced charge density in Na cluster, $N = 254$. [After Beck.[126]]

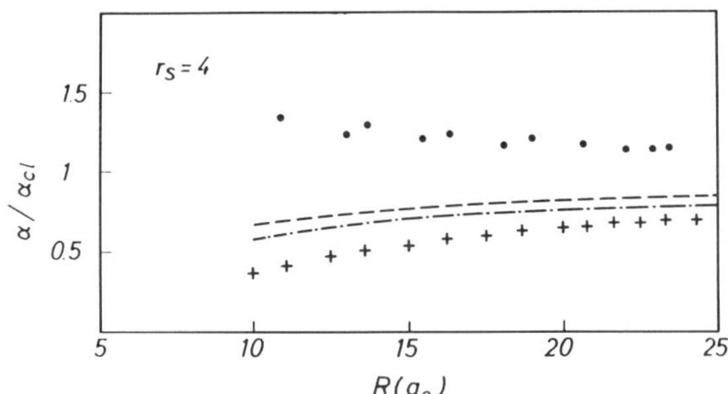

FIG. 29. Self-consistent calculation for sodium cluster polarizability (···). Examples of non-self-consistent calculations indicate weak screening: step density Thomas–Fermi calculation (–––); semiclassical approximation (—·—); quantum infinite barrier model (+ + +). [After Ekardt.[124]]

polarizability as a sum of the electronic and ionic response remains remarkably constant as the jellium stiffness is relaxed from perfect rigidity.

28. Theory for Open-Shell Clusters

A calculation of the polarizability for other than closed-shell clusters is difficult because of the lack of spherical symmetry. Nevertheless, the polarizability of open-shell clusters has been estimated using fractional occupation numbers to obtain spherically symmetric wave functions.[128] This calculation employs the jellium model with self-consistent density-functional methods. Figure 30 shows the results for lithium. The variations in polarizability are related to the electronic shell structure, but not in a simple way. For example, as a new electronic level fills, the polarizability increases because of the large spatial extent of the added wave function, but the polarizability decreases with the addition of further electrons whose wave functions are more localized.[125] However, there are exceptions to this rule. In these cases, the localization is suppressed,[125,128] and the polarizability increases as the level is filled.

29. Effects of Spheroidal Distortions

The specific effects of spheroidal distortions must be included. Assuming that screening for spheroids is essentially complete, like that for spheres, the polarizability follows from a calculation of the depolarizing factors, with the image plane a distance δ from the jellium edge. The polarizability is

$$\alpha_i = \alpha_0 D_0 / D_i \tag{29.1}$$

where D_0 is the depolarizing factor for a sphere, $(4\pi/3)$, D_i corresponds to the ith axis, and α_0 the polarizability of a spherical cluster with the same polarizability parameter δ.

Molecular-beam deflection experiments measure a polarizability averaged over all orientations. Therefore, the measured polarizability is expressed $\alpha = q\alpha_0$, where the depolarization parameter q is defined as

$$q = \left(\frac{1}{D_x} + \frac{1}{D_y} + \frac{1}{D_z}\right)\frac{D_0}{3} \tag{29.2}$$

The calculated values of q are given in Fig. 43 and are derived from depolarization factors[129] appropriate to the calculated[58] shapes of the clusters. The spheroidal distortions contribute up to 8% to the total polarizability.

[128] M. Manninen, R. M. Nieminen, and M. J. Puska, *Phys. Rev. B: Condens. Matter* [3] **33,** 4289 (1986).

[129] J. A. Osborn, *Phys. Rev.* **67,** 351 (1945).

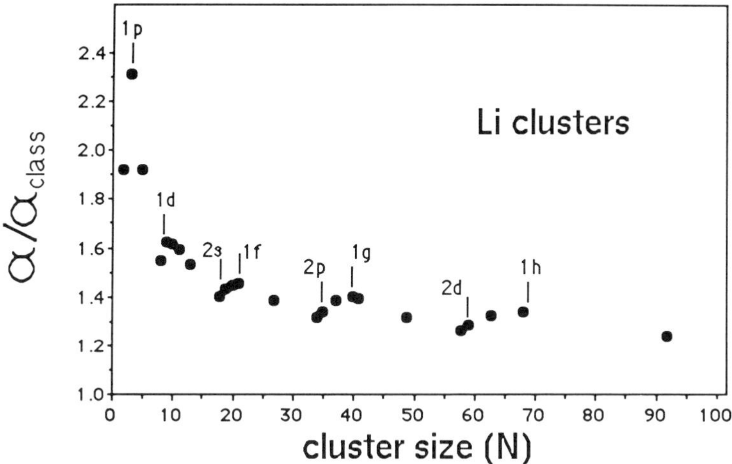

FIG. 30. Calculated polarizability of Li spheres. The beginnings of the shells are labeled. [After Puska *et al.*[125]]

The product of the depolarization factors and the calculated[128] polarizabilities is shown for Na in Fig. 31.

30. Experimental Polarizabilities

The experimental method involves molecular-beam deflection in an inhomogeneous electric field.[30,103,130,131] Polarizabilities per atom, normalized to the atomic polarizability, are plotted for Na and K clusters in Fig. 32. The bulk value, also normalized to the atomic polarizability, is indicated. The results for Na and K are the same within experimental error. The experimental values for sodium follow a general downward trend toward the bulk value and are approximately constant between $N = 20$–40. The two values for polarizability at $N = 20$ are derived from different experimental runs, and the lower polarizability correlates with conditions believed to produce cooler clusters. The experimental values for 20–40 may therefore reflect elevated temperatures for those clusters.

The experimental trend is actually not smooth but shows definite correlations with shell structure. Conspicuous local minima appear at $N = 2$, 8, and 18–20, corresponding to major shell closings, and no conspicuous features are seen above 20. The $1p$ shell shows variations which probably depend on the details of the individual cluster structures. It is also possible to

[130] R. W. Molof, H. L. Schwarz, T. M. Miller, and B. Bederson, *Phys. Rev. A* **10,** 1131 (1974).
[131] W. D. Hall and J. C. Zorn, *Phys. Rev. A* **10,** 1141 (1974).

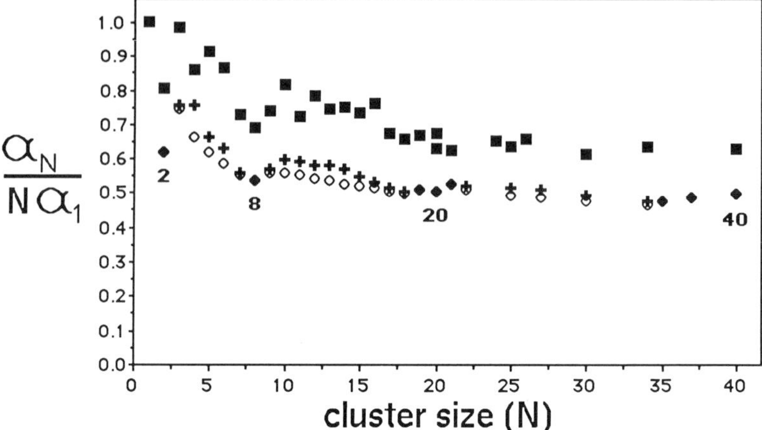

FIG. 31. Calculated polarizability per atom for Na clusters (○), including depolarization factors (+), compared with experimental values (■).

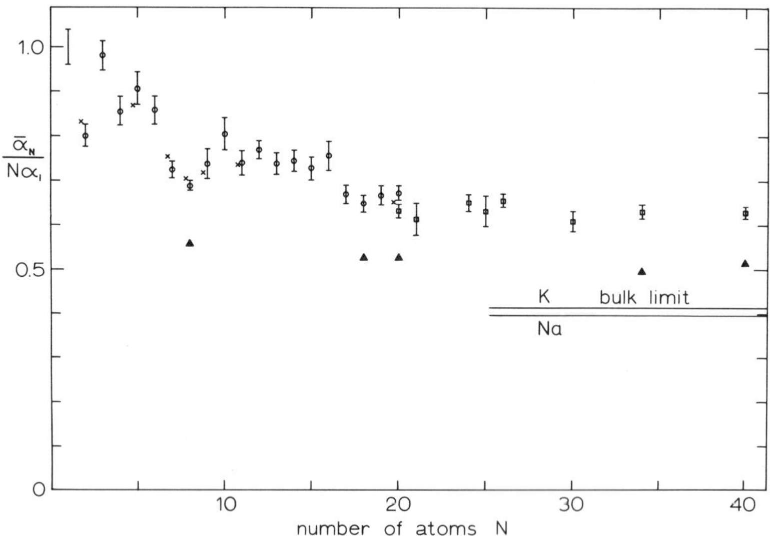

FIG. 32. Experimental polarizability compared with spherical theory (▲) (Ekardt). Experimental points: Na (○), K (×). [After Knight *et al.*[30]]

visualize two trends among the small clusters: the open- and closed-shell systems converge, respectively, toward the atom and dimer polarizabilities. In any case, the polarizabilities of the closed-shell clusters at 2, 8, and 18–20 are local minima, while the open-shell clusters in between show significantly higher polarizabilities.

31. Comparison of Theory and Experiment

The increase in experimental polarizability with decreasing cluster size suggests that the electronic screening is greater than that for classical particles and is qualitatively consistent with an increased effective size for the clusters resulting from charge spillout. The geometrical interpretation of the results is a convenient description based on more elaborate calculations involving detailed treatment of the energy levels, and we will not dwell on the details of what is implied about the radius or polarizability parameter.

We have seen correlations between a modified spherical theory[125] and shell structure. Shell effects are also reflected in the spheroidal distortion parameters, which, in turn, determine the depolarization factors. Combining these contributions, we find a preliminary value for the total theoretical polarizability (Fig. 31), which follows the experimental trend well. However, in general these and all recent theoretical values are 15–20% lower than the experimental ones.

Several effects may contribute to the discrepancy in average polarizability. These might include[103] thermally populated quadrupolar oscillations and centrifugal distortions, polarizability of the ion cores, and thermal expansion. Although inclusion of these factors in the theory might improve agreement, it is not known whether they would account for the difference. Perhaps more directly important is the question of whether the proper value of r_s for clusters indicates a contraction from the bulk value,[132] predicted to be $\sim 5\%$. This predicted contraction, borne out experimentally for supported clusters[6] but not for free clusters, would contribute a 15% correction to the present theory but in the wrong direction, which would double the disagreement.

The ion lattice correction[55] very likely makes an important contribution in the following way. The quantum-size effects operate through the level splittings. The calculation[123,124] ultimately depends on the level separations, which will certainly be reduced by the crystal-field splittings of the ion background, thereby increasing α. Inclusion of this factor will tend to improve agreement with experiment.

32. Conclusions

Considering the factors yet to be included in calculations of polarizability, we can conclude that experiment and theory may be in approximate agreement. However, the SCUJB theory is not expected to produce the detailed results which are required to describe systems as complex as real clusters. A complete self-consistent theory of ellipsoidal clusters including ion

[132] J. L. Martins, R. Car, and J. Buttet, *Surf. Sci.* **106,** 265 (1981).

background and depolarizing effects along with excited-state wave functions is needed. Such calculations will produce significant new physics.

The main static dipole response features of the cluster systems seem to be understood to first order. The alkali clusters exhibit metallic-type screening from the atoms to systems including 40 atoms. Experiments that follow the cluster behavior into the range of hundreds of atoms per cluster are yet to be done. The fact that the electron density at the cluster surface is comparable to the density of the bulk surface suggests that the basic operating physical mechanisms are understood, but more refined calculations are definitely needed. It is likely that detailed information about cluster surfaces may be acquired by further experiments on polarizability. Looking to the future, the dynamic response properties have been calculated,[133] and one may design the relevant plasma resonance experiments based on these predictions, which depend on the static dipole polarizability.

VI. Ionization Potential

33. Introduction

Interest in the ionization potential (IP) of alkali clusters was first recorded[5] about 20 years ago during the early development of techniques for making and studying clusters in beams. In these early experiments, it was found that the IP for small sodium clusters exceeded the bulk work function and varied with cluster size, then limited to $N \leq Na_8$. A second report[134] from the same group gave results for small clusters of other pure and mixed alkali metals. Later measurements[135] were extended to a series for Na ($N \leq 14$) and K ($N \leq 8$), including some mixed NaK systems. Sodium ($N \leq 8$) has also been studied[136] at higher resolution, with results similar to those of Ref. 135, including some post-threshold features. Later experiments with Na seeded beams[64,137] obtained IP values for selected larger clusters up to 65, and a variation of IP with $1/R$ (where R is the cluster radius) was reported.

Direct evidence for shell structure and the associated energy gaps should appear in the variation of IP with cluster size. For example, it is expected that the IP for a closed-shell cluster would be significantly larger than for clusters

[133] Cf. Refs. 123 and 124.

[134] P. J. Foster, R. E. Leckenby, and E. J. Robbins, *J. Phys. B* **2**, 478 (1969).

[135] A. Herrmann, E. Schumacher, and L. Wöste, *J. Chem. Phys.* **68**, 2327 (1978).

[136] K. I. Petersen, P. D. Dao, R. W. Farley, and A. W. Castleman, Jr., *J. Chem. Phys.* **80**, 1780 (1984).

[137] E. Schumacher, M. Kappes, K. Marti, P. Radi, M. Schär, and B. Schmidhalter, *Ber. Bunsenges. Phys. Chem.* **88**, 220 (1984).

of the next larger shell. Thus, one should observe features in the curve of IP versus N corresponding to the major features in the abundance spectra. This was first observed and reported[31] for potassium clusters between 2 and 30. Taken together with more recent measurements,[98] the results on K allow an examination of the behavior of the IP over a range which extends beyond the shell closing at 92.

In this section, we compare the various experimental and theoretical results in a framework based on Lang and Kohn's theory[55] of metal surfaces. This involves a self-consistent treatment of an inhomogeneous electron gas at the edge of the uniform positive jellium background. The calculation of the work function for the plane surface involves three principal energy terms: electrostatic, exchange correlation, and kinetic. We discuss these terms for the bulk material and for the clusters.

34. Ionization Threshold Measurements

In the following, we consider mainly results on sodium[64] and potassium. The latter comprises the range 3–30 potassium atoms per cluster,[31] for which a xenon lamp and monochromator were used, and from 26 to 101 potassium atoms,[138] for which a tunable dye laser was used. The combined results are shown in Fig. 33, where the IP has reached 116% of the bulk work function at $N = 101$. The curve is not smooth but rather displays discontinuities at 8, 18, 20, 40, 58, and 92, corresponding to the main features in the experimental abundance spectra (cf. Fig. 17). For the K clusters 3–8, the IPs are in satisfactory agreement with those of Refs. 135 and 139.

Since the resolution of the optical system was less than the ionization threshold width for the first series of measurements[31] and a laser was used for the second series,[138] spectrometer resolution is not an essential consideration in the analysis of the results. As was pointed out,[31] single step-function analysis does not apply, and determination of photoionization (PI) thresholds is accomplished by a linear extrapolation of the photoionization efficiency (PIE) profile to the energy axis. The measurements essentially give values for the appearance potential (AP). The AP was found[140] to differ from the IP by a few meV for K_2 because of small effects caused by the extracting electric field in the detector. In our discussion, we neglect the difference and use the term IP for convenience.

[138] W. A. Saunders, *Proc. SPIE—Int. Soc. Opt. Eng.* (1986).

[139] C. Brèchignac and P. Cahuzac, *Chem. Phys. Lett.* **117,** 365 (1985).

[140] M. Broyer, J. Chevaleyre, G. Delacrétaz, S. Martin, and L. Wöste, *Chem. Phys. Lett.* **99,** 206 (1983).

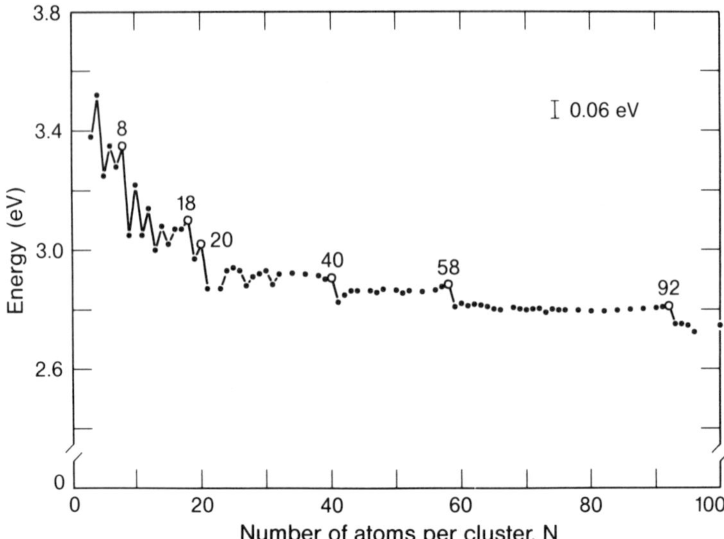

FIG. 33. Experimental K cluster ionization potentials. The bulk work function is 2.4 eV. [After Saunders.[98]]

With some notable exceptions (e.g., 3, 4, and 10), the K cluster thresholds are broad. While the laser PIE measurements often show only a part of the threshold edge of the PIE curve, the measurements of the series (3–30) give the PIE profiles to 1 eV beyond the peak. It was found (cf. Fig. 3 of Ref. 31) that the widths of the even and odd cluster thresholds average around $\frac{1}{4}$ and $\frac{1}{2}$ eV, respectively. Several of the odd thresholds show signs of unresolved structure.

Recent studies of the ionization thresholds for alkali dimers and trimers show definite temperature effects[141] (see Fig. 34). For the sodium trimer, it is tentatively concluded[142] that the approximately linear threshold profile is associated with a staircase function arising from unresolved vibrational spectra for a cluster system which possesses similar structures for both ionized and neutral states.[143] Evidence is also discussed for the effects of autoionizing Rydberg states and predissociation. For the larger clusters, one might expect similar effects to influence the ionization thresholds to some extent, which is so

[141] P. Fayet, J. P. Wolf, and L. Wöste, *Phys. Rev. B: Condens. Matter* [3] **33,** 6792 (1986).
[142] M. Broyer, J. Chevalyre, G. Delacrétaz, P. Fayet, and L. Wöste, *Chem. Phys. Lett.* **114,** 477 (1985).
[143] P. H. Guyon and J. Berkowitz, *J. Chem. Phys.* **54,** 1814 (1971).

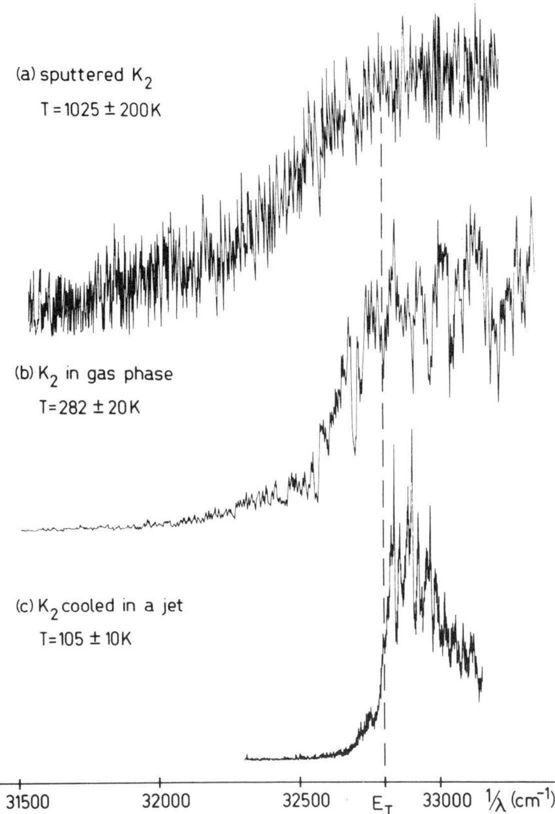

FIG. 34. Dependence of ionization threshold on temperature for sputtered potassium dimers. [After P. Fayet, J. P. Wolf, and L. Wöste, *Phys. Rev. B: Condens. Matter* [3] **33**, 6792 (1986).]

far not known. Better experimental resolution, which depends on low cluster temperatures,[65] will be required to investigate these questions.

The *shapes* of the PIE profiles show an odd–even alternation, which correlates with the measured IPs (cf. Figs. 1 and 3 of Ref. 31). The alternations, as well as the threshold widths, depend on the initial and final states involved in the ionization process. The fine structure of the energy levels of the neutral clusters is not well known, and less is known about the ion, although there are some theoretical calculations for trimers,[144] and a comparison of the level structure of Na_{60}^{+}[56] with the neutral cluster shows the two structures to be closely the same and just shifted relative to each other. In view of the lack of

[144] J. L. Martins, R. Car, and J. Buttet, *J. Chem. Phys.* **78**, 5646 (1983).

adequate data concerning both the neutrals and the ions, the interpretation of the IP thresholds is difficult. Since it is known that the resolution of the PIE profiles is not greatly improved by narrowing the bandwidth of the optical system, and that the internal cluster temperature is an important factor controlling resolution, in further developments it will be important to work with cold clusters.[65]

35. Theoretical Framework

We consider now the main features of the IP experimental data compared with theoretical predictions: theoretical framework, general trend of the IP with N and comparison of the extrapolated IP with the bulk work function (W), and discontinuities of the IP at shell closings.

Following the work of Lang and Kohn,[55] the work function is mainly the sum of three energy terms which depend on the Wigner–Seitz radius r_s: electrostatic, exchange correlation, and kinetic. The first, E_{es}, respresents the surface barrier resulting from the spilling of electron charge beyond the jellium boundary, and the second term μ_{xc}, written here as E_{xc}, is the exchange and correlation contribution of the chemical potential for a uniform electron gas. These two terms mainly determine the depth of the potential well. The kinetic energy term E_F is the bulk Fermi energy. The work function W is expressed

$$W = E_{es} + E_{xc} - E_F \tag{35.1}$$

where all contributions are taken here to be positive. Lang and Kohn calculate for bulk sodium, according to Eq. (35.1), $W = 0.91 + 5.28 - 3.13 = 3.06$ eV, compared to the quoted experimental value 2.7 eV. For potassium, calculated similarly, they give $W = 2.74$, compared to the experimental value 2.39 eV. Inclusion of the pseudopotentials decreases the calculated values by $\sim 10\%$ depending on the crystal face of the solid. We may conclude that the calculated work functions for the alkali metals are in reasonable agreement with experiment. Similar agreement was also obtained for several other metals.

The ionization potential of metal clusters in the uniform spherical jellium background model has been calculated by several authors.[52,56,57,132,145] The method is to calculate the energy difference between the ground states of the ion and its neutral parent. The variational calculations[145,146] do not show shell structure, while the others do show it. Agreement among the self-consistent calculations which show shell structure is quite good. However, it is interesting that in some of the calculations the shell structure was suppressed

[145] D. R. Snider and R. S. Sorbello, *Solid State Commun.* **47**, 845 (1983).
[146] M. Cini, *J. Catal.* **37**, 187 (1975).

in the final analysis or pronounced unphysical. The calculations were then interpreted mainly in terms of the smooth variation after the removal of the shell structure.

36. Ionization Potential and Cluster Size

Some experimental[137] and theoretical[57,145] treatments have emphasized the importance of the electrostatic contribution, and there has been considerable discussion of the appropriate form for the variation. Equation (36.1) gives the classical expression which has been widely used.

$$\mathrm{IP} = W + A(e^2/R) \tag{36.1}$$

The bulk work function is W, and R is the classical radius of a cluster. Various results of experiment and theory give values of A in the range of $\sim$0.3–0.5. There has been some question whether A might take on a value $A = \frac{3}{8}$ or alternatively $A = \frac{1}{2}$, which follows from the charging energy of a metal sphere. The $\frac{3}{8}$ factor arises from a consideration of the details of charge removal, including the image potential.[147,148]

Sufficient experimental and theoretical data, cf. the preceding references, are available to conclude that, while Eq. (36.1) represents an approximate and convenient form for fitting data, it is not an exact representation of the physical processes involved. It would certainly be natural to expand the IP function with $1/r$ as the leading term, which in many cases is a good approximation, although the addition of a higher-order term would in many cases improve agreement with experiment. Experimental data for sodium[137] indicate agreement with Eq. (36.1) with $W = 2.52$ eV compared to the experimental 2.7 eV. The curve which is plotted includes surface tension corrections for the atomic volumes. The agreement would be more satisfying if appropriate cluster radii were known independently and if the theoretical arguments for the $\frac{3}{8}$ factor were compelling. In fact, none of the more extended calculations arrives at this value.

As for the variation with cluster radius, the theoretical results are approximately of the form of Eq. (36.1), but there are significant differences.[57] It is particularly interesting to study the results of Ekardt for sodium,[56] in which he suggests, prior to experimental confirmation, that the calculated shell structure might be a real physical effect. He displays (Fig. 35) the electrostatic part of the IP plotted as a function of N, which reaches the proper asymptotic value 0.91 eV, and the total IP approaches 3.06 eV, as previously

[147] J. M. Smith, *AIAA J.* **3,** 648 (1965).
[148] D. M. Wood, *Phys. Rev. Lett.* **46,** 749 (1981).

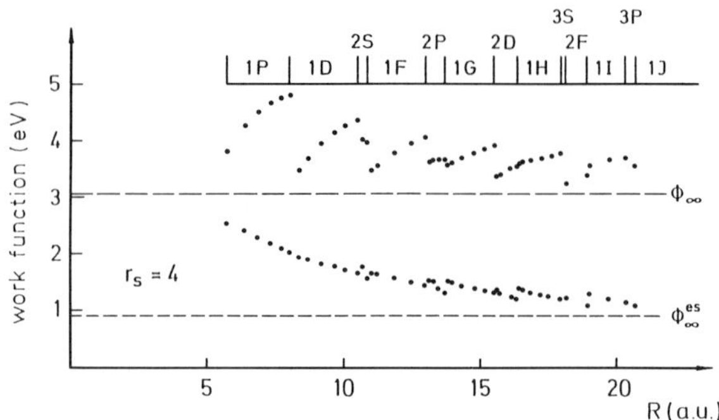

FIG. 35. Theoretical ionization potential calculation for Na clusters. Upper curve, total IP; lower curve, electrostatic contribution. The Lang–Kohn theoretical bulk work function and electrostatic term are, respectively, 3.06 and 0.91 eV. [After Ekardt.[124]]

calculated for the bulk by Lang and Kohn.[55] Ekardt[56] points out that by using Eq. (36.1), one overestimates the electrostatic contribution because of the spatial distribution of the missing charge over the whole sphere, and furthermore it does not vary as $1/R$. It should also be pointed out that in such calculations[56] the electrostatic term derives from the double layer at the surface and the "image force" is a part of the change in E_{xc} when an electron leaves the cluster, rather than part of the electrostatic energy.

Further experimental data are available.[138] In Fig. 33, we see the experimental IPs for potassium over the range 3–101. The same data are plotted versus $N^{-1/3}$ in Fig. 36, from which it may be inferred that a $1/R$ dependence is only approximate. By definition, we have

$$R = r_s N^{1/3} \tag{36.2}$$

With regard to Fig. 36, we see that a $1/N^{1/3}$ dependence is a fair approximation to the data if only the closed-shell clusters are included. In this case, the extrapolated value for the IP is 2.41 eV, in agreement with the experimental bulk work function $W = 2.4$ eV, and application of Eq. (36.1) gives $A = 0.33$. The inclusion of the open-shell clusters would change the above results, giving only a diffuse correspondence of experiment with a $1/R$ dependence and with $A \sim 0.30$. The extent of the wave functions of the open-shell clusters, discussed in Section V, and the spheroidal distortions probably operate similarly in giving reduced IPs between shell closings.

The jellium model calculation[52] (Fig. 37) for the IP of K_8 is 4.0 eV, compared to the experimental values 3.3[138] and 3.4 eV.[135,139] The difference is slightly larger than what is found in the case of the bulk work function. A similar difference is found for Na_8 with the experimental[135,136] IP = 4.1 eV, compared to the theoretical value 4.8 eV.[52,56,57] The theoretical values

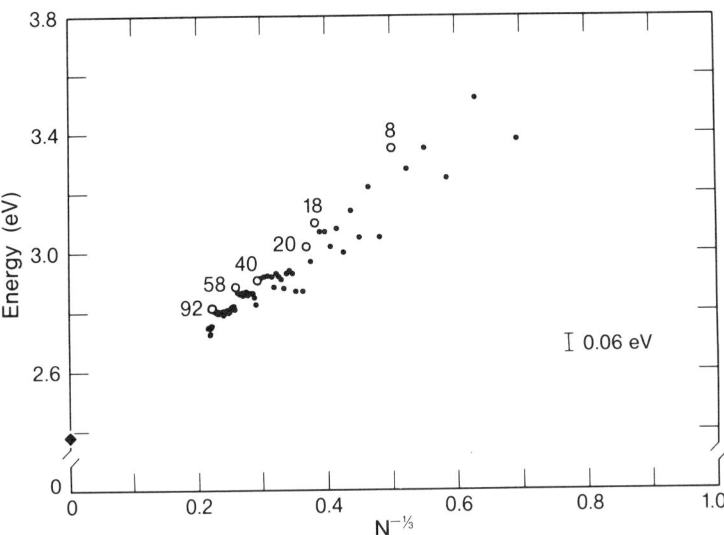

FIG. 36. Experimental K cluster ionization potential versus $N^{-1/3}$. Bulk work function is 2.4 eV. [After Saunders.[98]]

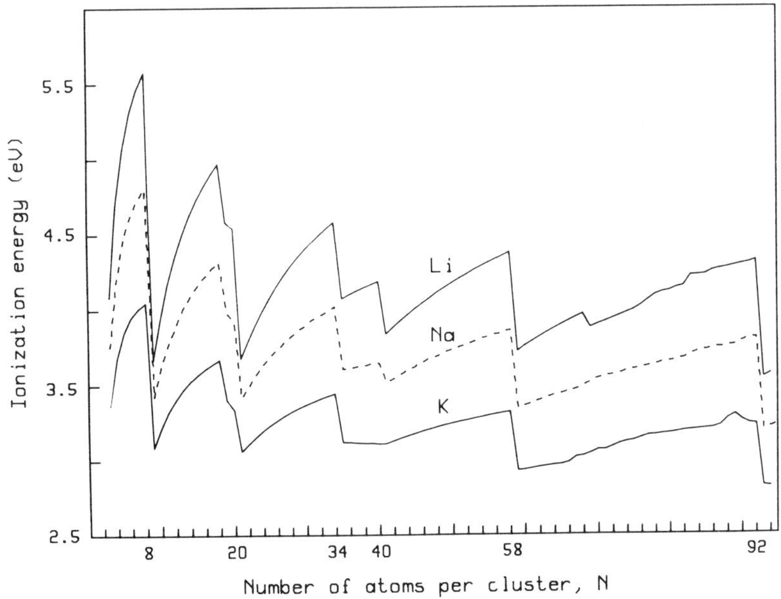

FIG. 37. Theoretical ionization potential for Li, Na, and K. The calculated work functions of the infinite half-plane jellium surface for Li, Na, and K are 3.37, 3.06, and 2.74 eV, respectively. [After Chou *et al.*[52]]

obtained from jellium calculations appear to be ~20% too high. It is expected that the introduction of a discrete atomic arrangement would bring the theoretical value closer to the experimental results. In fact, when the lowest energy structure is obtained for Na_8,[37] the resulting IP is found to be 4.3 eV.

37. Behavior of the Ionization Potential at Shell Closings

The interesting new features to be found in the data of Fig. 33 for potassium are the abrupt decreases in the IP at the shell closings. Qualitatively, this was expected, as we mentioned earlier. Quantitatively, agreement with calculations is poor. For example, the drop in potassium IP at 8, 40, and 58 is predicted[52] to be, respectively, 1.0, 0, and 0.4 eV, while the corresponding experimental values are 0.3, 0.08, and 0.08 eV. Thus, the physical effect is definitely observed, but the size is $\sim\frac{1}{3}$ what is predicted at 8, larger than predicted at 40, and $\frac{1}{5}$ what is predicted (Fig. 37) at 58.

In Fig. 38, one may see the displacements of the thresholds at the shell closings merely by superposing a series of profiles. Similarly, for the most recent[138] measurements on K, it may be seen that the profiles for the clusters just below a shell closing (e.g., Refs. 56 and 57) are virtually identical to the profile for the major feature.[58] However, they are distinctly different from the profiles for 59–62 (Fig. 39), whose IPs are discontinuously lower than the value for the previous shell. The same behavior is observed at 40 and 92. The fact that discontinuities in the IP show up at the shell closings previously identified with major features in the abundance spectra (Fig. 15) validates the

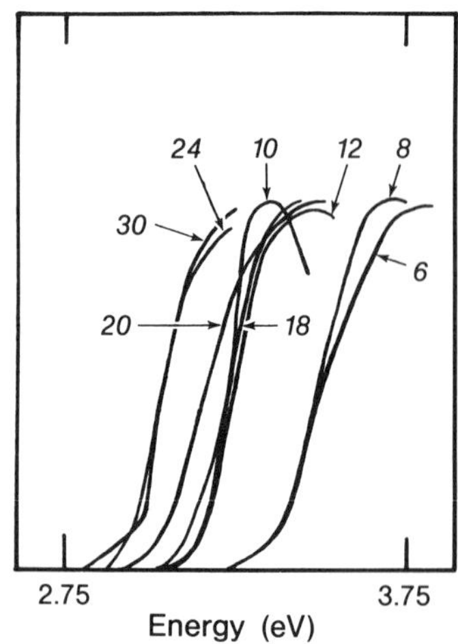

Fig. 38. Superposed ionization potential thresholds for K clusters. [After Saunders et al.[31]]

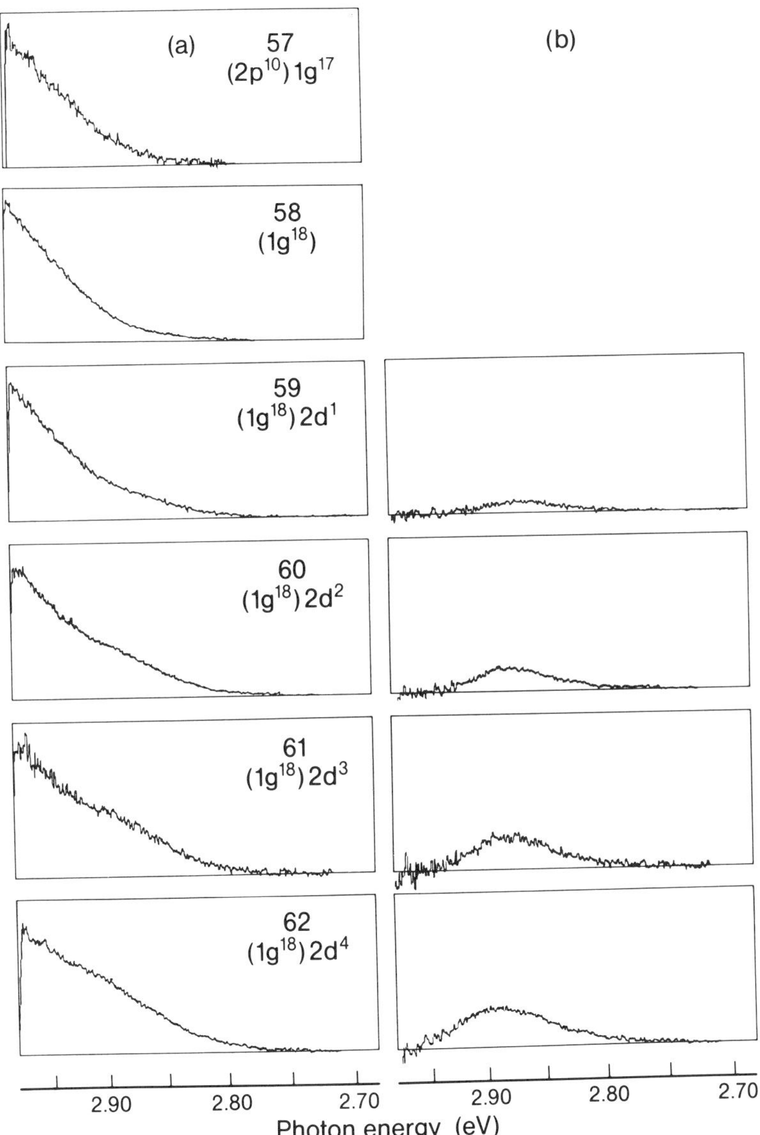

FIG. 39. Structure on PIE thresholds for K clusters: (a) profiles of ionization thresholds, (b) peaks remaining after subtraction of $N = 58$ profile. [After Saunders.[98,138]]

qualitative application of the shell model and gives a direct correspondence between energy levels and cluster behavior. The IP experiments constitute the beginning of a systematic spectroscopy of clusters, as is illustrated by the behavior of the potassium IP near 58, as discussed above.

The correspondence between IP profiles and the energy-level structures of the clusters may be appreciated by examining closely the series of clusters 58–62 (Fig. 39), in the last four of which a small peak of ~ 0.1 eV width appears superposed on the smooth threshold. It appears[138] that this peak, which grows linearly with N, represents the contributions of the $2d$ electrons to the ion yield. Similar behavior is observed qualitatively near 40 and 92.

Although the IP steps will depend on the details of the cluster wave functions and the individual energy levels, they are expected (cf. Appendix A) to scale according to

$$\Delta E \sim E_{\mathrm{F}}/N^{1/3} \tag{37.1}$$

derived from the degeneracies appropriate to the spherical symmetry. The experimental trend is in rough agreement. The small values of the steps may be the result of a small crystal-field splitting attributable to the small pseudopotentials which should have the effect of reducing the effective energy gaps.

Unfortunately, it is difficult to specifically separate the classical $(1/R)$ and quantum $(1/N^{1/3})$ effects contributing to the observed IPs. Further experiments and theoretical analysis are required to construct a complete picture. Ultimately, the important questions relate to the degeneracies inherent in the problem. The trend of the experimental energy gaps seems to be very roughly consistent with a $1/N^{1/3}$ dependence. The experimental errors are somewhat large to allow this as a firm conclusion. Certainly, however, at the other extreme, there is no indication that the Kubo [14] model with splitting of all degeneracies into a statistical manifold fits the experimental results at all.

38. Effects of Spheroidal Distortions

Assuming that the highest occupied levels of a series of neutral clusters establish the pattern for ionizing transitions, we should expect the subshell energies associated with the spheroidal distortions to be related to the cluster IPs. Qualitatively, a look at the Nilsson diagram (Appendix A) reveals fourfold sequences $(\Lambda \neq 0)$ leading to the filling of subshells at 18, 26, 30, and 34. Sequences are also seen in the IP curve which are consistent with these patterns. We may test the idea by extracting the cluster energies E_N for the neutral clusters from the Nilsson diagram and scaling the harmonic-oscillator frequency[103] as a function of N,

$$\hbar\omega_N \sim E_{\mathrm{F}}/N^{1/3} \tag{38.1}$$

The results, assuming that $E_N + \mathrm{IP}_N = \text{constant}$, are plotted in Fig. 40. The agreement (cf. Appendix A) is reasonable. One can also see odd–even alternations associated with the filling of the twofold-degenerate levels, with $\Lambda = 0$ (cf. Section IV) for $N = 2$–14, providing the cluster 12 is remarkable in the IP data as it was in the abundance spectra for not correlating with the Nilsson diagram. There is evidence[77,98] that 12 is distorted ellipsoidally with an increase in binding.

The question arises whether the observed odd–even alternation in the PIE curve follows from the level structure according to the Nilsson diagram or is a result of electron spin effects[112] or both. The fact that the experimental IP difference between 9 and 10 for K is significantly greater than calculated[31] from the theory of spin effects[112] suggests that the spin effects may be a part

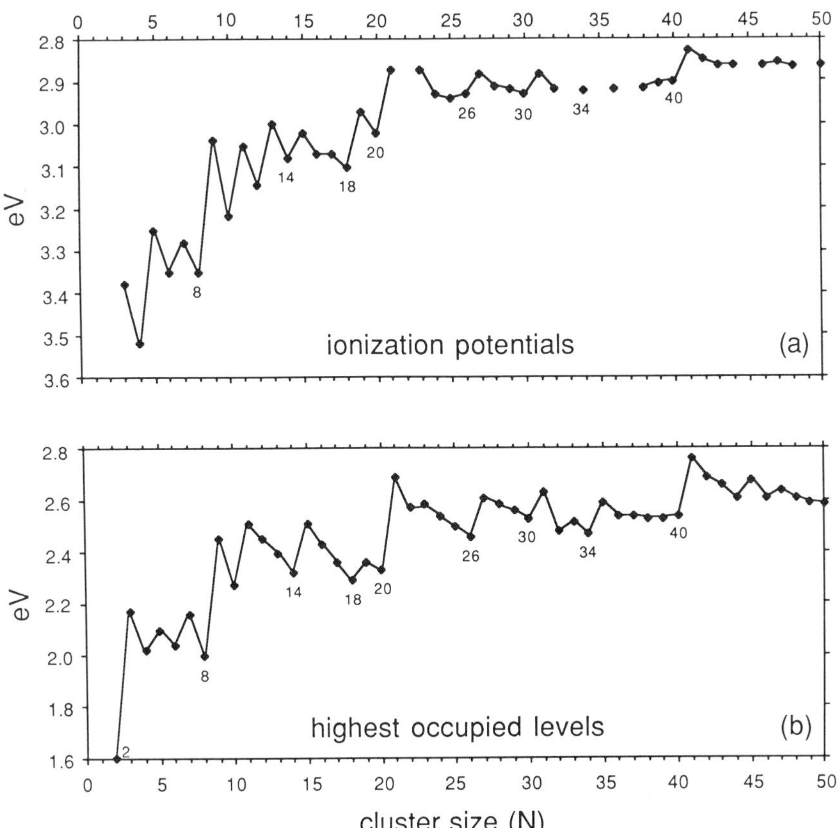

FIG. 40. (a) Experimental ionization potentials for K clusters; and (b) highest occupied levels derived from Nilsson diagram [c.f. procedures in Appendix A].

but only a part of the total effect. The spin effects are predicted[112] to die out for $N \sim 20$.

Another important contribution to the odd–even alternation in IP may be related to the odd–even alternation in cluster ion stabilities. It is known experimentally[110] and theoretically[108] that the "odd ions," i.e., those derived from odd neutrals, are the more stable and possess an even number of electrons lying at lower energies than their corresponding "even-ion" neighbors. This is a possible way of describing the final state in the ionization process; on this basis, the odd neutrals should have the lower IPs.

39. Conclusions

The prospects for performing resolved spectroscopy experiments for the larger clusters are promising. In particular, we note that, in addition to the peaks at the beginning of the $1g$, $2d$, and $3s$ shells (41, 59, and 93, as in Subsection 37 above), the earlier observations[31] found peaks at 3.8, 3.5, and 3.3 eV, which are associated, respectively, with the $1p$, $1d$–$2s$, and $1f$ shells. We expect these features in the ionization spectra to sharpen up and others to appear when significantly lower cluster temperatures are achieved.

We conclude that all of the gross features and much of the fine structure in the IP trends for the alkali metals correspond to shell structure predictions. The development of these measurements holds considerable promise for further exploration of the basic properties of clusters. Nevertheless, quantitative comparison between theory and experiment for the IPs requires further improvement beyond the present jellium model.

VII. Prospects for Metal-Cluster Physics

40. Introduction

Cluster physics is a fast-growing field with great promise for the production of new ideas and physical systems. The study of cluster structures is important not only for clusters and applied cluster systems but also for related problems in the physics of atoms, molecules, condensed matter, and transitions between these systems. The development of the field has seen an important increase in language compatibility and collaborations among experimentalists, theorists, physicists, and chemists. Like atomic and nuclear physics, the study of clusters has proceeded with the advantage of long sequences of homologous systems, intercomparisons among which make interpretations easier than would be true if only a small number of cases were available.

In this review, we emphasize that the electronic shell model is appropriate for analyzing the alkali metals, noble metals, and some simple divalent and trivalent metals. The model applies equally well to positive and negative ions and to neutrals. We have seen evidence for the metallic character even of small clusters and experimental evidence for specific energy-level structures corresponding to theoretical analysis. The foundation exists for further development of the model to apply more generally to other properties and other metals. Cluster production is understood in terms of a general thermodynamic argument based directly on the energy-level structure.

Looking to the near future, several lines of continued investigation seem clearly indicated by the foregoing discussions. These are set forth in the following sections. We continue to restrict the discussion to metallic systems.

41. Cluster Production, Structure, and Stability

An obvious first step in further work is to extend experimental and theoretical investigations to the other alkalis and other simple metals. The trends for sodium and potassium cover a wide range, and specific experimental data will be useful in refining the theory of energy-level development. The success of the simple shell theory in dealing with divalent metals and aluminum suggests that further application to other metals will be valuable. Extension of theoretical ideas to include spin and spin-orbit interactions, e.g., cesium and lead, seems promising. Studies of various mixed cluster systems will continue to be closely related to the pure clusters. The inclusion of single atom impurities and the stability of mixed clusters and elementary alloys is a challenging problem of theoretical interest with many practical applications.

The temperatures of clusters, still not well known except perhaps for dimers and trimers, affect the experimental measurement of properties. Early developments should include experiments designed to measure and control the temperatures of clusters systematically. In the seeded-beam method of making clusters, low temperatures are expected to be reached mainly by increasing the carrier gas pressures, which incidentally will increase total cluster production. Estimates made in connection with Section III and Appendix B suggest that heat capacity and stability against evaporation may be experimentally determined. These results should be a stimulus to related theoretical studies. A primary question is whether the clusters as investigated so far exist primarily in a liquid state and, if so, whether the development of a lattice structure will change measured properties.

Although fine structure in the mass spectra reflect quasiequilibrium conditions, the growth process is expected to be complicated. Actual experimental studies of the kinetics of cluster–cluster and cluster–carrier interactions may be studied by crossed-beam methods, which have been helpful in understanding chemical reactions. Experimental cross sections may be determined and compared with theoretical models.

The development of electron diffraction methods applied to small clusters will give direct information about the cluster structures, indicating the extent to which the ion background effects must be included in the jellium calculations.

Corresponding studies of cluster ions will be equally valuable. The development of SIMS sources has recently produced spectacular cluster results, and we can look forward to further development of cluster-ion sources in the near future. Cluster-ion systems are thought to be similar to corresponding neutral clusters, and in some respects, their study is easier because of the ease of manipulating ions and ion beams. Ion sources are prolific producers of clusters, and detection is perhaps three orders of magnitude better than for neutrals, which must be ionized with low cross sections in order to be detected. Comparative studies of neutrals and cluster ions will indicate the degree to which they are similar. Corresponding theoretical studies will develop. The properties of both positive and negative ions will be important.

Successful experiments[149,150] of chemical reactions with clusters suggest that this a profitable field of investigation, which should be applicable to the non-transition metals. The relation of reactivity to shell structure will certainly have fundamental as well as practical consequences. For catalysis, the spin states of the electrons can have a substantial influence.

The question of cluster stability and the stability of neutrals compared to cluster ions needs careful and systematic investigation. Electron- and photon-impact fragmentation studies are needed. Fragmentation spectroscopy will become an important branch of cluster studies. It is now known that fragmentation cross sections for metallic clusters are generally smaller than previously known typical values for molecules. This is no doubt associated with the nonlocalized bonding of the metallic electrons.

Although a rudimentary description of electonic shell structure is provided by the spherical jellium model, it is not expected that the present state of the model is adequate to describe details of the behavior of real clusters. *Ab initio* techniques are now successful in predicting structures and properties of extended systems. However, techniques such as Hellmann–Feynman force calculations have been attempted only for small clusters. The increasing collaborations of quantum chemists and condensed-matter theorists give some promise of developing new schemes for calculating total energies of

[149] R. L. Whetten, D. M. Cox, D. J. Trevor, and A. Kaldor, *Phys. Rev. Lett.* **54,** 1494 (1985).

[150] S. C. Richtsmeier, E. R. Parks, L. G. Pobo, and S. J. Riley, *J. Chem. Phys.* **82,** 3659 (1985).

large clusters using techniques for infinite systems. One approach is to use supercells to mimic clusters,[39] for which some excellent results have already been obtained.

With progress in techniques of calculation and in computer capacity, the properties of clusters can be studied in more detail and extended to yet larger clusters. The results will yield more information about the clusters and also indicate ways for improving the jellium model.

42. Cluster Properties

The shell closings are apparent in the polarizabilities of alkali clusters 8 and 20, but few data are available in the range between 20 and 40, and few features are observed. It is quite possible that temperature effects are important for the larger clusters and corresponding experimental studies are needed to evaluate them. Further developments await a self-consistent theoretical study of the distortions. The Nilsson analysis of the abundance spectra is in good agreement with the experimental results, and with proper depolarizing factors, the experimental trends in polarizability are reproduced. However, a more complete theory is needed. Since it is not known whether further features will be found in the curve of polarizability versus N, the range of the experiments must be extended. In view of the fact that polarizability depends on the energy-level splittings, which appear to be smaller than calculated in a simple jellium model, further calculations are needed, estimating crystal-field splittings and corrections to the local density approximation for comparison with experiment. Polarizability measurements should be extended to other simple metals and mixed clusters.

The expected shell effects in ionization potential exist, but the steps in IP at shell closings are $\frac{1}{3}$–$\frac{1}{2}$ the values predicted by the simple jellium model. In view of the fact that the calculations of bulk work function are improved by including the ion background contribution, it is reasonable to suppose that similar inclusions would improve the calculation of cluster IPs. In fact, the crystal-field splittings[55] would clearly tend to reduce the energy gaps. The discussion of the polarizability results suggested that agreement would be improved there, too, if the relevant energy gaps were smaller. The Nilsson analysis is applicable for IP studies and gives a pattern of results including fine-structure features in reasonable agreement with experiment. When the modified energy levels of the Nilsson model and the crystal-field splittings are included, agreement with experiment should be good. The IP calculation depends directly on the appropriate well depth, which, in turn, depends on the exchange-correlation contribution to the energy. Further theoretical study of the exchange-correlation interaction in clusters would seem appropriate.[118]

Further experiments on the temperature dependence of the IP threshold profiles and the energy gaps are needed to identify those features which are temperature dependent. Extended theoretical treatment is required to evaluate the effects of crystal-field splitting, structure of the final-state ions, transition probabilities, and the origin of the odd–even effects. A study of the second ionization potentials[100] should be undertaken experimentally and theoretically in connection with ion stabilities. Total ionization cross sections are known for alkali atoms and dimers. Experiments designed to follow the ionization cross sections over a series of clusters will be a real contribution to understanding the fundamental properties. Both experiment and theory are difficult.

No electron-impact ionization thresholds or cross sections are known for the alkali clusters. An experimental comparison with the corresponding data for photon ionization will be a contribution to knowledge of charge densities and energy-level structures of the clusters.

Experiments on the dynamical response properties are needed. Fortunately, an extended series of calculations by Ekardt[133] predict the location of the plasma resonance peak, the single-particle resonances and their coupling to the plasma resonance, and photoemission properties. This will be a fruitful area of experimentation.

43. Cluster Spectroscopy

In addition to the single-particle fine structure of PIE profiles, the vibrational structures, pre- and post-threshold phenomena, autoionization, and Rydberg states should be investigated. A theory of excited states is needed. Attempts should be made to observe collective rotational and vibrational modes of the cluster systems, analogous to nuclear collective modes. It is important to search for the predicted splitting of the plasma resonance associated with the spheroidal deformations.[58,103]

It is desirable to attempt the study of clusters that have been collected in a mass selective ion trap. This can be accomplished by ionizing neutral clusters or by directly trapping ions from an ion source. The latter is expected to give the larger concentration of ions with which one may expect to perform experiments over a wide spectral range on, for example, cluster energy gaps, photoionization, and photoemission phenomena. Transitions among excited states of cluster ions and plasma resonances may be observed and related to theoretical models. Theoretical calculations are needed to describe plasma resonances in cluster ions. Ultimately, it should be possible to work with single trapped cluster ions.

Experiments on inner-core excitations and interactions with the conduction band[151] are a promising beginning to deep-level spectroscopy of clusters. Photoemission spectroscopy experiments on clusters provide a promising probe of the energy levels and gaps.

44. New Materials

In the beginning, properties were mainly studied of clusters deposited in a matrix, and there were problems with identifying which clusters or interactions were associated with the observed spectral features. Pure "naked" cluster properties are now becoming sufficiently well known to inquire quantitatively into the measure of the cluster–surface interactions. New matrix isolation experiments should be done on mass-selected clusters designed to measure the change in a property (e.g., ionization potential, surface plasma resonance) of a cluster in a matrix compared to the naked cluster. The breaking of symmetries in the deposited clusters will be reflected in the surviving degeneracies. Together with Raman scattering experiments to examine the matrix, this will be a good method for doing surface physics.

Ion implantation methods can be used to create submicrometer arrays analogous to heterostructures with mass-selected cluster ions deposited regularly on surfaces. Clusters may be used to fabricate tunneling structures. These are but a few of many practical applications of cluster physics. If we adopt the view presented here that we can use a "giant-atom" model for the clusters, then there are *gedanken* experiments which might motivate further real experiments.

Molecular-beam epitaxy techniques have built solid structures proceeding atom by atom; similar techniques may allow a cluster-by-cluster fabrication of interesting systems. Deposition on substrates such as graphite could reveal the development of itinerant electronic states as the cluster density increases. The density dependence would be controlled by the excitation gap for the highest-energy electrons, and in turn this depends on the number of atoms in the cluster. Collective properties like superconductivity and magnetism will depend on intra- and intercluster interactions. These may be determined separately in experiments.

Localization arising from inter- and intracluster scattering in a deposited system can be examined. Disordered and ideal ordered substrates may allow controlled studies of the effects of disorder. The semiconducting, insulating, or metallic behavior of systems of this kind should depend on cluster size, the

[151] C. Brèchignac, M. Broyer, P. Cahuzac, G. Delacrétaz, P. Labastie, and L. Wöste, *Chem. Phys. Lett.* **120,** 559 (1985).

density of the clusters, temperature, and the disorder or order of the substrate. Collective and spin effects should also be sensitive to these parameters.

In general, the "giant-atom" picture supplies us with new building blocks whose properties depend on the constituent clusters. The resultant properties of aggregates of clusters will have an additional dependence on cluster density. There appear to be unlimited possibilities and every reason to believe that the present rapid growth of the field will continue to accelerate.

45. Conclusions

In retrospect, the future of clusters seemed uncertain when it depended on increasingly complex quantum chemical calculations for clusters larger than a dozen or two atoms. While this difficulty may be overcome to a large extent by the use of increasingly powerful computers, a solution based on shell structure has the advantage of elegance and simplicity. Such a scheme also possesses the power of predictability.

Appendix A. Applications of Nilsson Theory for Clusters

46. The Nilsson Diagram

The discussion of the Nilsson model of spheroidal distortions, introduced in Subsection 8, will be expanded here. This model permits us to determine the shape factors for the clusters and the single-particle energy-level structure as well. The model is basically simple, and it may be used to successfully predict properties such as variations in cluster stabilities and variations in the ionization potential as a function of N.

The Nilsson diagram (Fig. 41) is the graphical representation of the single-particle energy levels of the Nilsson Hamiltonian, Eq. (8.1). The energy levels (units of $\hbar\omega_0$) are calculated as a function of the distortion parameter η,[58] which is defined as

$$\eta = 2\left(\frac{z_0 - \rho_0}{z_0 + \rho_0}\right) = 2\left(\frac{\Omega_\perp - \Omega_z}{\Omega_\perp + \Omega_z}\right) \tag{46.1}$$

where z_0 and ρ_0 are the semiaxes of the spheroidal cluster. The strength of the anharmonic correction in Eq. (8.1) is determined by the parameter U, which is chosen so that the energy-level splittings correspond to those of the self-consistent jellium calculations.[152] The diagram as presented in Fig. 41 is drawn for the specific value $U = 0.04$.

[152] M. Y. Chou, A. Cleland, and M. L. Cohen, unpublished energy level calculations.

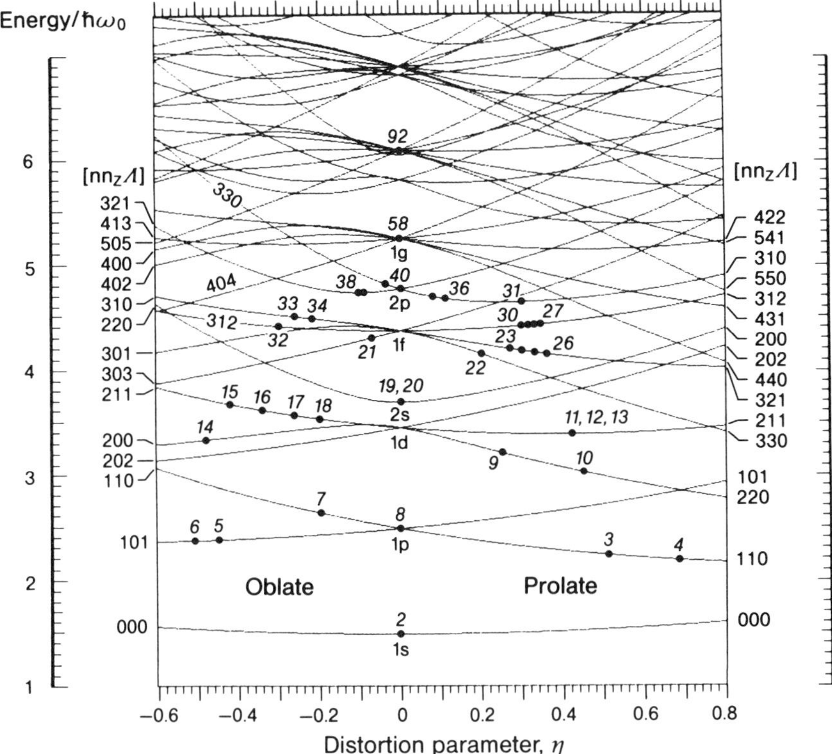

FIG. 41. The Nilsson diagram and spheroidal cluster distortions. [After Clemenger.[58]]

Assuming that the total energy of a cluster is given by $\frac{3}{4}$ the sum of the eigenvalues of the occupied levels,[27] the total energy of the cluster can be determined for each cluster as a function of η, as in Fig. 42. The energy minimum determines the value of η to be assigned to the cluster. These values appear as dots on the diagram on the respective highest occupied levels for the clusters. Thus, the distortion parameter for $N = 10$ is $\eta = +0.44$ at energy $3.0\hbar\omega_0$. The magnitude of η oscillates around 0 (spherical shell closings at $N = 2, 8, 20, 40, 58$, and 92) with maxima occurring approximately in midshell at $N = 4, 14, 26, 50$, and 74. The maxima for η decrease as N increases, signifying the tendency for the larger clusters to approach spherical shapes.

When the distortion parameter η and energy scale are known, the energy spectrum for a given cluster can be found from the intersections of a vertical line through η and the levels on the diagram. Two sets of quantum numbers are used to label the levels. These are indicated for the spherical clusters along the vertical axis through $\eta = 0$. The quantum numbers for distorted clusters

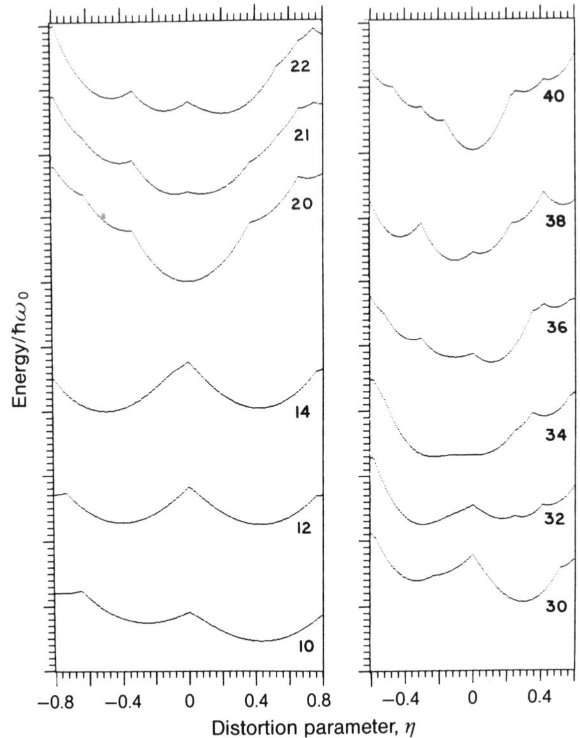

FIG. 42. Total electronic energy versus cluster distortion parameter for several even clusters. The number of atoms N is indicated for each curve. [After Clemenger.[103]]

are indicated at the left and right borders of the diagram. The number of oscillator quanta is n, the number of quanta along the symmetry axis is n_z, and the projection of the electronic orbital angular momentum along the z axis is Λ. The levels are either two- or fourfold degenerate, respectively, as $\Lambda = 0$ or $\neq 0$.

The effect of the anharmonic term is most evident in those regions for $\eta = 0$, where energy gaps exist between, e.g., the $1d$–$2s$ levels and the $1f$–$2p$ levels. For $U = 0$, those gaps vanish since the Hamiltonian reduces to that for the three-dimensional isotropic harmonic oscillator.

47. Energy Scales

The energy scale may be calibrated by assuming that the highest occupied level lies at the bulk Fermi energy, E_F. For a spherical harmonic oscillator, the number of levels is $\sim(3N)^{1/3}$, so that

$$\hbar\omega_0 = E_F/(3N)^{1/3}$$

We will refine this estimate below following arguments[103] which were adapted from the treatment of de Shalit and Feshbach.[153]

In the spherical harmonic oscillator, for a system whose lowest n levels are filled, the mean-square radius of the electron distribution is

$$\langle r^2 \rangle = 3\hbar(3N)^{1/3}/4m\omega_0 \tag{47.1}$$

The mean-square radius of the electron distribution can be estimated by assuming[103] a constant electron density within a radius $R + t$, where R is the jellium radius and t is a measure of the extent of electron spillout.[55,118] (See also discussions in Sections V and VI.) This approximation, combined with Eq. (47.1), gives

$$\hbar\omega_0 = 0.98\ E_F N^{-1/3}(1 + t/R)^{-2} \tag{47.2}$$

where $R = N^{1/3}r_s$, and the parameter t is taken to be 1.4 a.u.[103] for Na, corresponding to the energy scale in the self-consistent jellium calculations.[56,152]

48. Predictions of the Nilsson Model

Several important properties may be calculated from Nilsson theory scaled as in the previous section. The shape corrections to the cluster polarizabilities may be found from depolarization factors based on the distortion parameters. The depolarization parameters q, defined in Eq. (29.2), are displayed in Fig. 43, where effects of shell structure may be seen. Details are given in Section V.

Figure 44 shows the calculated energies of the highest occupied levels of sodium. Large steps are seen at the major shell closings (bold-faced numbers), while weaker steps are seen at the subshell closings associated with the spheroidal distortions. In Section VI, the experimetal ionization potentials are compared with the energies of the highest occupied levels for potassium (Fig. 40). The correspondence verifies both the spheroidal distortions and the level structure.

The enhanced stabilities of the closed-shell structures as inferred from the cluster abundances (Section IV) can be associated with cluster energies. Figure 45 may be used to derive relative electronic energy values for clusters N and $N - 1$. A low value of this difference indicates that the last added atom is relatively tightly bound. For example, it is useful to compare clusters

[153] A. de Shalit and H. Feshbach, "Theoretical Nuclear Physics," Vol. 1. Wiley, New York, 1974.

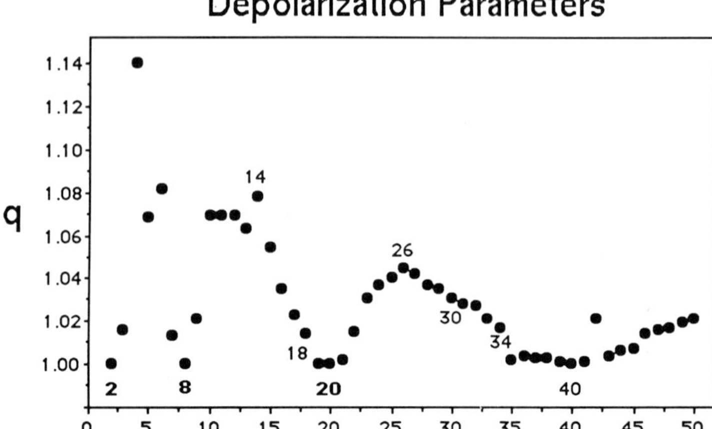

FIG. 43. Depolarization parameters q derived from the distortion parameters on the Nilsson diagram.

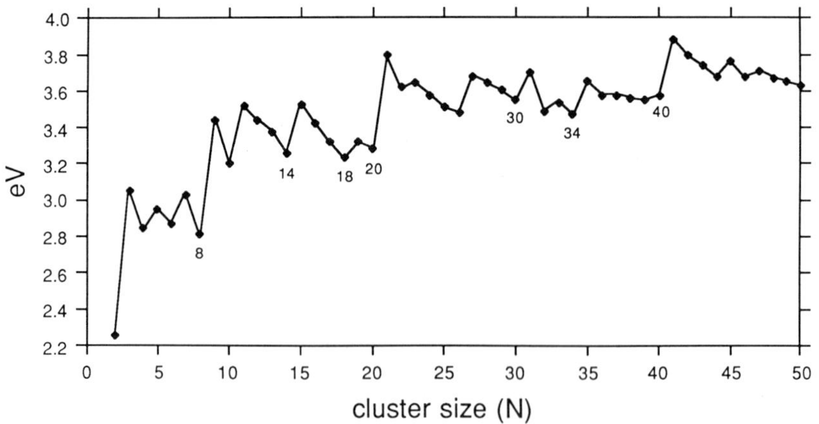

FIG. 44. Highest occupied levels derived from the Nilsson diagram for Na.

7–8 and 8–9 in Fig. 45. The observed shell structure reflects the structure of the Nilsson diagram, as was seen in Fig. 19.

Although a close relation between binding energies and cluster abundances is indicated, quantitative comparisons require the application of thermodynamic arguments. In Appendix B, we formulate a theory for cluster production based on frozen equilibrium distributions, which permits quantitative comparison between Nilsson theory and the experimental abundance spectra.

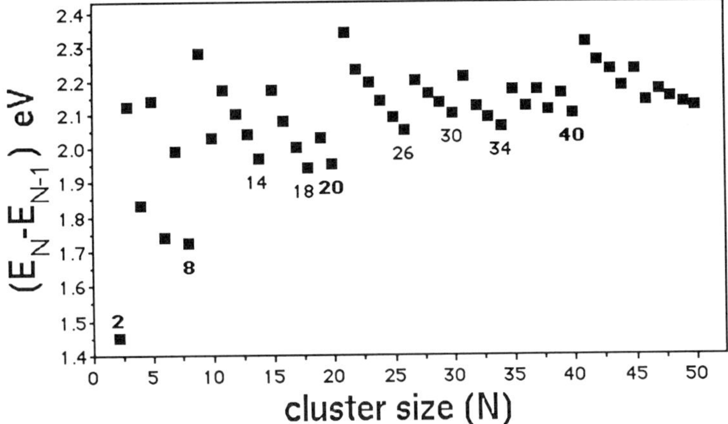

FIG. 45. Electronic energy differences $E_N - E_{N-1}$, derived from the Nilsson diagram for Na.

Appendix B. Fine Structure in Cluster Abundance Spectra

The fine structure in the mass spectra (Section IV) is clearly related to the stabilities of the clusters. Therefore, the production processes must be sensitive to the binding energies. In this appendix, we examine cluster distributions in thermodynamic equilibrium. We show that the fine structure in the abundance spectra from nozzle sources closely follows the predicted thermodynamic equilibrium distributions in the nozzle channel with some variations which can be explained as resulting from evaporation of unstable species. In the following, evaporation refers to the release of an individual atom or a dimer. Fragmentation means the release of a larger fragment. It is found that the significant processes involve mainly atoms and dimers.

49. Equilibrium Distributions

The equilibrium distributions of clusters can be calculated by examining all possible reactions of the form

$$X_I + X_J + \cdots + X_K \leftrightarrow X_L + X_M + \cdots + X_N$$

In thermodynamic equilibrium at temperature T and pressure P in volume V, the cluster concentrations ρ_N can be found from the law of mass action,[154] which in this case gives

$$(f_N/\rho_N V)^{1/N} = K(P, T) \tag{49.1}$$

where f_N is the N-mer partition function, evaluated below, and $K(P, T)$ does not depend on N.

The N-mer partition function is defined as $\sum \exp(-E_n/kT)$,[154] where the sum is over all the states of the N-mer. In order to calculate the partition function, we assume a model with the following features:

(1) The cluster ($N > 2$) has $3N - 6$ vibrational modes, which all have the same frequency ω_0 following an Einstein-type model.[54]
(2) The clusters are assumed to be spherical.
(3) The ground-state binding energy BE_N can be approximately by

$$\mathrm{BE}_N = Nb + E_N \tag{49.2}$$

where E_N is the total electronic energy, and b is a smoothly varying function of N.

The partition function is factored in the usual way

$$f_N = f_{\mathrm{tr}\,N} f_{\mathrm{rot}\,N} f_{\mathrm{vib}\,N}$$

and the terms become for $N > 2$

$$\begin{aligned} f_{\mathrm{tr}\,N} &= N^{3/2} V/V_q \\ f_{\mathrm{rot}\,N} &= N^{3/2}(8\pi/5)^{3/2} R_N^3/V_q \\ f_{\mathrm{vib}\,N} &= \frac{e^{-\mathrm{BE}_N/kT}}{(1 - e^{-\hbar\omega_0/kT})^{3N-6}} \\ V_q &= \left(\frac{2\pi\hbar^2}{mkT}\right)^{3/2} \end{aligned} \tag{49.3}$$

where m is the atomic mass and R_N is the radius of the N-mer, which we take to be $N^{1/3}r_s$. The spheroidal distortions may also be incorporated in $f_{\mathrm{rot}\,N}$, but such refinements do not significantly affect the results.

As an example, we first calculate the equilibrium density of atoms and clusters. From Eq. (49.2) we find that

$$\left(\frac{f_N}{\rho_N V}\right)^{1/N} = \left(\frac{f_1}{\rho_1 V}\right)$$

so that for large N

$$\rho_1 = \left(\frac{1}{V_q}\right)(1 - e^{-\hbar\omega_0/kT})^3 e^{\mathrm{BE}_N/NkT} \tag{49.4}$$

In the limit of large N, $-\mathrm{BE}_N/N$ is the bulk cohesive energy, and $P = \rho_1 kT$ is the bulk vapor pressure. For sodium, the cohesive energy is 1.13 eV, and it is

[154] F. Reif, "Fundamentals of Statistical and Thermal Physics." McGraw-Hill, New York, 1965.

found that if $\hbar\omega_0/k = 110$ K, Eq. (49.4) predicts the sodium vapor pressure to within 30% of the measured value for $300 < T < 1000$ K, spanning over 13 decades in the vapor pressure.

Absolute cluster concentrations may be calculated similarly by solving Eq. (49.2) for all N, and in order to treat supersaturated vapors, appropriate boundary conditions may be imposed. Since we are only interested in fine-structure features, we will consider relations between the densities of clusters of neighboring sizes. Within this model, Eqs. (49.2) and (49.3) give

$$\ln \frac{\rho_N^2}{\rho_{N+1}\rho_{N-1}} = -\frac{2E_N - E_{N+1} - E_{N-1}}{kT} = \frac{\Delta_2(N)}{kT} \tag{49.5}$$

This expression involves the electronic contribution to the binding energy, which will be calculated with the help of the Nilsson model. In order to compare with experiment, the relation between detected intensities, I_N, and the densities in the source must be established. If variations in the detection efficiency with cluster size are small, ρ_N may be replaced by I_N in Eq. (49.5), giving

$$\ln \frac{I_N^2}{I_{N+1}I_{N-1}} = \frac{\Delta_2(N)}{kT} \tag{49.6}$$

The left-hand side of Eq. (49.6) represents the experimental intensities as from Fig. 15. The right-hand side represents values for E_N derived from the Nilsson diagram (Fig. 41). A test of the applicability of Eq. (49.6) to cluster production processes can be made by plotting the results of the left- and right-hand sides of Eq. (49.6) and comparing as in Fig. 46. A best fit of the two plots is found for temperature $T = 1030 \pm 400$ K, which is in rough agreement with the measured nozzle channel temperature of 800 K.

Equation (49.5) predicts peaks at *8*, 10, 14, 18, *20*, 26, 30, 34, *40*, 46, 50, 54, *58* (Fig. 46a). Of these, all but the peak at 18 are observed experimentally, (Fig. 46b), and additional ones are seen at 12, 17, 23, and 43. The spheroidal Nilsson model does not predict a peak at 12, although distortion along the third axis lowers the energy[77,98] of 12 sufficiently to produce a peak in $\Delta_2(N)$.

The extraneous odd peaks appear to result from evaporations. For example, if 10% of the equilibrium 18-mers evaporated an atom, the expected peak at 18 would be moved down one notch in the sequence. Similarly, the peaks at 23 and 43 may arise by contrast from relatively low abundances for 22 and 42. In a further discussion of evaporation in the next section, it will be found that 22 and 42 are both minimally stable with respect to the evaporation of a dimer (which itself is usually stable), resulting in enhanced abundances for the closed-shell peaks at 20 and 40.

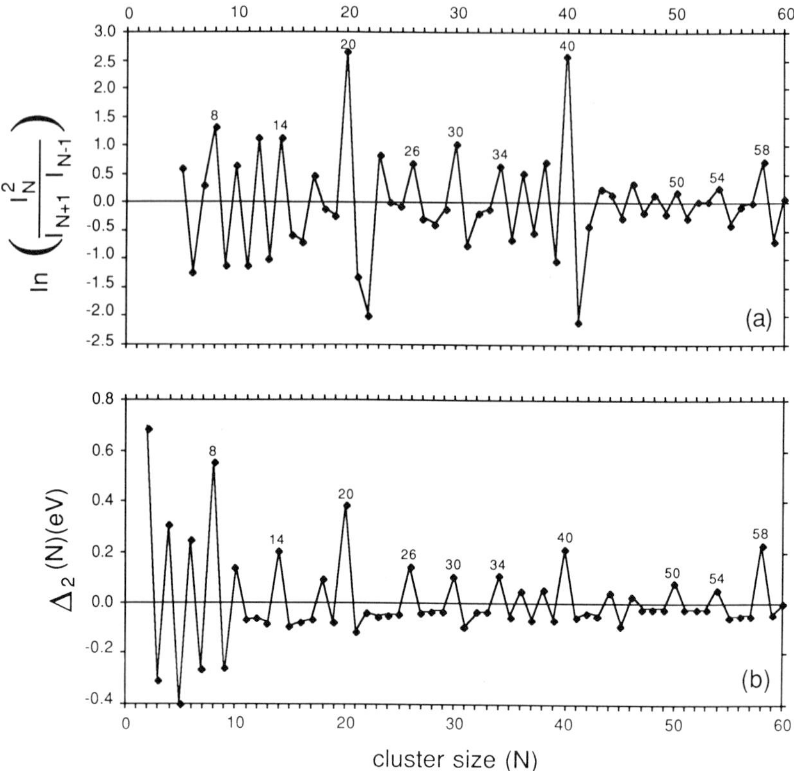

FIG. 46. (a) Measured neighboring cluster abundances, compared with (b) total energies from the Nilsson diagram, according to Eq. (49.6).

50. Evaporation and Fragmentation

At elevated temperatures, a cluster may have enough internal energy to evaporate or fragment significantly, reducing the internal temperature. For example, assuming a binding energy of 1 eV per atom and a heat capacity of $3Nk$, we estimate temperature drops of 200 and 250 K, respectively, for the evaporation of an atom or a dimer. The evaporation rate depends on the temperature and on the stability. We now calculate the evaporation and fragmentation rates according to the model of section 49 and compare with experiment.

In thermodynamic equilibrium, the formation rate of a cluster equals its fragmentation rate. The formation rate will be estimated using equilibrium densities and assuming classical geometrical formation cross sections. For the process

$$X_{N-M} + X_M \rightarrow X_N$$

the normalized formation rate may be written as

$$W_{N-M,M} = \frac{1}{\rho_N}\left(\frac{d\rho_N}{dt}\right) = \sigma v'\left(\frac{\rho_{N-M}\rho_M}{\rho_N}\right) \tag{50.1}$$

where $\sigma \sim \pi(R_{N-M} + R_M)^2$, and v' is the mean relative speed of interacting particles. The density ratio can be calculated from Eq. (49.1) and the partition functions. However, we must treat separately the cases $M = 1$, $=2$, and >2 according to the respective partition functions. We find

$$\begin{aligned}
\frac{\rho_{N-M}\rho_M}{\rho_N} &= g(M, T)V_q^{-1}\left(\frac{(N-M)M}{N}\right)^4 e^{-\mathrm{DE}/kT}(1 - e^{-\hbar\omega_0/kT})^3 \\
g(1, T) &= 1 \\
g(2, T) &= 2\pi R^2 V_q^{-2/3}\left(\frac{1 - e^{-\hbar\omega_2/kT}}{1 - e^{-\hbar\omega_0/kT}}\right)^3 \\
g(M > 2, T) &= \left(\frac{8\pi}{5}\right)^{3/2}\left(\frac{r_s^3}{V_q}\right)(1 - e^{-\hbar\omega_0/kT})^3
\end{aligned} \tag{50.2}$$

where DE is defined as

$$\mathrm{DE} = (\mathrm{BE}_{N-M} + \mathrm{BE}_M - \mathrm{BE}_N)$$

R is the dimer bond length and ω_2 is the dimer vibrational frequency. From Eqs. (50.1) and (50.2), the equilibrium formation and fragmentation rates can be determined.

We will investigate the evaporation rates for sodium clusters as a specific example. In Figs. 45 and 47 we show the values of $(E_N - E_{N-1})$ and $(E_N - E_{N-2})$. It is clear from Eqs. (49.2) and (50.2) that large values for these functions identify clusters which are relatively unstable for evaporation of an atom or a dimer, respectively. Sodium clusters 22 and 42 are the least stable with respect to dimer evaporation. Quantitative estimates of the evaporation rates can be made if we assume that the binding energy of the clusters per atom is slightly less than that of the bulk, and that variations in this quantity from one cluster to another are given by the Nilsson model. Hence for $15 < N < 25$ the quantity b in Eq. (49.2) is taken to be -3.2 eV.

We assume the bulk value for $\hbar\omega/k = 110$ K. Furthermore, the dimer bond length = 5.8 a.u., $\hbar\omega_2/k = 220$ K, and $\mathrm{BE}_2 = -0.73$ eV.[132] Using the above

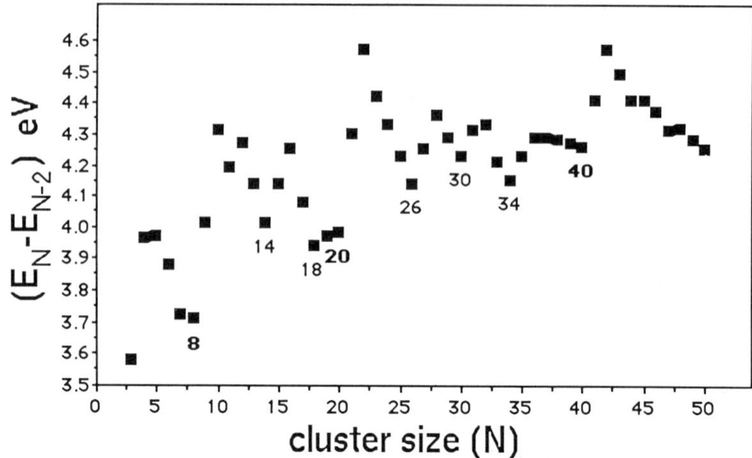

FIG. 47. Electronic energy differences $E_N - E_{N-2}$ derived from Nilsson diagram for Na.

values, we find that at 400 K the dimer evaporation rate for $N = 20$ is $\sim 2/s$, while that of $N = 22$ is $10^6/s$. In order to be stable for at least 1 ms, the temperature of $N = 22$ can be 345 K at most. The upper limit for $N = 20$ is 550 K.

In cases where fragmentation is severe when, for example, the clusters are not cooled in the nozzle expansion rapidly enough to stabilize, even the most stable ones may need to cool by evaporating one or two atoms. Applying the above model, the probability of evaporating units larger than the dimer is insignificant. The resulting effect on the abundance spectra will be to move edges in the spectra down by one or two notches. Such effects are, in fact, observed in the K mass spectra when the nozzle temperature is too high (Fig. 11).

ACKNOWLEDGMENTS

We intended this review to be selective rather than exhaustive, and we regret any omissions of relevant work.

We thank the following colleagues for stimulating and productive discussions: Walter Ekardt, J. L. Martins, L. Wöste, Keith Clemenger, Winston A. Saunders, and Andrew Cleland. We are grateful to the following colleagues for providing unpublished information: I. Katakuse, L. Wöste, and K. H. Meiwes–Broer, and M. Manninen. The following were kind enough to provide original photographs for reproduction in this paper: I. Katakuse, D. Beck, K. Sattler, and L. Wöste. We are grateful to D. Blackman for assisting in the construction of computerized drawings, and to Sari Wilde for her excellent work in typing the manuscript.

This work has been supported in part by the Materials Research Division of the U.S. National Science Foundation under grants DMR84-17823 and DMR83-19024, and by the Director, Office of Energy Research, Office of Basic Energy Sciences, Materials Sciences Division of the U.S. Department of Energy, under Contract No. DE-AC03-76SF00098.

NOTE ADDED IN PROOF

A self-consistent calculation[155,156] of the electronic structure of Na_6Mg and Na_8Mg has recently been made in the local density approximation. The charge density is found to be delocalized with no evidence for directional bonding. A central Mg atom in a Na cluster primarily affects the s states and modifies the set of shell closing numbers, with the result that both Na_6Mg and Na_8Mg are predicted to be closed shell systems. The experimental results for K_6Mg and K_8Mg[115] appear to be consistent with this prediction.

[155] S. B. Zhang and M. L. Cohen, *Phys. Rev. B* (in press) (1987).

[156] W. D. Knight, W. A. de Heer, W. A. Saunders, K. Clemenger, M. Y. Chou, and M. L. Cohen, *Chem. Phys. Lett.* **134,** 1 (1987).

Stage Ordering in Intercalation Compounds

S. A. SAFRAN

Exxon Research and Engineering,
Corporate Research Science Laboratories,
Annandale, New Jersey

I. Staging: Modulated Structure and Effects on Properties

1. INTRODUCTION

Systems with modulated structures are interesting because they exhibit structure on length scales that can be much longer than typical atomic dimensions. The large-length-scale structure can be the result of a balance of competing interactions. For example, in some magnetic systems, the competition between near-neighbor ferromagnetic and further neighbor antiferromagnetic interactions yields long-period structures.[1] Similarly, the competition between adsorbate–adsorbate and adsorbate–substrate interactions is responsible for incommensurate lattices in epitaxial layers and in some quasi-two-dimensional intercalation compounds.[2] Modulated structures can

[1] J. von Boehm and P. Bak, *Phys. Rev. Lett.* **42,** 122 (1979).
[2] P. Bak, *Rep Prog. Phys.* **45,** 587 (1981).

also arise in systems where the interaction range is large. For example, the long-range strain interactions of hydrogen atoms in metals yield macroscopic density waves of hydrogen; these systems have been reviewed in Ref. 3. The long-range nature of these modulated structures results not only in unusual ground-state structures, but also in complex and subtle phase transitions at finite temperatures.

Very often, modulated structures are seen in systems where a "guest" species has been introduced into a host lattice. In adsorbed layers, the guest atom or molecule resides on the host surface, while in intercalation compounds, the guest resides between the layers of the host material. The properties of such guest–host systems are not only of scientific interest but of technological importance, since they allow for materials engineering on the microscopic level. The structure of such systems is the crucial step for the understanding and control of these materials. Electronic, thermal, and transport properties are extremely sensitive to structure, especially in systems of reduced dimensionality, such as adsorbed layers or intercalates.

Current interest in intercalation compounds originally stemmed from the realization that the conductivity (electronic or ionic) of the guest species was relatively large and controllable.[4,5] Intercalation compounds occur in layered host materials where the interlayer bonding is weak in comparison with the intralayer binding. Two common examples of host materials are graphite, which shows a large enhancement of its electronic conductivity, and transition-metal dichalcogenides, which are of interest in battery applications based on their ionic conductivity. Although the original motivation for recent studies of the physical properties of intercalation compounds was technological, it was soon realized that these materials had a rich variety of novel structural properties. In particular, the in-plane structure of a single layer of intercalant has been shown to include quasi-two-dimensional solid, liquid, gas, and incommensurate solid phases, along with their associated phase transitions.[6] These phenomena afford the researcher a look at quasi-two-dimensional behavior in a bulk material! However, intercalation compounds also exhibit a modulated structure that is unique, namely *staging*.

The staging phenomenon is characterized by a periodic sequence of intercalate layers in the host matrix, as shown schematically in Fig. 1. The stage number n refers to the number of host layers separating two intercalant layers. Staging is characteristic of intercalation compounds; its existence does not seem to depend on the details of the order or disorder of the in-plane atomic arrangements. Although staging was originally discussed in terms of a perfectly periodic one-dimensional structure of infinite intercalate layers,

[3] H. Wagner and H. Horner, *Adv. Phys.* **23,** 587 (1974).

[4] F. L. Vogel, *J. Mater. Sci.* **12,** 982 (1977).

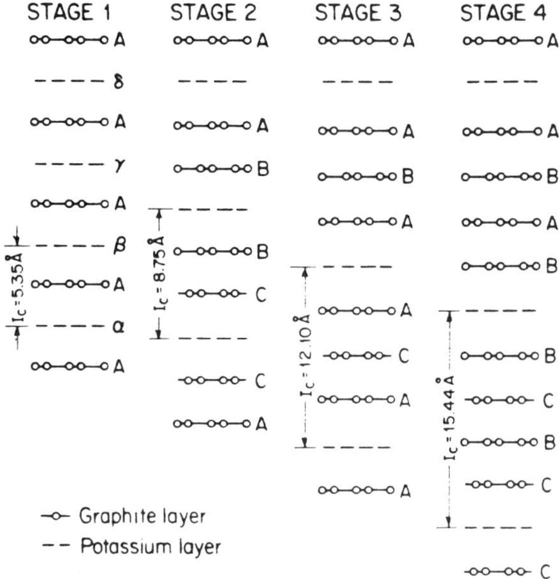

FIG. 1. Schematic diagram illustrating the staging phenomenon in graphite–potassium compounds for stages $1 \leq n \leq 4$. The potassium layers are indicated by (---) and the graphite layers by (—○—) and indicating schematically a projection of the carbon atom positions. The ...ABC... denotes the stacking order of the carbon layers, and the ...$\alpha\beta\gamma\delta$... denotes the stacking order of the potassium layers. For each stage, the distance I_C between adjacent intercalate layers is indicated. [From M. S. Dresselhaus and G. Dresselhaus, *Adv. Phys.* **30,** 139 (1981).]

recent experiments have shown that this periodicity is not a fixed chemical property of the intercalation compound, but is rather a physical consequence of the balance between the energy and entropy of the interacting intercalate layers.[7] Thus, different stages of intercalation represent different phases of the material; phase transitions between stages can be induced by changes in temperature and/or pressure.[8,9] In addition, stage ordering is not perfect; a given sample may consist of "packages" of intercalate and host that fluctuate in size.[10,11] Whether this fluctuation is a result of kinetics or a more intrinsic phenomenon is a subject of current study.[12,13]

[5] J. E. Fischer, *Phys. Chem. Mater. Layered Struct.* **6,** 481 (1979).
[6] S. A. Solin, *Adv. Chem. Phys.* **49,** 455 (1982).
[7] S. A. Safran, *Phys. Rev. Lett.* **44,** 937 (1980).
[8] J. E. Fischer, C. D. Fuerst, and K. C. Woo, *Synth. Met.* **7,** 1 (1983).
[9] R. Clarke and C. Uher, *Adv. Phys.* **33,** 469 (1984).
[10] G. Kirczenow, *Phys. Rev. Lett.* **52,** 437 (1984).
[11] M. E. Misenheimer and H. Zabel, *Phys. Rev. Lett.* **54,** 2521 (1985).
[12] G. Kirczenow, *Phys. Rev. Lett.* **55,** 2810 (1985).
[13] P. Bak and G. Forgacz, *Phys. Rev. B: Condens. Matter* [3] **32,** 7535 (1985).

In this article the staging phenomenon is analyzed. While the point of view is theoretical, connections and applications to experiment are emphasized. Although staging is observed in many types of intercalation compounds, the most numerous experiments and models have focused on graphite intercalation compounds. While some of the theory depends on particular electronic properties of graphite, the staging phenomenon is discussed in general terms. Most of the experimental examples are drawn from studies of intercalated graphite with a few examples from measurements on intercalated layered dichalcogenides.[14–16] In the last ten years (1975–1985) excellent reviews by Ebert[17] and Herold[18] have appeared on the synthesis and structure of intercalated graphite. In addition, the electronic and transport properties have been discussed by Fischer.[5] Dresselhaus and Dresselhaus[19] have reviewed a broad range of topics from the basic chemistry, physics, and materials science, to engineering applications. Solin[6] and Clarke[9] have recently summarized work on structural phase transitions and high-pressure effects.

Almost all of these recent reviews of the properties of intercalated graphite have emphasized the experimental consequences of the staging phenomenon. Here, the focus is on the conceptual and theoretical issues: the microscopic origin of the interactions responsible for staging (Section II), the phenomenon of staging phase transitions (Section III), and the kinetics of staging and intercalation (Section IV). The discussion of these topics is preceded by a brief introduction to intercalation compounds (with an emphasis on intercalated graphite) and an outline of the effects of stage ordering on the lattice, electronic, and transport properties of intercalated graphite.

2. Intercalation Processes and Staging

Intercalation of a layered host material occurs when the host contacts a reservoir of guest atoms or molecules. The guest species intercalates the host if the chemical potential difference between the reservoir (e.g., a gas of alkali atoms) and the guest atoms in the host is positive. Intercalation then proceeds and reaches equilibrium when the chemical potential difference is zero. Control of the chemical potential of the guest species is accomplished

[14] J. R. Dahn and W. R. McKinnon, *J. Phys. C* **17,** 4231 (1984).
[15] D. Kalarchichi, R. F. Frindt, *Phys. Rev. B: Condens. Matter* [3] **28,** 3663 (1983).
[16] K. K. Bardhan, G. Kirczenow, and J. C. Irwin, *J. Phys. C* **18,** L131 (1985).
[17] L. B. Ebert, *Rev. Mater. Sci.* **6,** 181 (1976).
[18] A. Herold, *Phys. Chem. Mater. Layered Struct.* **6,** 323 (1979).
[19] M. S. Dresselhaus and G Dresselhaus, *Adv. Phys.* **30,** 139 (1981).

experimentally by varying the temperature of the reservoir (e.g., the two-zone method of intercalation,[20,21] where the host is maintained at one temperature and the reservoir at another) or the electric potential in an electrochemical cell. In the former case, for a reservoir consisting of a dilute gas of intercalant, the chemical potential is proportional to $k_B T \log P$, where P is the gas pressure. In the latter case, the chemical potential is proportional to the voltage difference in the cell. In graphite intercalation compounds, the two-zone, thermal method is commonly used, although several groups have reported electrochemical data as well.[22,23] For intercalation in transition-metal dichalcogenides—e.g., TiS_2, TaS_2, $NbSe_2$—the electrochemical method predominates.[14–16]

If there are no sharp minima in the free energy of the guest species, the chemical potential at equilibrium is a smooth function of the concentration of intercalant in the host. Any structure in a plot of intercalant concentration x versus chemical potential μ usually indicates the existence of ordered phases which choose a preferred stoichiometry. The presence of stage ordering is thus observed as a series of plateaus in the $(x-\mu)$ plane. A given stage exists for a certain range of chemical potential. Outside of this range, the concentration drops sharply to that characterizing a higher or lower stage, as shown[6,20] in Fig. 2. The finite slope of the line connecting regions of two different stages signifies that the transition between them is not sharp; there can be intermediate phases with continuously varying stoichiometries.[11,24] In some cases, these phases can be due to kinetic effects and will result in hysteresis in the $(x-\mu)$ plots.

The chemical potential differences between different stages reflect their different energies. However, for the intercalation process to proceed, it is necessary that the free energy of the guest atom or molecule be reduced compared with its free energy in the reservoir phase. If the free energy reduction due to the host environment overcomes the free energy needed to expand the host structure, intercalation takes place. (Of course, there are corrections to this picture due to cooperative interactions between intercalants; these interactions are discussed below.) In host materials such as graphite or layered transition-metal dichalcogenides, the bonding is weak (van der Waals) between layers, but is covalent within a single layer. Thus, intercalation takes place through the separation of the host layers without

[20] A. Herold, *Bull. Soc. Chim. Fr.*, p. 999 (1955).
[21] J. G. Hooley and M. Bartlett, *Carbon* **5,** 417 (1967).
[22] S. Aronson, F. J. Salzano, and D. Bellafiore, *J. Chem. Phys.* **49,** 434 (1968).
[23] E. McRae, A. Metrot, P. Willman, and A. Herold, *Physica B (Amsterdam)* **99B,** 489 (1980).
[24] R. Clarke, N. Caswell, and S. A. Solin, *Phys. Rev. Lett.* **42,** 61 (1979).

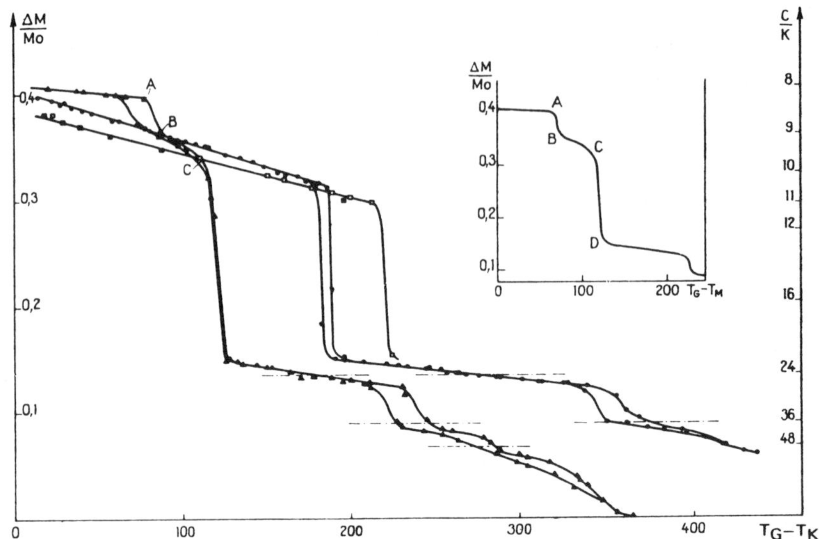

FIG. 2. Adsorption/desorption isobars for the potassium–graphite system. The carbon-to-potassium atomic ratio C/K corresponding to the relative mass change $\Delta M/M_0$ is shown on the right. The intercalant temperature T_i is fixed at 298°C. Closed circles refer to absorption, and open circles refer to desorption of intercalate. [From A. Herold, *in* "Physics of Intercaltion Compounds" (L. Pietronero and E. Tosatti, eds.), p. 7. Springer-Verlag, Berlin and New York, 1981.] T_M = 298 (▲), 492 (●), and 600°C (□).

interrupting their in-plane structure. The energy to separate the host interlayer bonding is usually compensated by an exothermic chemical reaction. The intercalant is ionized, partly or fully, when inside the host.

Intercalation compounds are classified according to whether the charge is transferred to or from the host. In the former case, a donor compound is formed by a transfer of electrons from the guest (e.g., alkali-metal atom) to the host. Acceptor compounds, which generally involve the intercalation of a molecular species, result in a transfer of electrons from the host layers to the guest layers.

Most studies of transition-metal dichalcogenide hosts have focused on intercalation with alkali metal atoms. Graphite intercalation compounds have included both donor (e.g., Li) and acceptor (e.g., $SbCl_5$) intercalants. More than one hundred such compounds have been synthesized. This number is expected to dramatically increase since ternary intercalation compounds can be synthesized (i.e., two guest species).[25,26] The chief motivation for studying

[25] B. R. York, S. K. Hark, and S. A. Solin, *Phys. Rev. Lett.* **50,** 1470 (1983).

[26] A. Herold, *in* "Physics of Intercalation Compounds" (L. Pietronero and E. Tossati, eds.), p. 7. Springer-Verlag, Berlin and New York, 1981.

these compounds lies not only in staging, but in the fine tuning of in-plane properties that is possible through the combination of intercalants with different chemical and/or physical properties. For example, two-dimensional analogs of metal–ammonia systems have been synthesized.[27] These can be used for studying quasi-two-dimensional metal–insulator transitions.

The quasi-two-dimensional nature of even binary (guest–host) intercalation compounds provides a vast variety of structures and phase transitions. The existence of staging allows the synthesis of compounds with a high in-plane density of intercalants. Since these planes can be separated by large distances (pure stages as high as 8 have been reported[28]), these materials are suitable for studying two-dimensional liquids, epitaxial ordering and incommensurability, and quasi-two-dimensional magnetic systems.

An example of the reduction of dimensionality by the separation of intercalant layers due to staging can be seen in Fig. 3, where the diffraction pattern of $SbCl_5$ intercalated in graphite is shown.[29] For stage 1, Homma and Clarke[29] found a coherent stacking sequence of intercalant layers with energetically equivalent structures (that differ only by translations), resulting

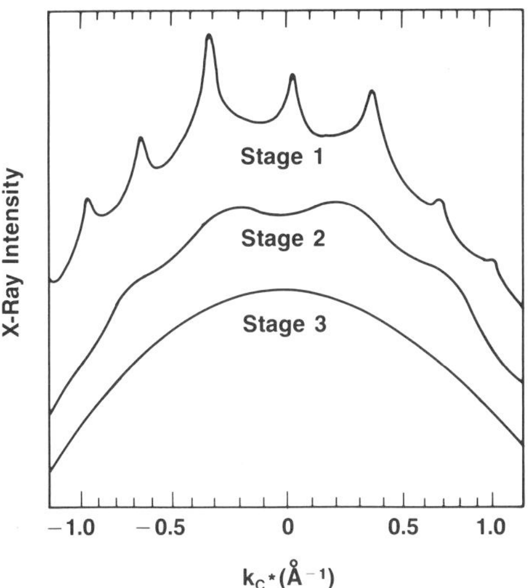

FIG. 3. Ambient temperature ($h0\ell$) x-ray diffractometer scans probing the $h = 1.11$ Å$^{-1}(\sqrt{7})$ intercalant c-axis correlations in $SbCl_5$ graphite intercalation compounds. k_c^* is the out-of-plane component of the diffraction vector. [From H. Homma and R. Clarke, *Phys. Rev. B: Condens. Matter* [3] **31,** 5865 (1985).]

[27] X. W. Qian, D. R. Stump, B. R. York, and S. A. Solin, *Phys. Rev. Lett.* **54,** 1271 (1985).
[28] C. Underhill, T. Krapchev, and M. S. Dresselhaus, *Synth. Met.* **2,** 47 (1980).
[29] H. Homma and R. Clarke, *Phys. Rev. B: Condens. Matter* [3] **31,** 5865 (1985).

in the diffraction peaks shown. As the stage is increased, the structure rapidly becomes two dimensional; the stacking order exists only over a short distance ($\sim$20 Å) in the direction perpendicular to the layers the $\hat{c}$ axis). The property of quasi two dimensionality is important not only in the study of structural in-plane phase transitions, but also in the analysis of magnetic properties of intercalation compounds. For example, $NiCl_2$[30] and $FeCl_3$[31] intercalated into graphite show nearly ideal two-dimensional magnetic behavior, with possible corrections for the finite size of the intercalation domains.

To account for both the staging and the in-plane density in these quasi-two-dimensional materials, it is common to characterize the compounds as GH_{qn}, where G represents the guest species (atomic or molecular) and H represents the host species. The stage number is n; it represents the number of host layers which separate two neighboring intercalant layers. The in-plane stoichiometry of the guest species is characterized by the index q, which is the ratio of host to guest atoms (or molecules) in a single plane. For example, in graphite, the stage 1, heavy alkali metals have an ideal stoichiometry MC_8 ($M = $ K, Rb, Cs). For stages $n \geq 2$, the ideal stoichiometries are MC_{12n}.

While the stoichiometric notation implies the existence of a single-stage compound (n) with a unique in-plane stoichiometry (q), both observation and theoretical analysis show that departures from ideal staging exist in thermodynamic equilibrium at finite temperature.[32] In some cases, the ideal stoichiometry is perturbed due to the presence of thermally generated vacancies[7]; the intercalant layer may have a range of continuously variable in-plane stoichiometries.[33] The statistical-mechanical description of this process and its implications for the integrity of stage ordering are discussed below. There it is shown that staging transitions induced by temperature occur as a result (1) of vacancies in layers that were completely filled and (2) of extra atoms in layers that were completely empty at zero temperature. These defects disturb the zero-temperature sequence of filled and empty layers and induce stage disorder, as shown schematically in Fig. 4a. We term this type of disorder, vacancy induced stage disorder. Experimental evidence for this phenomena has been presented by Caswell and Solin[24] and by Woo *et al.*[33] and is also discussed below.

While these considerations are relevant for infinite intercalant layers in thermodynamic equilibrium, there is considerable evidence that the intercalant layers are distributed inhomogeneously between every pair of graphite

[30] M. Suzuki and H. Ikeda, *J. Phys. C* **14,** L923 (1981).

[31] Y. S. Karimov, A. V. Zvarykina, and Y. N. Novikov, *Sov. Phys.—Solid State (Engl. Transl.)* **13,** 2388 (1972); *Fiz. Tverd. Tela* **13,** 2836 (1971).

[32] W. Metz and D. Hohlwein, *Carbon* **13,** 87 (1975).

[33] K. C. Woo, H. Mertwoy, J. E. Fischer, W. A. Kamitakahara, and D. S. Robinson, *Phys. Rev. B: Condens. Matter* [3] **27,** 7831 (1983).

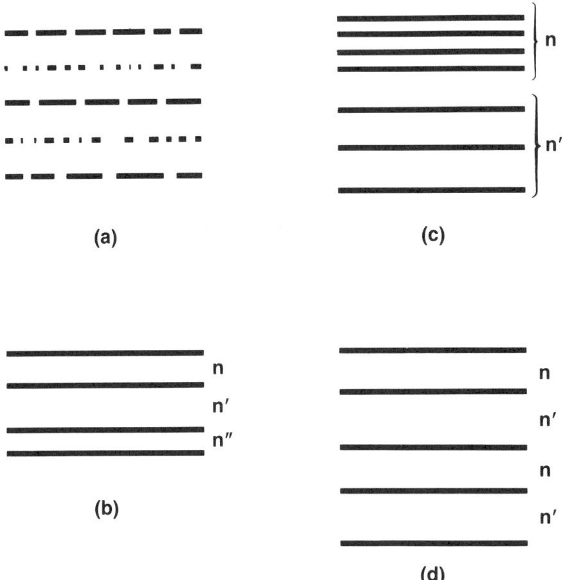

FIG. 4. Types of stage ordering in intercalation compounds. (———) represent the intercalate layers with n and/or n' host layers between them. (---) indicate layers with a lower density of intercalate atoms than the saturation density; the spacing between the dashes represents the proportion of vacancies in those layers. (a) Vacancy-induced stage disorder: Disorder is due to partial filling of layers. (b) Random staging: Disorder is due to randomness in the interlayer spacings; all layers are at their saturation densities. (c) A mixed phase; stages n and n' are mixed on a macroscopic length scale. (d) A fractional stage; this is a periodic state which is a *microscopic, ordered* mixture of stages 1 and 2.

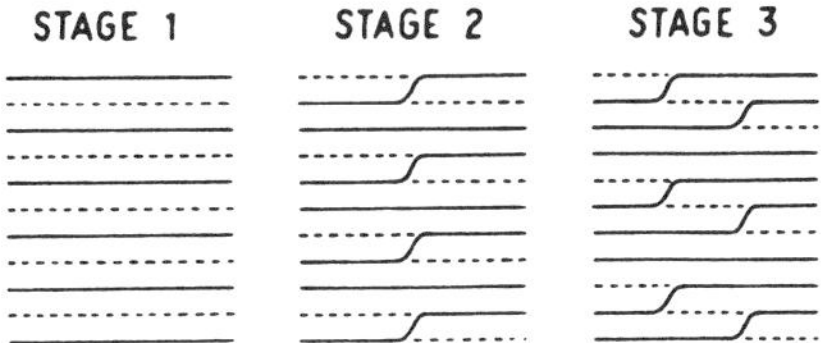

FIG. 5. Staging in intercalation compounds in the domain structure suggested by Daumas and Herold. (———) are the graphite layers, and (---) are the intercalate layers.

layers, as shown in Fig. 5. Locally, the intercalant domains or islands are arranged to form an ordered, staged structure.

This picture was first suggested by Daumas and Herold[34] as a step in understanding the kinetics of intercalation. If the intercalant layers were infinite in lateral extent, entire macroscopic planes of intercalant would have

[34] N. Daumas and A. Herold, *C. R. Hebd. Seances Acad. Sci., Ser. C* **286,** 373 (1969).

to move when a sample responds to a change in the intercalant chemical potential by a change of stage. The kinetics of such a process would seem to be prohibitively long, in comparison to the time scale of intercalation, which is typically several hours. In the domain picture shown above, only smaller-scale rearrangements of the intercalant islands are needed. Other experimental evidence exists which implies that staging transitions can occur as temperature or pressure is changed with all the intercalant remaining in the host. For example, Daumas and Herold heated stage-2 KC_{24} in a carbon monoxide environment. If whole layers of intercalant had to exit the sample, the potassium vapor would have reacted with the CO, and no change of stage would have been observed. The fact that they observed a change of stage from stage 2 (KC_{24}) to stage 3 (KC_{36}) was taken as evidence that the sample could restage with only internal movements of the intercalant domains. Similar conclusions were reached by Caswell and Solin[24] in their thermal experiments and by Clarke,[35] Solin,[36] and Fischer[37] in restaging transitions that they observed as a function of applied pressure.

The existence of islands or finite domains of intercalant presents the interesting possibility of disorder in the staging sequence. For infinitely large intercalant layers, there is no staging disorder (in the absence of in-plane vacancies). This is because the entropy gain from a random reshuffling of the intercalant layers lowers the free energy by an amount proportional to $k_B T \log N_l$, where N_l is the number of layers. The energy loss, due to the proximity of layers that would be separated in a perfectly staged structure, is proportional to the number of atoms *in* the layer, N_0. In the thermodynamic limit the energy term dominates (N_0 is proportional to the area of the sample, N_l to its length), and perfect stage order is maintained. This argument, however, breaks down if the layers are of finite extent. Kirczenow[10] has shown how the entropy gain of the rearrangement of finite-size layers in intercalation compounds can lead to stage disorder without atomic vacancies. We term this form of stage disorder—where the domain size induces the disorder—random staging. A schematic illustration is shown in Fig. 4b. This form of stage disorder is distinct from a mixed-phase compound which contains large domains of pure stage n and n', as illustrated in Fig. 4c. Mixed phases of pure-stage materials are a natural consequence of concentration constraints of almost any model of staging; they are absent if the intercalation compound is held at constant chemical potential (except at special values of μ). Recent experiments of Fuerst *et al.*[38] and of Misenheimer and Zabel[11] show evidence

[35] R. Clarke, N. Wada, and S. A. Solin, *Phys. Rev. Lett.* **44,** 1616 (1980).

[36] N. Wada and S. A. Solin, *Physica B (Amsterdam)* **105,** 268 (1981).

[37] J. E. Fischer and H. J. Kim, *Synth. Met.* **12,** 137 (1985).

[38] C. D. Fuerst, J. E. Fischer, J. D. Axe, J. B. Hastings, and D. B. McWhan, *Phys. Rev. Lett.* **50,** 780 (1983).

of random staging (apparently due to domain disorder), especially in high-stage materials. The theories of stage disorder and their experimental applications will be discussed in detail in Sections III and IV.

Although staging is a form of modulated structure—with the attendant phase transitions and disorderings discussed above—modulations of stage ordering itself are also possible. While the previous discussion has assumed the existence of only one periodicity—i.e., a pure stage—as the low-temperature state of an intercalation compound at fixed chemical potential, more complex structures have been discussed theoretically by Safran,[7,39] Millman and Kirczenow,[40] and Hawrylak and Subbaswamy.[41] These structures have been observed experimentally by Fuerst and co-workers[38] and by Kim *et al.*[42] Such arrangements, illustrated in Fig. 4d, have been termed "fractional" stages, since they correspond to microscopically ordered sequences of layers of intercalant with more than $\frac{1}{2}$ of the host interlayer positions (or galleries) occupied. The stage is then generally defined as the reciprocal of the fraction of occupied galleries; a stage $\frac{3}{2}$ would then consist of a periodic sequence of two filled layers followed by an empty layer followed by two filled layers, followed by an empty layer.... The existence of such states is a result of competing interactions, analogous to those discussed in the magnetic problem by von Boehm and Bak,[1] and will be treated below.

3. Staging Effects on Host Properties

In this section, the consequences of pure-stage ordering on the electronic, lattice, and transport properties of intercalated graphite are examined. The discussion is limited to the properties of intercalated graphite, since it is by far the best studied example of a staged intercalation compound. Only those properties that are relevant to the ensuing discussion on the energetics of staging are discussed in detail. For complete reviews of experimental data on intercalated graphite, the reader should consult the aforementioned reviews.[5,6,9,19]

a. Electronic Structure

Measurements of the optical and transport properties of graphite after intercalation show that the intercalation reaction is accompanied by a charge transfer from the intercalant to the host. The most direct experimental

[39] S. A. Safran, *Synth. Met.* **2,** 1, (1980).

[40] S. E. Millman and G. Kirczenow, *Phys. Rev. B: Condens. Matter* [3] **26,** 2310 (1982).

[41] P. Hawrlyak and K. R. Subbaswamy, *Phys. Rev. B: Condens. Matter* [3] **28,** 4851 (1983).

[42] H. J. Kim, J. E. Fischer, D. B. McWhan, and J. D. Axe, *Phys. Rev. B: Condens. Matter* [3] **33,** 1329 (1986).

evidence for this charge transfer comes from nuclear magnetic resonance measurements of the Knight shift of the intercalant NMR lines. The Knight shift is proportional to the electron density at the nucleus and thus indicates the degree of ionization of the intercalant species. Early work in this area was done by Carver,[43] who concluded that the first-stage C_8Cs compounds have a partially ($\sim 50\%$) ionized Cs ion, while the second-stage $C_{24}Cs$ compound consists of completely ionized alkali metal. The degree of charge transfer from or to an intercalant atom or molecule is denoted by f, with $f = 1$ representing complete charge transfer. For monovalent donors, this corresponds to complete ionization of the valence electron. More recently, measurements of the Knight shift of ^{13}C NMR lines have been performed[44] by Conrad *et al.* These experiments are more difficult to interpret since the Knight shift is mostly sensitive to the *s*-electron density and not to the *p*-like orbitals out of which the conduction bands of graphite are formed. The ^{13}C measurements are in qualitative agreement with the results of Carver; stage-1 compounds show $f < 1$, while stage $n > 1$ materials show $f \approx 1$. The measurements are too imprecise to show the details of the stage dependence of f for large values of n.

Since there is a charge transfer from the intercalant to the host, information about the distribution of this charge among the various host layers in a staged compound can be obtained from studies of the electronic properties of the host material. The calculation of this distribution has been the focus of recent studies,[45–51] since experiments such as reflectivity, electron spectroscopy, and magneto-optical studies all require band models for their interpretation.[19] A crucial input to these band models is the potential of each of the graphite layers. Two limits of the charge distribution—and hence the potential—can be envisioned. In the first, all of the transferred charge is assumed to reside on the graphite layers adjacent to the intercalant (the bounding layers); the remaining graphite layers in the intercalant–graphite–intercalant sandwich (the interior layers) have graphitic properties. The second possibility is a uniform distribution of charge among all the graphite layers. This model

[43] G. P. Carver, *Phys. Rev. B: Solid State* [3] **2,** 2284 (1970).

[44] J. Conrad, H. Estrade, P. Lauginie, H. Fuzellier, G. Furdin, and R. Vosse, *Physica B (Amsterdam)* **99,** 457 (1980).

[45] L. Pietronero, S. Strässler, H. R. Zeller, and M. Rice, *Phys. Rev. Lett.* **41,** 763 (1978); *Solid State Commun.* **30,** 399 (1979).

[46] S. A. Safran and D. R. Hamann, *Phys. Rev. B: Condens. Matter* [3] **22,** 606 (1980).

[47] S. A. Safran and D. R. Hamann, *Phys. Rev. B: Condens. Matter* [3] **23,** 565 (1981).

[48] S. Shimamura and A. Morita, *J. Phys. Soc. Jpn.* **51,** 502 (1982).

[49] T. Ohno and H. Kamimura, *J. Phys. Soc. Jpn.* **52,** 223 (1983).

[50] N. A. W. Holzwarth, S. Louie, and S. Rabii, *Phys. Rev. B: Condens. Matter* [3] **28,** 1013 (1983).

[51] S. A. Safran, N. A. W. Holzwarth, and D. R. Hamann, *in* "Physics of Intercalation Compounds" (L. Pietronero and E. Tossati, eds.), p. 138. Springer-Verlag, Berlin and New York, 1981.

suggests a rigid-band picture of an intercalation compound which consists of the energy bands of pure graphite with a Fermi level that is raised or lowered by the addition of electrons (donors) or holes (acceptors). Such models were studied by McClure[52] and by Dresselhaus *et al.*[53]

Since the true charge distribution in a staged intercalation compound is probably inhomogeneous, but with a non-negligible fraction of charge located on the interior layers, more recent models of the electronic properties have taken into account the variation of the electrostatic potential in their calculations. Phenomenological approaches have been taken by Dresselhaus and Leung,[54] who computed a three-dimensional band structure, based on the Slonczewski–Weiss–McClure[55] model of the graphite energy bands. To account for the nonuniformity of the charge distribution, they introduced a potential difference between the bounding and interior layers. This shift was estimated to be ~0.3 eV based on experiment. In a similar vein, Blinowski and Rigaux[56] studied the band structure of third- and fourth-stage acceptor compounds. In their calculation, Blinowski and Rigaux used a simplified model for the band structure of coupled graphite layers. This model enabled them to extract analytic forms for the energy bands and band gaps, for comparison with optical absorption experiments. Although both of these studies did differentiate between the potentials of the bounding and interior layers, they did not focus on a self-consistent model for the charge distribution in high-stage compounds.

Physical insight into the nature of this charge distribution was greatly advanced by Pietronero *et al.*[45] with a simple Thomas–Fermi model for the screening of the ionized intercalant layers. They considered a single layer of ionized intercalant with its charge donated to an infinite continuum of graphite layers. Using a simple model for the graphite energy bands, they showed that the screening of the intercalant follows a power-law dependence with a potential $\phi(z) \sim (z + z_0)^{-2}$, where z is the distance along the c axis from the intercalant layer. Due to this power law dependence, the screening is anomalously long ranged; in a simple metal the screening is exponential, with a screening length of 1–3 Å. This approach was extended by Safran and Hamann[46] to examine the problem of a general configuration of intercalant layers and to derive an expression for the total energy as a function of stage (see Section II).

[52] J. W. McClure, *Phys. Rev.* **119,** 606 (1960).

[53] M. S. Dresselhaus, G. Dresselhaus, and J. E. Fischer, *Phys. Rev. B: Solid State* [3] **15,** 3180 (1977).

[54] G. Dresselhaus and S. Y. Leung, *Solid State Commun.* **35,** 819 (1980).

[55] J. C. Slonczewski and P. R. Weiss, *Phys. Rev.* **109,** 272 (1958).

[56] J. Blinowski and C. Rigaux, *J. Phys. Orsay, Fr.* **41,** 667 (1980).

The nonlinear Thomas–Fermi equations describing the screening can be approached using a continuum, energy-density formalism. Since the interest here is to see the effects of staging on the charge distribution, the neglect of the discrete nature of the graphite layers should be a good approximation for high stages, where the distance between intercalant layers is much larger than the 3.35 Å which separates adjacent graphite layers. Consistent with this approximation, the small c-axis band dispersion of the graphite layers is neglected and the energy bands are expanded about their values near the K point of the two-dimensional graphite Brillouin zone. Figure 6 shows the two-dimensional structure of the graphite lattice and the corresponding two-dimensional Brillouin zone. For donor intercalants, the electrostatic and band energy is written in a Hartree approximation as

$$E[n] = \frac{1}{2}\int dz\,[V_1(z) - V_2(z)][n(z) - \rho(z)] + \int dz\, n(z)\varepsilon_b[n(z)] \qquad (3.1)$$

In Eq. (3.1), $n(z)$ and $\rho(z)$ are the electron and ion charge distributions, respectively; the charge is assumed to be homogeneously distributed perpendicular to the c axis. The total band energy per electron due to the in-plane, graphite conduction π-band dispersion is ε_b. (Graphite has $2s$ and $2p$ valence electrons. The $2s$ and $2p_x$, $2p_y$ orbitals combine to make in-plane σ bonding orbitals. The p_z electron is in a valence π band; the donated charge resides in conduction π orbitals.) The first term in Eq. (3.1) is the electrostatic energy, and

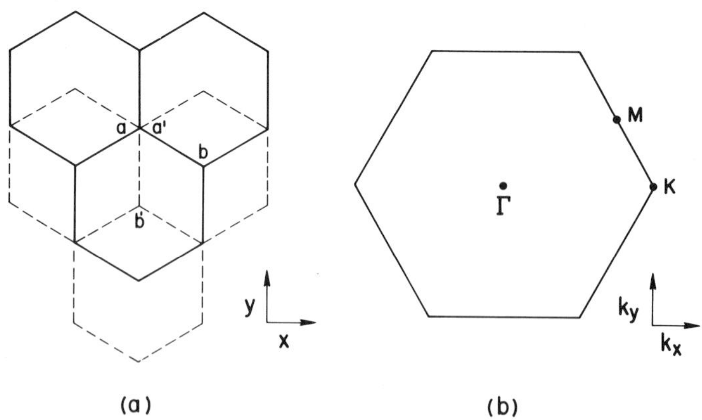

FIG. 6. (a) Real-space lattice for graphite layers in AB stacking. (———) is the A layer, and (---) is the B layer displaced along the c axis by c_0. The two atoms in the unit cell are shown for the A (unprimed) and B layers. (b) Two-dimensional Brillouin zone for a single graphite layer. The symmetry points Γ, K, and M are located at $(0, 0), (2\pi/a), (\frac{2}{3}, 0)$ and $(2\pi/a)(\frac{1}{2}, 1/2\sqrt{3})$, respectively, where a is the graphite in-plane lattice constant.

the potentials V_1 and V_2 satisfy

$$\frac{\partial^2 V_i}{\partial^2 z} = -\frac{4\pi e^2}{\varepsilon}\theta_i, \qquad i = 1, 2 \tag{3.2}$$

where $\theta_1 = n(z)$ and $\theta_2 = \rho(z)$. The graphite c-axis dielectric constant is ε, which accounts for the screening of the intercalant ions by the bonding π electrons and σ electrons that are not explicitly considered in this calculation.

For small shifts of the Fermi level, the energy bands are linear in the two-dimensional wave vector for a single layer of graphite; hence $\varepsilon_b[n(z)] = 2\beta n^{1/2}/3$, where for a single graphite layer with q intercalant atoms per carbon and a charge transfer of f electrons per intercalant, $\beta n^{1/2} = \gamma_0(\pi\sqrt{3}\, qf)^{1/2}$. Here, $\gamma_0 \sim 3$ eV[55,57] is the tight-binding matrix element between nearest-neighbor carbon atoms in a single plane. Minimizing the energy of Eq. (3.1) with respect to the electron charge distribution $n(z)$, and using Eq. (3.2), one finds that within one unit cell "sandwich" of intercalant and n-graphite layers, Poisson's equation can be written in dimensionless form as

$$\frac{\partial^2 \phi}{\partial \zeta^2} = \phi^2(z) - \tilde{\rho}(z) \tag{3.3}$$

The dimensionless distance is $\zeta = z\tau$, where

$$\tau^2 = 4\pi e^2(\sigma/c_0)^{1/2}/\varepsilon\beta \tag{3.4}$$

is related to the ratio of Coulomb to band energies, with $\sigma = \sigma_C/6$, where σ_C is the planar density of carbon atoms and c_0 is the spacing between adjacent carbon planes ($c_0 = 3.35$ Å). The dimensionless electron and ion densities are $\tilde{n} = n(z)c_0/\sigma$ and $\tilde{\rho} = \rho(z)c_0/\sigma$. From the energy minimization, one finds the nonlinear relationship between the potential and the charge

$$\phi^2(z) = \tilde{n}(z) \tag{3.5}$$

which arises from the zero density of states at the Fermi level for a single graphite layer. These equations can be solved for a *general* configuration of intercalant and graphite layers as described by Safran and Hamann.[46] Here it is noted that the solution is given in terms of the minimum value of the potential in the region between two intercalant layers, which varies approximately as

$$\phi_0 \approx (12)^{1/2}\pi^2(c + d)^{-2} \tag{3.6}$$

where c is the distance between intercalant layers (in the dimensionless units described above) and $d = 2(24/p)^{1/3}$, where p is the charge density of the intercalant layer normalized to σ/c_0. The potential and charge distribution

[57] N. A. W. Holzwarth, *Phys. Rev. B: Condens. Matter* [3] **21,** 3665 (1980).

[see Eq. (3.5)] are thus described by power laws as opposed to exponential decays, due to the nonlinearity of the Thomas–Fermi equations. In Section II, an expression for the energy associated with staging will be derived from this formulation.

To examine the effects of the discrete nature of the graphite layers as well as the modifications introduced by *interlayer* hopping matrix elements, Safran and Hamann[47] calculated a *self-consistent* band structure for a model system consisting of a thin film of n graphite layers bounded by two partially ionized intercalant layers. The hybridization between the graphite and intercalant layers is ignored. Similarly, effects due to the σ electrons are lumped into a dielectric constant for the graphite layers. The quantum mechanics of the electrons in the graphite layers is described using a linear combination of atomic orbitals (LCAO) graphite π-band Hamiltonian, modified to describe the thin film.[57] The in-plane spatial variation of the electrostatic potential due to the charged intercalant and graphite layers is ignored; both are treated as charged sheets. These assumptions lead to a self-consistent set of equations for the coefficients in the LCAO expansion, where the exact wave function of the interacting system is written $\psi_p^k(\mathbf{r})$, where $\mathbf{k}$ is the two-dimensional wave vector defined in the Brillouin zone of Fig. 6, and p is a band index. The wave function is expanded in a basis of orthonormal orbitals $\Phi(\mathbf{r} - \mathbf{R}_{\alpha i})$ located on the carbon atom in layer i ($i = 1, \ldots, n$). The index $\alpha = a, b$ denotes the two atoms in the unit cell of a single graphite layer (see Fig. 6), and

$$\psi_p^k(\mathbf{r}) = \sum_{\alpha i} c_{\alpha p i}(\mathbf{k}) \Phi(\mathbf{r} - \mathbf{R}_{\alpha i}) \tag{3.7}$$

The total energy per carbon atom (U_n) of the n-graphite layer is the sum of Hamiltonians describing the n-graphite layers (without Coulomb terms due to charge transfer) and the sum of the Coulomb terms for the electron–electron, electron–ion, and ion–ion energies. The average charge *per carbon atom* in the ith layer is q_i, which is given by

$$q_i = \sum_{pk\alpha} f_p(\mathbf{k}) |c_{\alpha p i}(\mathbf{k})|^2 - 1 \tag{3.8}$$

where $\tilde{f} = \sum q_i$, with $\tilde{f}/n$ the average charge transfer per carbon atom in the n layers. Note that $\tilde{f}$ is related to f, the charge transfer per intercalant, by the number of intercalant atoms per carbon in the intercalant layer; i.e., for $C_{12n}M$, $\tilde{f} = f/12$. The factor $f_p(\mathbf{k})$ is the product of the Fermi function and an energy normalization, as described in Ref. 47. The energy U_n can then be written:

$$U_n = \frac{1}{n}\left[\sum_{pk} f_p(\mathbf{k}) \sum_{ij\alpha\beta} c^*_{\alpha p i}(\mathbf{k}) c_{\beta p j}(\mathbf{k}) E_{ij\alpha\beta}(\mathbf{k}) \right.$$
$$\left. - \frac{V_0}{2}\left(\sum_{ij} |i - j| q_i q_j + \frac{\tilde{f}^2}{2n}[(n-1) + 2\eta]\right)\right] \tag{3.9}$$

In Eq. (3.9) $V_0 = 2\pi\sigma_0 e^2 c_0/\varepsilon$ and η is the ratio of the carbon–intercalant to adjacent carbon–carbon c-axis distances. The first term represents the band energy, where the matrix $E_{ij\alpha\beta}(\mathbf{k})$ is given in Ref. 47 and includes interlayer matrix elements. It is equivalent to the Hamiltonian of Leung and Dresselhaus[58] if second-layer interactions are set to zero. The second term is the electron–electron interaction of the charged sheets, while the last term is the sum of the electron–ion and ion–ion interactions. The coefficients $c_{\alpha pi}(\mathbf{k})$ are determined by the minimization of U_n with the constraint that the wave functions be orthonormal. These conditions yield

$$\sum_{\beta j} [E_{ij\alpha\beta}(\mathbf{k}) + \tilde{V}_i \delta_{\alpha\beta}\delta_{ij}] c_{\beta pj}(\mathbf{k}) = \varepsilon_p(\mathbf{k}) c_{\alpha pi}(\mathbf{k}) \tag{3.10}$$

where $\varepsilon_p(\mathbf{k})$ is the eigenvalue and where $\tilde{V}_i$ is the self-consistent potential at layer i due to the other charged layers. In the charged sheet approximation

$$\tilde{V}_i = -V_0 \sum_j q_j |i - j| \tag{3.11}$$

with a starting guess for the $\{q_i\}$, the eigenvalue equation is diagonalized and the new charge density is determined from the values of $c_{\alpha pi}(\mathbf{k})$. The process is iterated to self-consistency. This procedure yields the energy bands $\varepsilon_p(\mathbf{k})$ as well as the charge distribution and potential. The results are compared with the Thomas–Fermi calculation in Table I, where good agreement is seen. (Note that the Thomas–Fermi results quoted here use a larger value of the graphite dielectric constant than used in the band calculations.[47]) If the same value of the dielectric constant is used, the band calculation results[47] in a *more* homogeneous charge distribution than the Thomas–Fermi results, probably due to the effects of c-axis hopping. The resulting expressions for the stage dependence of the energy are discussed in Section II.

More sophisticated calculations of the self-consistent charge distribution and band structure were performed by Shimamura and Morita,[48] who considered the effects of explicitly including all of the $2s$ and $2p$ valence electrons of carbon. The previous calculations had considered only the $2p_z$ valence and conduction states. By including a much larger basis of valence states occupied by the electrons present in pure graphite, a more accurate measure of the dielectric screening of the Coulomb interactions could be obtained, and the need for a phenomenological dielectric constant is obviated. However, the calculation of Shimamura and Morita did *not* include interlayer c-axis hopping. As noted above, this hopping tends to make the charge distribution more homogeneous. A nonempirical calculation of the band structure and charge distribution—still based on the thin-film approximation—was carried out by Ohno and Kamimura.[49] Thus, they neglect any hybridization between the graphite and intercalant orbitals as well as

[58] S. Y. Leung and G. Dresselhaus, *Phys. Rev. B: Condens. Matter* [3] **24,** 3490 (1981).

TABLE I. COMPARISON OF CHARGE DISTRIBUTION IN STAGED COMPOUNDS FROM VARIOUS CALCULATIONS [a]

Layers	All bands thin film[b]	π bands only: Dielectric[c]	π bands only: σ bands[c]	Thomas–Fermi[e]	*Ab initio*[f]
$n = 3, f = 1$					
1	0.493	0.438	0.478	0.420	0.40
2	0.013	0.124	0.045	0.160	0.10
$n = 6, f = 1$					
1	0.485	0.417	0.469		
2	0.012	0.059	0.024		
3	0.003	0.024	0.007		
$n = 3, f = 0.3$					
1	0.447	0.411			
2	0.105	0.177			

[a] The stage is denoted by n and the charge transfer per intercalant is f. The table entries are normalized to the total charge in the n layers.
[b] T. Ohno and H. Kamimura, *J. Phys. Soc. Jpn.* **52,** 223 (1983).
[c] S. A. Safran and D. R. Hamman, *Phys. Rev. B: Condens. Matter* [3] **23,** 565 (1981).
[d] S. Shimamura and A. Morita, *J. Phys. Soc. Jpn.* **51,** 502 (1982).
[e] L. Pietronero, S. Strässler, H. R. Zeller, and M. Rice, *Phys. Rev. Lett.* **41,** 763 (1978).
[f] N. A. W. Holzwarth, S. Louie, and S. Rabii, *Phys. Rev. B: Condens. Matter* [3] **28,** 1013 (1983); the numbers here are the charges for the conduction electrons only (see Ref. 53).

hopping between graphite layers separated by an intercalant layer. In addition, the intercalant layers are treated as charged sheets. The calculation is performed self-consistently, using a numerical-basis-set LCAO method within the local-density approximation for the exchange and correlation energy; the calculations of Safran and Hamann and of Shimamura and Morita included only Coulomb terms. The 1s, 2s, and 2p orbitals of carbon were used as a numerical basis set by Ohno and Kamimura.

Ohno and Kamimura found, in agreement with the phenomenological treatments of the band structure, that most of the donated electrons occupy the lowest two conduction π bands, with wave functions mostly localized on the graphite bounding layers. In Table I, their results for the inhomogeneous charge density along the c axis is presented and compared with the continuum Thomas–Fermi result[45] and the thin-film calculations of Safran and Hamann[47] and of Shimamura and Morita.[48] Ohno and Kamimura[49] found that the finite spatial extent of the graphite orbitals has the effect of attracting the charge more strongly to the bounding layer; their results show the strongest screening when compared with the previous calculations.

The thin-film calculations have thus evolved from the simple Thomas–Fermi picture of power-law screening to more detailed treatments which include more details of the graphite band structure and Coulomb interactions. However, all these treatments ignore the details of the intercalant potential, whose effects may be as important as the fine tuning of the graphite bands included in the most recent calculations. In Table I the results for the charge density and layer potentials derived from first-principles, self-consistent pseudopotential calculations for LiC_{18} are also shown. These calculations, by Holzwarth *et al.*,[50,51] include the Li potential and the full three-dimensional band structure. The results show that in stage 3 the charge is localized on the bounding layer with only 6% of the *total* charge-density difference associated with the interior layer. However, the conduction-electron contribution (as defined by the density of electrons in partially occupied bands) is much more delocalized, with 19% of the conduction charge associated with the interior layer. This seems to indicate that the charge distribution is more homogeneous than the calculation of Ohno and Kamimura[49] would indicate.

Since there are no direct experimental probes of the π-electron charge density, comparison of theory and experiment is not straightforward. Most studies of Fermi surface structure, optical absorption, and magneto-optical properties are more directly related to band structure rather than to the charge distribution. The interpretation of these measurements is thus highly model dependent; the implications for the screening problem are not yet clear. The reader is referred to the aforementioned reviews for a survey of band-structure studies. Here, it is noted that a simple interpretation of the C 1*s* binding energy as measured in x-ray photoelectron spectroscopy (XPS) has been presented by DiCenzo *et al.*[59] and that the *c*-axis charge distribution has been estimated. The 1*s* electron energy depends on the charge associated with the individual carbon atoms and can be related to the shifts of the band filling in the individual layers of a staged compound. Using a two-dimensional model of the band structure of these layers, DiCenzo *et al.* fit their XPS lines. Their results are in good qualitative agreement with the theoretical results of Ref. 47. However, the XPS data seem to indicate a total charge transfer of $f \approx 0.05$–0.6 for KC_{48} and KC_{60}, in contrast to Knight-shift data discussed above which show[43,44] $f \approx 1$; the comparison between the XPS measurements and theory are thus only semiquantitative.

b. Lattice Properties

The changes in the host lattice structure that occur upon intercalation have a marked effect on the phonon spectrum of intercalation compounds. For

[59] S. B. DiCenzo, G. K. Wertheim, S. Basu, and J. E. Fischer, *Phys. Rev. B: Condens. Matter* [3] **24,** 2270 (1981).

[60] W. D. Ellenson, D. Semmingsen, D. Guerard, D. Onn, and J. E. Fischer, *Mater. Sci. Eng.* **31,** 137 (1977).

example, the enlarged unit cell along the c axis (consisting of the intercalate layer and n host layers) is responsible for a large number of phonon modes for wave vectors along the c axis. These "zone-folding" effects, along with one-dimensional models for phonons in graphite intercalation compounds, have been studied by Ellenson *et al.*,[60] Rossat-Mignod,[61] Zabel and Magerl,[62] and Dresselhaus *et al.*[63]

Another manifestation of stage ordering on the lattice properties is the change in the (in-plane) carbon–carbon bond length. Although this is a very small effect, on the order of 1%, it shows a systematic $1/n$ dependence on the stage number and may be the source of a long-range *attractive* interaction between intercalant layers, as discussed below. Nixon and Parry[64] were the first to observe the expansion of the graphite lattice constant upon intercalation. Their results, along with those of Guerard *et al.*,[65] are shown in Fig. 7 along with data for acceptor intercalants,[66] where the graphite lattice constant contracts upon intercalation. The microscopic mechanism for the change in bond length was discussed by Pietronero and Strässler,[66] who showed that the change in the carbon–carbon bond length u (measured in Å) is related to the charge transfer f by the formula

$$u = u_1\tilde{f} + u_2|\tilde{f}|^{3/2} + u_3\tilde{f}^2 \tag{3.12}$$

where $u_1 = 0.157$ Å, $u_2 = 0.146$ Å, and $u_3 = 0.236$ Å. The charge transfer $\tilde{f}$ is the charge transfer *per carbon* ($\tilde{f} = f/q$ for $C_{qn}X$). The linear term comes from the change in matrix elements due to the change in the atomic potential. The second term in Eq. (3.12) comes from the added band energy of the donated electrons, while the last term arises from the additional electrostatic energy. The linear term is responsible for the observation of expansion of the graphite lattice for donors and contraction for acceptors. These effects are probably also responsible for the changes in the Raman spectrum of the in-plane graphite modes upon intercalation.[67,68] It should be noted that a detailed interpretation of experiments using Eq. (3.12) should be performed with caution. This formula for the change in the bond length is only valid in a

[61] J. Rossat-Mignod, D. Fruchart, M. J. Moran, J. W. Milliken, and J. E. Fischer, *Synth. Met.* **2,** 143 (1980).

[62] A. Magerl and H. Zabel, *Phys. Rev. Lett.* **46,** 444 (1981); H. Zabel and A. Magerl, *Phys. Rev. B: Condens. Matter* [3] **25,** 2463 (1982).

[63] G. Dresselhaus, R. Al-Jishi, J. D. Axe, C. F. Majkrzak, L. Passell, and S. K. Satija, *Solid State Commun.* **40,** 229 (1981).

[64] D. E. Nixon and G. S. Parry, *J. Phys. C* **2,** 1732 (1969).

[65] D. Guerard, C. Zeller, and A. Herold, *C. R. Hebd. Seances Acad. Sci., Ser. C* **283,** 437 (1976).

[66] L. Pietronero and S. Strässler, *Phys. Rev. Lett.* **47,** 593 (1981).

[67] C. Underhill, S. Y. Leung, G. Dresselhaus, and M. S. Dresselhaus, *Solid State Commun.* **29,** 769 (1979).

[68] S. A. Solin, *Physica B (Amsterdam)* **99B,** 443 (1980).

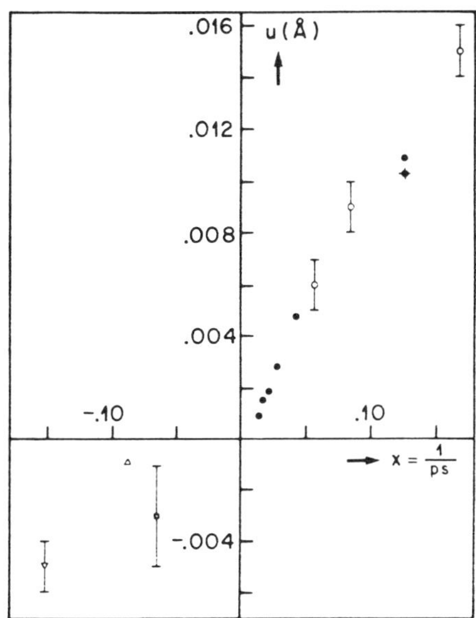

FIG. 7. Experimental values for the bond-length change u (in Å) for various intercalation compounds plotted as a function of the ratio of the number of intercalate molecules to the number of carbon atoms in a given compound. The nominal formula of a compound is CA_x, where $x = 1/ps$, with s being the stage and p being the in-plane stoichiometry. [From L. Pietronero and S. Strässler, *Phys. Rev. Lett.* **47,** 593 (1981).] (○), G–Li; (●), G–K; (+), G–Cs; (Δ), $C_{11.3}(NiCl_{2.13})$; (∇), G–$FeCl_3$; (□), G–AsF_5

2D rigid-band approach. The values of the coefficients can be modified by hybridization, although the stage dependence of the effect should be unchanged.

Of relevance to the problem of staging is the observation that the change in the graphite in-plane lattice constant decreases approximately as $1/n$ as the stage increases. The charge transfer f or $\tilde{f}$ is not expected to show such a drastic change. A simple explanation for this decrease in u was first suggested by Safran and Hamann,[46] who wrote the change in energy as

$$\Delta U = q(nC\epsilon^2 - 2\Delta_0\epsilon) \tag{3.13}$$

for a compound with stoichiometry $C_{qn}X$. In Eq. (3.13), $C \approx 68$ eV,[69] while the strain $\epsilon = u/u_0$, where u_0 is the in-plane carbon–carbon bond length in pure graphite. The largest force occurs on the bounding layer (which has the most charge) and is proportional to Δ_0 (the microscopic origin of Δ_0 is discussed in Ref. 66). The factor n appears in Eq. (3.13) since it is assumed that all n graphite

[69] D. L. Blakslee, D. G. Proctor, E. J. Seldin, G. B. Spence, and T. Weng, *J. Appl. Phys.* **41,** 3373 (1970).

layers maintain the same in-plane lattice constant; the misfit dislocations that would arise if the layers had different lattice constants would cost too much energy. The energy is then minimized by $\epsilon = \epsilon_0/n$, where $\epsilon_0 = C/\Delta_0$, in agreement with the stage dependence of the observed lattice constant.

c. Transport Properties

Since a primary focus of interest in intercalated graphite is the enhanced in-plane conductivity, there have been many measurements on the stage dependence of both the in-plane and *c*-axis conductivity.[19] The in-plane conductivity decreases approximately as $1/n$ for high stages,[23] since most of the conduction is due to the enhanced charge on the bounding host layers, whose volume fraction decreases with increasing stage as $1/n$. The *c*-axis resistivity, ρ_c, for donor compounds has been measured[70] as a function of temperature, as shown in Fig. 8. The change in slope of $\rho_c(T)$ as a function of stage was suggested to arise from an instability in the *c*-axis charge distribution; however, no further quantitative comparisons have been made. Although detailed studies of this phenomenon could possibly elucidate the nature of the inhomogeneous *c*-axis charge distribution, it is possible that the anomalous temperature dependence is linked to changes in the in-plane structure as well.

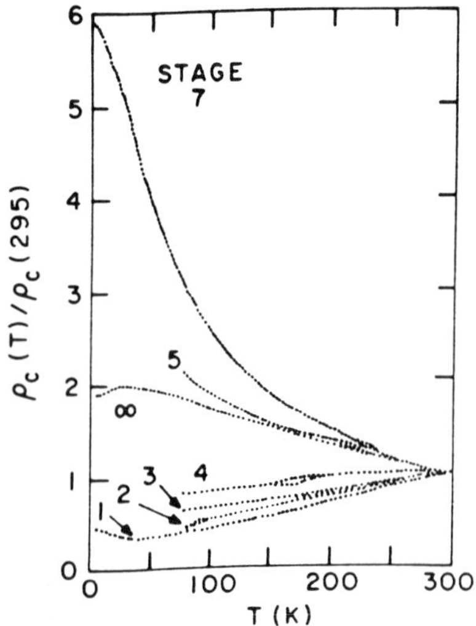

FIG. 8. Normalized *c*-axis resistivity versus temperature for stages 1–5 and 7, potassium–graphite intercalation compounds; data for pure graphite (stage ∞) are included for comparison. [From K. Phan, C. D. Fuerst, and J. E. Fischer, *Solid State Commun.* **44**, 1351 (1982).]

[70] K. Phan, C. D. Fuerst, and J. E. Fischer, *Solid State Commun.* **44**, 1351 (1982).

Another physical property that has been shown to be extremely sensitive to the inhomogeneous *c*-axis charge distribution is the orbital magnetic susceptibility of graphite intercalation compounds. DiSalvo *et al.*[71] were the first to measure the anisotropy of the magnetic susceptibility of donor compounds. They observed that the susceptibility was *paramagnetic* (pure graphite shows a large *diamagnetism*) for the magnetic field parallel to the *c* axis. Furthermore, the ratio of the *c*-axis susceptibility (χ_c) to the estimated Pauli (spin-only) contribution to the susceptibility (χ_P) increased rapidly as a function of stage, varying from ≈ 2 in stage 1 (KC_8) to 17 in stage 4 (KC_{48}). More recent experimental estimates of the spin and orbital susceptibility in both donors and acceptors have been made by Ikehata *et al.*[72] and by Yoshida *et al.*[73]

Safran and DiSalvo[74] attributed the paramagnetic nature of the susceptibility to the unusual band structure of graphite, which shows a saddle point in the density of states for large charge transfer. Near the saddle point, the *x*- and *y*-direction effective masses are opposite in sign. This effect changes the orbital susceptibility, which in the Landau–Peierls approximation[75] is proportional to the inverse of the product of the two effective masses, from diamagnetic to paramagnetic. The complete calculation for the susceptibility of a single graphite layer as a function of band filling was presented in Ref. 74, where a peak in the orbital susceptibility was predicted for small band filling. Since interior layers have small amounts of charge, and hence locally small changes in the band filling, they should contribute to an even larger paramagnetic susceptibility. However, analytic calculations by Blinowski and Rigaux[76] showed that the numerical results of Safran and DiSalvo[74] were inaccurate at small values of the band filling and that there is no peak in the susceptibility of a *single* graphite layer; Blinowski and Rigaux *did* confirm the paramagnetic nature of the orbital susceptibility.[76]

A more recent calculation by Safran[77] of the orbital susceptibility for two coupled graphite layers showed that indeed there is a paramagnetic peak for small values of the Fermi energy, which can account for the observed increase of the orbital susceptibility with stage. The peak is associated with the interlayer coupling. Saito and Kamimura[78] have calculated the orbital

[71] F. J. DiSalvo, S. A. Safran, R. C. Haddon, J. V. Waszczak, and J. E. Fischer, *Phys. Rev. B: Condens. Matter* [3] **20,** 4883 (1979); F. J. DiSalvo and J. E. Fischer, *Solid State Commun.* **28,** 71 (1978).

[72] S. Ikehata, H. Suematsu, and S. Tanuma, *Solid State Commun.* **50,** 375 (1984).

[73] Y. Yoshida, Y. Saito, I. Iye, and S. Tanuma *J. Phys. Soc. Jpn.* **54,** 2635 (1985).

[74] S. A. Safran and F. J. DiSalvo, *Phys. Rev. B: Condens. Matter* [3] **20,** 4889 (1979).

[75] J. Callaway, "Quantum Theory of the Solid State," Student ed., p. 499. Academic Press, New York, 1976.

[76] J. Blinowski and C. Rigaux, *J. Phys. Orsay, Fr.* **45,** 545 (1984).

[77] S. A. Safran, *Phys. Rev. B: Condens. Matter* [3] **30,** 421 (1984).

[78] R. Saito and H. Kamimura, *Phys. Rev. B: Condens. Matter* [3] **33,** 7218 (1986); *Synth. Met.* **12,** 295 (1985).

susceptibility of graphite intercalation compounds for stages $n = 1-5$ using a Hamiltonian which reproduces the more detailed band-structure calculations discussed above.[49] They find that both the interlayer hopping matrix elements as well as the shifts in potential due to the inhomogeneous c-axis charge distribution are important in quantitatively determining the stage dependence of the orbital susceptibility. The susceptibility has a peak for donors at stage 4, in agreement with experiment. However, the absolute magnitude of the calculated susceptibility is smaller than the measured value. Thus, while the susceptibility is *in principle* a sensitive measure of the inhomogeneous c-axis charge distribution (since layers with *small* numbers of electrons contribute the most), the theoretical interpretation is not yet complete.

II. Mechanisms of Stage Ordering

In the previous section the effects of stage ordering on the electronic, lattice, and transport properties were considered. Here, the microscopic origins of the effective interactions between intercalate atoms are considered, and their consequences on the energetics of staging are compared with experiment. Section III analyzes the phase transitions between stages in terms of these interactions, while Section IV treats the consequences of finite-size intercalant islands or domains.

4. In-Plane Interactions

Although the origin of stage ordering is the effective repulsive interaction *between* layers of intercalate atoms, the nature of the interactions *within* a single layer is also relevant, as discussed in Section III. Current theoretical models[7,40,41,79,80] of staging in intercalation compounds all make use of a lattice-gas model to describe the intercalant. Thus, the hard-core and short-range (screened) Coulomb repulsions between the intercalant atoms are taken into account by the lattice-gas construction. Of course, the highly ordered commensurate phases, which have been observed in some compounds,[81,82] indicate the presence of longer-range repulsions. These can be taken into account in more detailed lattice-gas models.[83,84] Their relevance to staging

[79] J. R. Dahn, D. C. Dahn, and R. R. Haering, *Solid State Commun.* **42,** 179 (1982).

[80] D. P. DiVincenzo and T. C. Koch, *Phys. Rev. B: Condens. Matter* [3] **30,** 7092 (1984).

[81] G. S. Parry, *Mater. Sci Eng.* **31,** 99 (1977).

[82] P. Lagrange, D. Guerard, M. El Makrini, and A. Herold, *C. R. Hebd. Seances Acad. Sci., Ser. C* **287,** 179 (1978).

[83] P. Bak and E. Domany, *Phys. Rev. B: Condens. Matter* [3] **20,** 2818 (1979).

[84] C. Lee, H. Aoki, and H. Kamimura, *J. Phys. Soc. Jpn.* **49,** 870 (1980).

lies in the preference for particular values (or ranges) of in-plane stoichiometries corresponding to these commensurate structures, which are favored because of the corrugation energy due to the host potential. This corrugation energy is related to the energy difference between an intercalant atom located above the center of a carbon hexagon compared with one in a position over the C–C bond. Detailed microscopic calculations by DiVincenzo and Mele[85,86] estimate this energy difference to be about 1 eV for Li, but only 0.1 eV for the heavier alkali-metal compounds.

In addition to the effective hard-core repulsion of the intercalant atoms, there are also longer-range attractive interactions between them. One source of these attractions is the elastic deformation of the graphite lattice by the intercalant atoms which separate the graphite planes. To model these interactions, the intercalant atoms are treated as pairs of "elastic dipoles"[3,87,88,89] which force the layers apart. In the simplest treatments[87,88] these dipoles only have diagonal z components. Since each one deforms the lattice, it is favorable for two dipoles in the same plane to be adjacent to each other (see Fig. 9a), implying an attractive interaction. Consider the interaction between two isolated intercalant atoms located at the origin and at (x, y, z), respectively. In an infinite graphite medium, the elastic interaction is long-ranged and has the form

$$U_e = -\frac{P^2}{4\pi C_{44}}\left(\frac{a_3\rho^2 - 2z^2}{(a_3\rho^2 + z^2)^{5/2}}\right) \tag{4.1}$$

where C_{ij} is the elastic constant,[87] $a_3 \approx c_{33}/c_{44} \approx 9$, P is the dipole strength (typically 1–3 eV), and $\rho^2 = x^2 + y^2$. For $z = 0$ (i.e., two atoms in the same layer) this interaction is attractive. The elastic energy per intercalant atom for a layer of intercalate is estimated to be lowered by 0.13 eV compared with the energy of isolated atoms.

The elastic energy of a homogeneous intercalant layer is linear in the density of intercalate atoms, as calculated above. It is thus applicable to a *dilute* layer of intercalate. For more concentrated densities, more than two-body interactions must be considered. The physical origin of this correction is the saturation of the graphite layer separation by a finite density of intercalant. This is shown[79] in Fig. 9b, where the c-axis lattice constant of Li_xTi_2 is plotted as a function of the in-plane Li volume fraction x. This intercalation compound does not show any *long-range* staging; so to a first approximation x

[85] D. P. DiVincenzo and E. J. Mele, *in* "Intercalated Graphite" (M. S. Dresselhaus, G. Dresselhaus, J. E. Fischer, and M. J. Moran, eds.), p. 123. North-Holland Publ., Amsterdam, 1983.

[86] D. P. DiVincenzo and E. J. Mele, *Phys. Rev. Lett.* **53,** 52 (1984).

[87] S. A. Safran and D. R. Hamann, *Phys. Rev. Lett.* **42,** 1410 (1979).

[88] S. A. Safran and D. R. Hamann, *Physica B (Amsterdam)* **99B,** 462 (1980).

[89] S. Ohnishi and S. Sugano, *Solid State Commun.* **36,** 823 (1980).

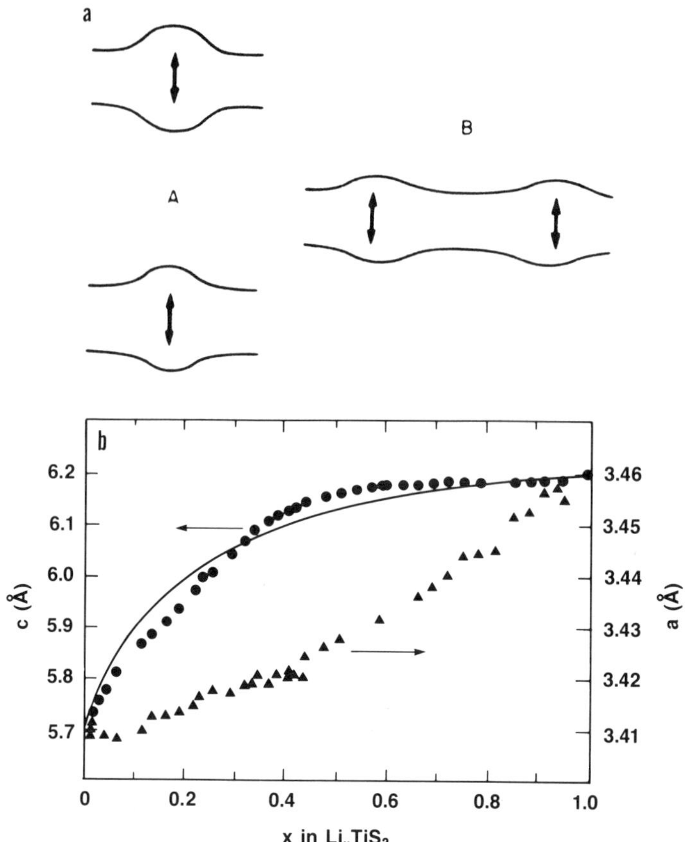

FIG. 9. (a) Pairs of elastic dipoles (arrows) in an infinite, anisotropic medium. The dipoles in configuration A repel, while those in B attract. (b) Measured lattice-parameter variations for Li_xTiS_2 [from J. R. Dahn and R. Haering, *Solid State Commun.* **40**, 245 (1981)]. The solid curve is a fit to the c-axis expansion using the model described in the text. [From J. R. Dahn, D. C. Dahn, and R. Haering, *Solid State Commun.* **42**, 179 (1982).]

corresponds to the in-plane density of intercalant. For small x, the expansion of the c axis is linear in x, in accord with the elastic model presented above. For large values of x, the layers are already separated to a point where the bonding is very weak. Dahn *et al.*[79] estimated this effect, by equating the forces between the expansion of the host layers due to the intercalant (which is modeled by a spring of equilibrium length c_L and strength k) and the separation of the host layers due to their intrinsic interactions (equilibrium spacing c_0 and spring strength K). They assume that the host–host bonds are never disrupted, and that the intercalant merely adds additional bonding to the adjacent host layers. If the in-plane ratio of intercalant to available host sites is x, the force balance equations dictate

$$K[c(x) - c_0] = xk[c_L - c(x)] \tag{4.2}$$

where $c(x)$ is the average separation of the host layers. They find that

$$\frac{c(x) - c_0}{c_L - c_0} = \frac{x}{\alpha + x}, \qquad \alpha = \frac{K}{k}$$

and they estimate that α is small (0.2). This simple model shows the saturation of the c-axis spacing which goes to c_L as x becomes large. The resulting energy per intercalant is

$$U_e = J\frac{x}{\alpha + x} \tag{4.3}$$

which has terms to all orders in x.

It is to be noted that this calculation does not account for the anharmonicity of the host bonding, which should contribute strongly to the saturation of the c-axis expansion as a function of x. This anharmonicity implies that the intercalant has completely weakened the host–host bonds once there is local separation of the adjacent layers. One way to account for this effect is to reduce the number of host–host bonds by the fraction of intercalant. Thus, the factor of K in Eq. (4.2) would be replaced by $K(1 - x)$. Similarly, the factor α in Eq. (4.3) would be replaced by $\alpha(1 - x)$. This form for the elastic energy has the correct limits as $x \to 0$ and $x \to 1$, in contrast to the simple form of Eq. (4.3) above.

These results are important because they indicate the importance of "many-body" interactions; the expressions for the energy contain terms higher than quadratic in the in-plane density. In Section III, the implications of these many-body interactions on the phase diagrams is discussed. A more complete treatment of these interactions would include the anharmonic nature of the c-axis bonding in the host material in a more detailed manner.

In addition to the elastic terms, there are other in-plane energies that result in many-body effective interactions. For example, the total energy of a single layer of intercalant—with all of the charge transferred to the graphite bounding layer—can be derived from the Thomas–Fermi model discussed above.[46] The energy per intercalant is proportional to[46] $qp^{5/3}$, where $p = \rho c_0/\sigma$, where there are q carbon atoms per intercalant. The intercalant charge density is ρ, and 6σ is the planar density of carbon atoms. This energy thus depends in a nonquadratic fashion on the intercalant density and represents another correction to the picture of simple, two-body intercalant-intercalant interactions.

In addition to these "many-body" effects, which are all calculated in the approximation of a uniform layer of intercalant, there are additional contributions to the in-plane energy which reflect the details of the host and intercalant layer periodicities. The intercalant lattice has a tendency to lock in to the periodicity of the host lattice. A particularly simple model for this "commensurability" energy was discussed by DiVincenzo and Koch[80] in a treatment of Li-intercalated graphite. The phenomenological form of this potential is given in Ref. 80. Here it is noted that the effective in-plane alkali potential is the sum of a two-body attractive term and an additional function of the in-plane density which has a minimum as the density corresponding to the 2×2 in-plane superlattice. The denser $\sqrt{3} \times \sqrt{3}$ superlattice structure is accounted for by constraining the maximum in-plane density to correspond to this value. In lattice-gas terms, the $\sqrt{3} \times \sqrt{3}$ superlattice is taken as the basic intercalant lattice.

Even more detailed calculations involving the Coulomb interactions, intercalate–graphite atom–atom interactions, and total electron energy have been performed by Miyazaki *et al.*[90] as a function of the *c*-axis distance between the intercalant and bounding graphite layers. They find that LiC_6 is the favored stoichiometry for lithium intercalates. However, heavy alkali intercalates favor an MC_{14} (in-plane) structure at ambient pressure and MC_8 structure at higher pressures.

5. Interlayer Interactions

At low temperatures, stage ordering depends on the existence of relatively long-ranged *interlayer* interactions. If these interactions are attractive, they lead to phase separation—i.e., the coexistence of regions of high intercalate density and low intercalate density. If the interlayer interactions are repulsive, the intercalate layers form a periodic structure along the *c* axis at a fixed value of the chemical potential—i.e., stage ordering results.

Two possible mechanisms for the long-range, repulsive interaction in intercalation compounds have been analyzed theoretically: (1) elastic interactions[87–89,91,92] and (2) electrostatic repulsions (including screening by the donated electrons) between the intercalant layers.[46,47]

The physical origin of the interlayer elastic repulsion can be seen from Fig. 9a. The elastic interaction of two simple (*z*-component) elastic dipoles is attractive in the same plane, but repulsive between planes. Ohnishi and Sugano[89] have shown that even the interplanar components of this interaction can sometimes be attractive if one takes into account the shearing of the

[90] H. Miyazaki, Y. Kuramoto, and C. Horie, *J. Phys. Soc. Jpn.* **53**, 1380 (1984).
[91] G. Kirczenow, *Phys. Rev. Lett.* **49**, 1853 (1982).
[92] S. E. Ulloa and G. Kirczenow, *Phys. Rev. Lett.* **55**, 218 (1985).

carbon planes adjacent to the intercalant. However, although the elastic interaction of two *isolated, intercalant atoms* does result in long-range repulsion [see Eq. (4.1)], it may be of little relevance in the ideal case of fully intercalated graphite crystals. This is because the physical origin of these interactions is the strain induced in the graphite layers which reduces their *c*-axis spacing in between the intercalant layers in order to be coherent with the rest of the sample. For a uniformly intercalated layer, there is no bending of the graphite around the layer. There is some residual strain around each intercalant atom or molecule which does give rise to interlayer repulsions (and which is responsible for the stacking order observed in some intercalation compounds[93–95]). However, the linear nature of the elasticity equations implies that these strains (and interlayer interaction strengths) decay as e^{-Gz}, where G is the first reciprocal lattice vector corresponding to the in-plane intercalant structure.

For intercalant islands or domains, these coherency strains do exist and drive a randomly mixed-stage sample to pure stage ordering in the low-temperature limit. Kirczenow[10,91,92] has recently shown how the competition between these strain-mediated interactions and the entropy of disordering *finite-size* intercalant domains can even result in stage disorder at finite temperatures. The existence of and interactions between intercalant domains is a subject of current study and will be examined in Section IV. Here the discussion is limited to the case of repulsions between infinite intercalant layers, where the long-range coherency strains due to intercalation vanish for uniformly intercalated materials. The discreteness of the intercalant atoms can give rise to a residual interlayer repulsion, which may be important because (1) the interaction strength is large, even though the range is short (range $z \sim G^{-1}$), and (2) linear elasticity is not valid on these short length scales.

There is, however, another mechanism that *does* give rise to a long-range, interlayer elastic interaction in intercalation compounds. Unfortunately, it results in an *attractive* effective interaction between intercalant layers and cannot be used to understand the origin of simple staging which is due to *repulsions*. This additional elastic interaction comes from the uniform in-plane expansion of the n host layers in a single intercalant–host–intercalant sandwich, as discussed in Section I,3,b. From Eq. (3.13), it is seen that the elastic energy due to this expansion (which is uniformly distributed among the n host layers in the absence of misfit dislocations) is attractive and proportional to $1/n$, with a proportionality constant which is typically -0.05 eV. This energy is lowest for stage 1 and hence can be termed an

[93] J. B. Hastings, W. B. Ellenson, and J. E. Fischer, *Phys. Rev. Lett.* **42,** 1552 (1979).
[94] H. Zabel, S. C. Moss, N. Caswell, and S. A. Solin, *Phys. Rev. Lett.* **43,** 2022 (1979).
[95] H. Miyazaki, C. Horie, and T. Wantanabe, *J. Phys. Soc. Jpn.* **53,** 3843 (1984).

"antistaging" interaction between intercalant layers. Although it is weak in magnitude compared with the electrostatic repulsions (discussed below), it is *very long ranged.* The presence of this $1/n$-type energy may be responsible for limiting the maximum stage obtainable in intercalation. When the attractions dominate the stronger, but shorter-ranged repulsions, staging is no longer the lowest-energy state of the intercalation compound. This implies the existence of a maximum stage, which is estimated below.

For infinite intercalant layers, the longest-range *repulsive* interaction that seems to exist in graphite intercalation compounds is electrostatic in origin. The long-range nature of the effective electrostatic repulsion between intercalant layers arises from the anomalously long-range nature of the screening of the intercalant layers, as discussed in Section I,3,a. In the Thomas–Fermi approximation, the charge density of the innermost layer was shown to decrease as $1/n^4$ (in the asymptotic limit of $n \to \infty$). In Ref. 46 it is shown that the total band and electrostatic energy is approximately given by

$$\tilde{E}[n] = E_\infty\left(1 + \alpha_0 \frac{1}{(1 + n/\alpha_1)^{-5}}\right) \tag{5.1}$$

where $\tilde{E}$ is the energy per intercalant for a compound with $C_{12n}X$ stoichiometry. For unit charge transfer ($f = 1$), $R_\infty \approx 1.1$ eV, $\alpha_0 \approx 0.96$, and $\alpha_1 \sim 4.1$. Smaller values of f would decrease the value of E_∞ but increase the range of the interaction (α_1), since the charge density is more homogeneous, as seen in Table I. It is important to notice that the stage dependence of $\tilde{E}_n$ (which represents a repulsion, since it is smaller at large n), is not approximately n^{-5} for the stages of interest ($1 < n < 15$), but is a much slower function of n.

The stage dependence of the total electrostatic and band energies has also been calculated in the self-consistent, thin-film calculations discussed above.[47,49,96] From Eq. (3.9) the total energy per intercalant can be calculated once the self-consistent band structure and charge density are determined. For stoichiometry $C_{qn}X$ the energy per intercalant, $\tilde{E}(n) = qU_n$, where U_n is the energy per carbon of Eq. (3.9). The total thermodynamic potential per intercalant is $\bar{E}_n$, given by

$$\bar{\tilde{E}}_n = -\mu/n + q(U_n - U_\infty) \tag{5.2}$$

where μ is the chemical potential for the intercalant atoms. The chemical potential is defined to include all *stage-independent* interactions. This expression does describe the stage dependence of μ for constant q.

At zero temperature the phase boundaries ($\mu_{n,n'}$) between stages n and n' are given by

$$\mu_{n,n'} = q\left(\frac{U_n - U_{n'}}{n^{-1} - n'^{-1}}\right) \tag{5.3}$$

[96] T. Ohno and H. Kamimura, *Physica B + C (Amsterdam)* **117B + C,** 611 (1983).

Assuming a simple sequence of stages 1,2,3,..., as μ is lowered, the range of stability of a given stage $\Delta\mu_n$ is

$$\Delta\mu_n = \mu_{n,n-1} - \mu_{n,n+1} = n(\bar{E}_{n+1} + \bar{E}_{n-1} - 2\bar{E}_n) \tag{5.4}$$

The range of stability as a function of stage is shown in Fig. 10 for both the Thomas–Fermi calculation[46] and the model, thin-film band calculation[47] discussed above. The nonempirical, self-consistent band calculation of Ohno and Kamimura[96] also yields total energies and hence estimates for $\Delta\mu_n$. Since the charge distribution of these calculation is more *inhomogeneous* compared with the results of Safran and Hamann,[47] the range of stability of the various stages is substantially smaller (see Fig. 10). In contrast to Safran and Hamann, who concluded that the electrostatic repulsions are large enough to account for a sizeable part of the binding energy of staged compounds, Ohno and Kamimura claim that this contribution is smaller and that the stronger, but shorter-ranged elastic repulsions discussed above may dominate.

Comparison of these calculations with experiment may be one way to resolve the question of whether the long-range electrostatic/band energy is responsible for stabilizing staged structures. In Table II the results of Salzano and Aronson[22] for the thermodynamics of intercalation of $C_{12n}K$ have been reinterpreted in terms of the range of stability $\Delta\mu_n$. Also shown are more

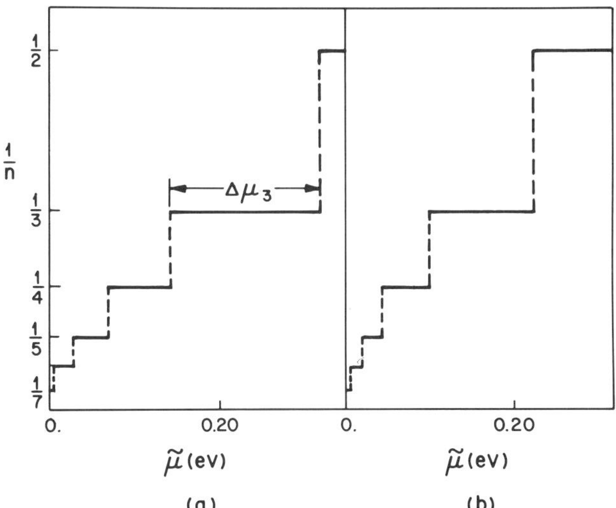

FIG. 10. Stage dependence of the chemical potential $\tilde{\mu} = \mu - \mu_0$, where μ_0 is chosen so that $\tilde{\mu}_{7,8} = 0$. The calculation is for a graphite intercalation compound with stoichiometry $C_{12n}X$ and the charge transfer $f = 1$. The quantity of physical importance is $\Delta\mu_n = \mu_{n,n-1} - \mu_{n,n+1}$. (a) Self-consistent band calculation with dielectric constant $\varepsilon = 3$. (b) Continuum Thomas–Fermi model. [From S. A. Safran and D. R. Hamman, *Phys. Rev. B: Condens. Matter* [3] **23**, 565 (1981).]

TABLE II. RANGE OF STABILITY OF STAGES 2–6 FROM BOTH THEORY AND EXPERIMENT (IN eV)

	$\Delta\mu_2$	$\Delta\mu_3$	$\Delta\mu_4$	$\Delta\mu_5$	$\Delta\mu_6$	
Experiment[a]	0.150	0.070	0.038	0.02		$T_g = 387°C$
	0.162	0.069	0.041	0.02	0.01	$T_g = 485°C$
	0.168	0.070	0.040	0.02	0.018	$T_g = 541°C$
Theory[b]		0.029	0.019	0.007		$f = \frac{1}{4}$
		0.069	0.036	0.024		$f = \frac{1}{2}$
		0.176	0.072	0.044		$f = 1$

[a] Experimental errors are all ± 0.005 eV. [From R. Nishitani, Y. Uno, and H. Suematsu, *Phys. Rev. B: Condens. Matter* [3] **27,** 6572 (1983).]

[b] Theory calculated for dielectric constant $\varepsilon = 3$ and $T = 0$. [From S. A. Safran and D. R. Hamman, *Phys. Rev. B: Condens. Matter* [3] **23,** 565 (1981).]

recent results of Nishitani *et al.*,[97,98] which combined *in situ* measurements of the chemical potential with x-ray data. Both of these sets of experiments are in good agreement with the calculations of Safran and Hamann[47] if the value of f is set to be $\frac{1}{2}$. However, this is not in agreement with other studies,[53] which show that $f \approx 1$. The reason for this discrepancy is not yet clear, and could be related to the assumed value of the dielectric constant in the calculations of Ref. 47. It is important to note that the strong screening found in the nonempirical calculations of Ohno and Kamimura[49,96] yield values for $\Delta\mu_n$ that are about an order of magnitude smaller than those observed. Calculations of $\Delta\mu_n$ for the short-ranged elastic repulsion have not yet been performed. Although the microscopic origin of the staging interactions seems to be consistent with long-range screening and electrostatic repulsion, several discrepancies still exist, and the problem remains unresolved.

III. Staging Transitions

The previous section surveyed the variety of interactions in intercalation compounds. The repulsive interplanar interaction wa shown to be responsible for stage ordering. Estimates of the zero-temperature range of stability of perfectly staged states were made for both elastic and electronic interactions and compared with experiment. In this section the energetic analysis of staging is combined with the thermal effects which lead to stage disorder. The stage, and hence the stoichiometry of an intercalation compound, is therefore temperature dependent. Staging is thus viewed not as a chemical property, but

[97] R. Nishitani, Y. Uno, and H. Suematsu, *Phys. Rev. B: Condens. Matter* [3] **27,** 6572 (1983).

[98] R. Nishitani, Y. Uno, and H. Suematsu, *Synth. Met.* **7,** 13 (1983).

rather as a *phase* of an intercalation compound. Phase transitions between stages can give rise to stage disorder. These transitions can be induced by either temperature or external pressure.

6. Phase Diagrams for Staging Transitions

Staging phase transitions are best described by phase diagrams which show the stage (and in-plane stoichiometry) as a function of intercalant concentration, temperature, and/or pressure. To calculate these phase diagrams, the statistical mechanics of staging must be understood. The current state of both theory and experiment suggests that mean-field descriptions of stage disordering are sufficient at the present time. These mean-field theories *do* take into account the presence of long-range stage ordering.

In mean-field theory the *free energy* of the intercalant is given by the sum of the intercalant energy (discussed in Section III) and the configurational entropy of the intercalant atoms. Since the intercalant is confined to the regions between host layers, a lattice-gas model can be used to describe the *c*-axis configurations. It is convenient, although not necessary, to use a lattice-gas model to describe the in-plane structures as well. This procedure is rigorously correct for intercalant layers that are commensurately ordered with respect to the host lattice. Since the focus here is on staging, the details of the in-plane structure of the intercalant are probably not crucial.

The simplest lattice-gas model to describe staging transitions was discussed by Safran.[7] This zeroth-order model shows stage ordering at low temperatures, staging phase transitions (vacancy-induced stage disorder) at elevated temperatures, and even fractional staging. Above a critical temperature, equilibrium states consist of stage-1 materials with continuously variable in-plane stoichiometry. This result accounts for the fact that some intercalation compounds show good stage ordering, while others may show *no* long-range order. However, the model did *not* predict random staging (associated with domains) where the intercalant layers are at their close-packed in-plane density, but show stacking disorder (see Fig. 4). This form of stage disorder requires *finite* intercalant layers, and disappears in the thermodynamic limit. In addition, the elementary model, which was based solely on *two-body* interactions between intercalant (including effective interactions mediated by the host), predicted a phase diagram which was symmetric with respect to intercalant atoms and "holes"—i.e., symmetry about 50% of the close-packed intercalant density.

The statistical mechanics of staging can be analyzed with phenomenological models. However, these models must be related to the underlying physical mechanisms so that the correct dependence of the interactions on the intercalant density, interlayer spacing, etc. is included. Safran's original model assumes *infinite* intercalant layers. The free energy includes the long-ranged interlayer interaction, which was modeled as a power law, an attractive interaction between intercalant atoms in the same plane, and a chemical potential term. The microscopic origins of the long-range interlayer interactions and in-plane attractions were discussed in Section II. There it was noted that the power-law form for the interplanar interaction was only correct asymptotically for $n \to \infty$. More recent calculations by Kirczenow[10] have used a longer-ranged interaction, consistent with this result.

In this and later versions of the lattice-gas model, staging phase transitions are coupled to changes in the in-plane density of intercalant. At finite temperatures, vacancies are formed in layers that were completely saturated at $T = 0$, while a finite density of intercalant exists in interlayer regions that were completely empty at $T = 0$. The physical origin of this change in in-plane density is the entropy gained by disordering the intercalant layers. In the thermodynamic limit of intercalant layers that are infinite in their lateral extent, there is no significant entropy gained by merely disordering the stacking order of completely saturated layers. (This disorder would contribute a term proportional to $\log N_1$, where N_1 is the number of host layers. Since the repulsive interaction varies as N_0, the number of atoms *within* a layer, the entropy term is negligible because $N_0 \sim N_1^2$ in the thermodynamic limit.)

The variable in-plane intercalant density is denoted by $\sigma_i(\mathbf{r})$, which varies from 0 to a maximum value of 1; i.e., $\sigma_i(\mathbf{r})$ is normalized to the close-packed in-plane density. The index i labels the host layer position, while $\mathbf{r}$ labels the in-plane coordinate of the intercalant atom. The fractional occupation of the ith layer is $\sigma_i = \int d\mathbf{r}\, \sigma_i(\mathbf{r})/A_0$, where A_0 is the area of the layer. Mean-field theory analyzes the free energy as a function of the $\{\sigma_i\}$. The mean-field free energy F is written

$$F = T\sum_i [(1-\sigma_i)\log(1-\sigma_i) + \sigma_i \log \sigma_i] + \sum U(\sigma_i) + \sum_{ij} V_{ij}\sigma_i\sigma_j - \mu \sum_i \sigma_i \tag{6.1}$$

The first term represents the contribution of the entropy of mixing within a layer, and the last term is the chemical potential of the intercalant. The in-plane effective interactions of the intercalant atoms is included in the term $U(\sigma_i)$, which in general may be a complicated function of the in-plane density. The mean-field form for the effective interlayer repulsion is quadratic in the average layer densities and is given by the term proportional to V_{ij}. As discussed in Section II, the effective interlayer repulsion is approximately a power law, due to the anomalous screening of the electrostatic interactions of the charged intercalate atoms—i.e., $V_{ij} = V_0|i-j|^{-\alpha}$, where $\alpha \to 5$ as the stage $n \to \infty$.

The simplest form for the in-plane interactions was approximated by Safran[7,39] as a quadratic function of the in-plane density, $U(\sigma) = -U_0\sigma^2$. The free energy is then minimized with respect to the $\{\sigma_i\}$. As noted above, the free energy is symmetric about $x = \frac{1}{2}$, where x is the average (overall, three-dimensional) concentration, normalized to the saturation concentration. The phase diagrams are then calculated as a function of (1) the chemical potential and (2) the intercalant concentration. These two cases represent the two physical situations (1) of an intercalation compound in equilibrium with a reservoir of intercalant, where the pressure (and corresponding μ) of the reservoir is fixed, and (2) of an isolated intercalation compound with a fixed total number of intercalant.

The resulting phase diagrams are shown in Figs. 11 and 12, for the case $\alpha = 4$ and $V_0 = U_0/2$. The high-stage phases all terminate in a dilute, stage-1 phase ("gas") whose concentration vanishes as $T \to 0$. The transitions from stage to stage for $n > 2$ are all first order; the concentration ranges over which these states are stable are within 6% of their $T = 0$ concentrations, as shown in Fig. 12. On the other hand, the stage-2 to stage-1 transition is second order for $T > T_t$, with a large range of stoichiometries predicted for the stage-2

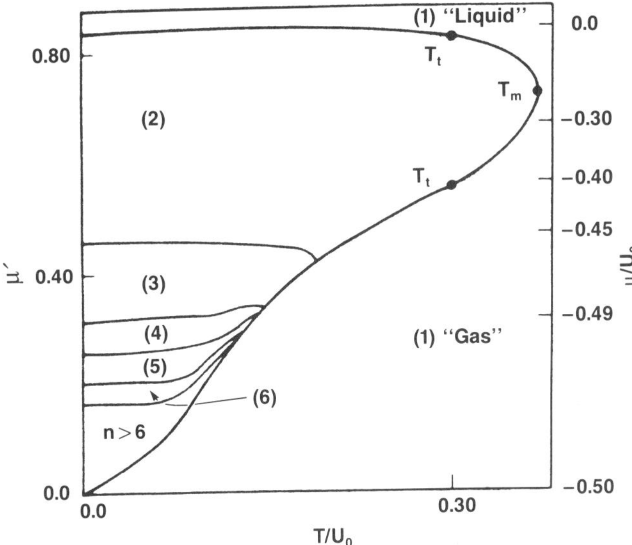

FIG. 11. Phase diagram for $V_0 = U_0/2$ and $\alpha = 4$ plotted as a function of the temperature T and the chemical potential μ, both normalized to U_0. For clarity, the phase diagram is given as a linear function of $\mu' = (\mu/U_0 + \frac{1}{2})^{1/\alpha}$, leading to a nonlinear μ scale on the right-hand side of the figure. The integers in parentheses are the stable, pure-stage phases. Fractional stages are not shown. T_m and T_t are the maximum and tricritical temperatures, respectively. [From S. A. Safran, *Phys. Rev. Lett.* **44**, 937 (1980).]

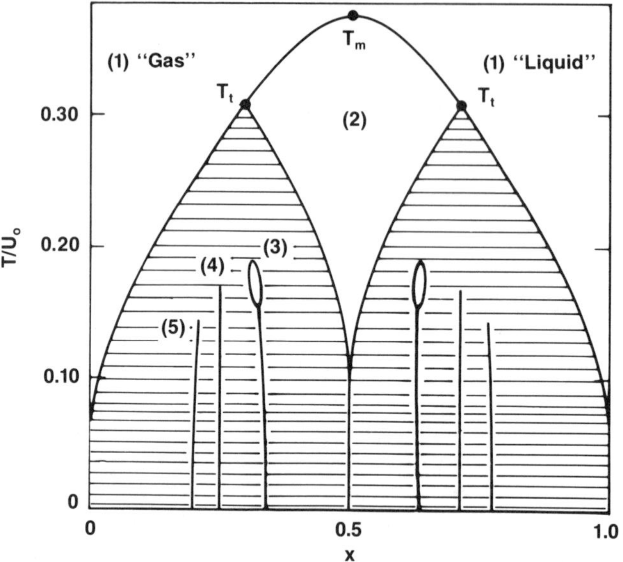

FIG. 12. Same phase diagram as in Fig. 11 plotted as a function of the normalized temperature T/U_0 and normalized concentration x. The integers in parentheses are the single-phase pure-stage states, while the cross-hatched regions denote two-phase regions. The fractional stages are not shown, but are predicted by the theory. They can be found by reflecting the phase diagram about $x = \frac{1}{2}$. For clarity, only stages 1–5 are shown. [From S. A. Safran, *Phys. Rev. Lett.* **44**, 937 (1980).]

phase at high temperatures. Furthermore, since the stage-2 to stage-1 transition is isomorphic to that of a metamagnet in a magnetic field,[99] the maximum temperature at which stage 2 is stable is T_{m}, where

$$T_{\mathrm{m}} = \tfrac{1}{4}(U_0 - V_a + V_b) \tag{6.2}$$

while the tricritical point (T_{t}) is given by

$$T_{\mathrm{t}} = T_{\mathrm{m}}\{1 - V_b/[3(U_0 - V_a)]\} \tag{6.3}$$

Here $V_a = V_0 2^{-\alpha}\xi(\alpha)$, and $V_a + V_b = V_0\xi(\alpha)$, where $\xi(\alpha) = \sum_{p=1}^{\infty} p^{-\alpha}$. For $T < T_{\mathrm{m}}$, the stage-1 phase consists of both "gas" and "liquid" phases, each having a restricted range of composition, while for $T > T_{\mathrm{m}}$, the composition range is unrestricted, and no staged ($n > 1$) states are possible. Finally, the existence of fractional stages, corresponding to periodic structures with multiple staging periodicities, is also found.

The three most striking predictions of this simple model are: (1) the existence of low-density, stage-1 phases which are reached from the ordered,

[99] D. Furman and M. Blume, *Phys. Rev. B: Solid State* [3] **10**, 2068 (1974).

high-stage compounds by an increase in temperature, (2) the second-order nature of the stage-2 to stage-1 transition, and (3) the occurrence of fractional stages and the symmetry of the model about $x = \frac{1}{2}$. Features (1) and (2) arise because the disordering of the stages is coupled to changes in the in-plane density of intercalant. In the simple model discussed above, the *change* in the in-plane energy is linear in the intercalant density, permitting a rather wide range of densities to be stable. This is not the case for the more sophisticated models of $U(\sigma)$ discussed below. However, experiments indicate that there is a finite *range* of concentrations at which a given stage is stable, in qualitative agreement with prediction (1). This was first pointed out in the experiments of Clarke *et al.*,[24] who presented x-ray evidence which showed that the staging transition is accompanied by melting in Cs–graphite intercalation compounds.

An extensive study of the temperature/concentration phase diagram has been performed by Woo *et al.*[33] and by Fischer *et al.*[8] for Li–graphite intercalation compounds. The phase diagram is shown in Fig. 13. The experimental phase diagram is in agreement with the simple lattice model as

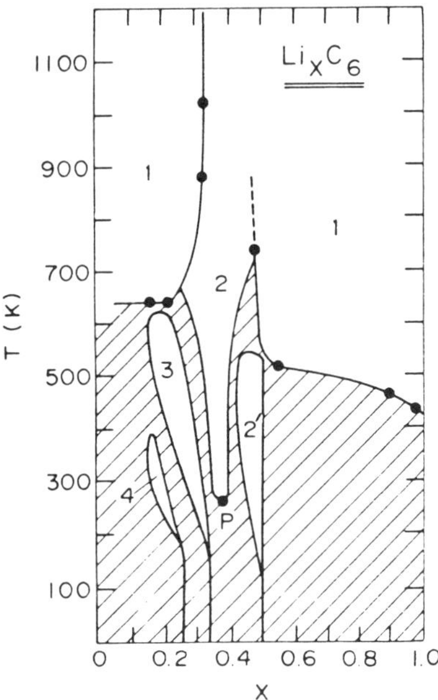

FIG. 13. The experimental concentration–temperature phase diagram for Li–intercalated graphite. The concentration of intercalant normalized to its saturation value is x. Dilute stage 2 is denoted as 2, while 2′ denotes a dense stage-2 phase. [From J. E. Fischer, C. D. Fuerst, and K. C. Woo, *Synth. Met.* **7**, 1 (1983).]

far as the temperature-driven destaging transitions. At high temperatures, all the phases of the material (except for $x = \frac{1}{3}$) are unstable to a dilute stage-1 compound. Both high- and low-density phases of this stage-1 compound exist ("gas"/"liquid" phases discussed above). In addition, stage 2 has the largest range of stable densities at high temperatures, in agreement with theory. However, no continuous stage-1 to stage-2 transition is observed; the Li–graphite system is chemically stable only for $T < 750$ K.

The existence of a dilute stage-1 phase in the Li–graphite system argues for the importance of the entropy of mixing in the free energy discussed above. It also suggests that there is no long-range stage ordering in the Li_xTiS_2 compounds because the intercalation temperature is greater than the appropriate T_m.[14,79,100] The value of T_m depends on the strength of the in-plane interaction, which is related to the band structure and may be quite different for C and TiS_2. There is x-ray evidence for the existence of at least low ($n = 2, 3$) stages in Li_xNbSe_2[101] and Ag_xTiS_2[102] (see below), which indicates that for these materials T_m is greater than the intercalation temperature.

While the experimental phase diagram for Li–graphite shows many of the features predicted by the simple lattice-gas model, it shows neither fractional stages nor the symmetry about $x = \frac{1}{2}$ predicted by the two-body (quadratic) form for the in-plane interaction U. There have been several attempts to modify the lattice-gas model in order to make it more realistic and physically plausible.

The major modification to the simple lattice model is the modeling of the in-plane density dependence of both the intralayer $[U(\sigma)]$ and interlayer interactions (V_{ij}). Deviations of these interactions from forms that are quadratic in the in-plane density yield phase diagrams that are *not* symmetric about $x = \frac{1}{2}$. In addition, they may suppress the stability of the fractional stages.

The first such modification was analyzed by Dahn *et al.*,[79] who included the energetics of the separation of the host layers (see Section II), resulting in an expression for $U(\sigma)$ which gives terms to all orders in σ, $U(\sigma) = J\sigma/(\alpha + \sigma)$, with $\alpha = K/k$, where K and k are the spring constants of the host material and of the Li–host bonds, respectively. As pointed out above, this expression does not really account for the anharmonicity of the host bonds, which are severely disrupted for large σ; in the limit $\sigma \to 1$, the sandwich thickness $c(x)$ [see Eq. (4.2)] does not approach the equilibrium host–intercalant c-axis bond length c_L (as it must). The resulting expression for the in-plane energy is thus only suggestive of the effects of nonquadratic terms. A more accurate expression

[100] A. Thompson, *Phys. Rev. Lett.* **40,** 1511 (1978); *Physica B (Amsterdam)* **99B,** 100 (1980).
[101] J. Dahn and R. Haering, *Solid State Commun.* **44,** 29 (1982).
[102] K. Bardhan, G. Kirczenow, and J. Irwin, *J. Phys. C* **18,** L131 (1985).

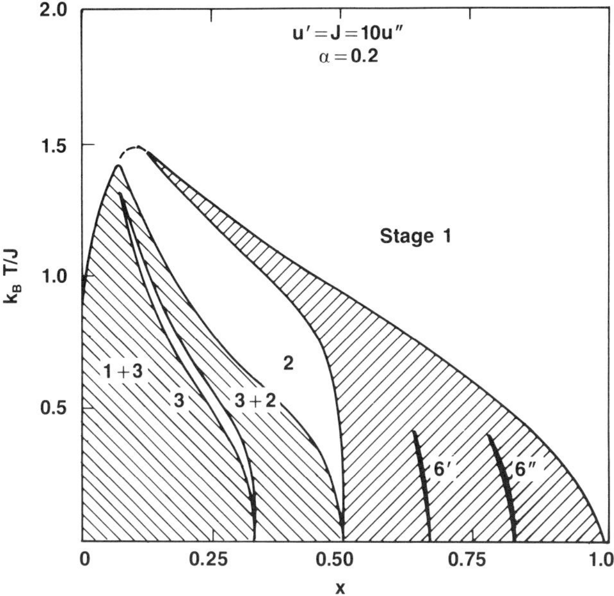

FIG. 14. Phase diagram for $\alpha = 0.2$ as a function of the normalized intercalant concentration x and T/J, where J characterizes the in-plane elastic energy. Note that the phase diagram is not symmetric about $x = \frac{1}{2}$. [From J. R. Dahn, D. C. Dahn, and R. R. Haering, *Solid State Commun.* **42**, 179 (1982).]

was suggested with $\alpha \rightarrow \alpha(1 - x)$ to approximately account for the anharmonicity. The resulting phase diagram, with V_{ij} extending only over nearest- and next-nearest-neighbor layers is not symmetric about $x = \frac{1}{2}$, as shown in Fig. 14.

In a series of papers,[40,103,104] Millman and Kirczenow performed extensive calculations of the phase diagrams of staged intercalation compounds. To account for the nonlinear elastic effects which are important in the intraplanar energy and which arise from the energy cost for separating the host layers, they wrote $U(\sigma) = -\varepsilon Z\sigma^2 + \gamma\Theta$. The first term is the standard two-body, effective, in-plane intercalant–intercalant interaction introduced above. The second term comes from the assumed elastic rigidity of the host layers. The energy cost for separating the host layers is γ (per intercalant site), *independent* of the density of intercalant, as long as the density is nonzero. Thus, $\Theta = 1$ if $\sigma > 0$ and $= 0$ for $\sigma = 0$. This energy (which overestimates the elastic energy for small σ) breaks the symmetry of the phase diagrams about $x = \frac{1}{2}$. As shown in

[103] S. E. Millman and G. Kirczenow, *Phys. Rev. B: Condens. Matter* [3] **28,** 3482 (1983).
[104] S. E. Millman, G. Kirczenow, and D. Solenberger, *J. Phys. C* **15,** L1269 (1982).

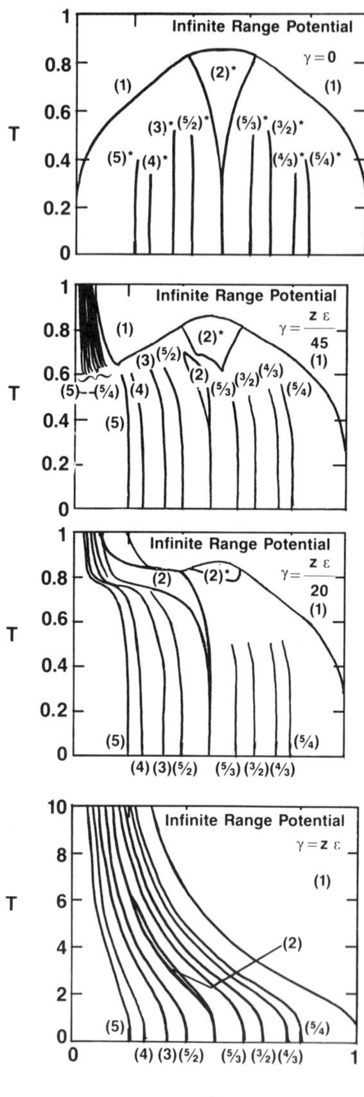

FIG. 15. Temperature–density phase diagrams for various values of the nonlinear, in-plane, elastic energy, γ. Results are shown for the repulsive interplanar interactions $V_{ij} = Z\varepsilon|i - j|^{-5}/6$. The single-phase regions are denoted by lines or loops, except for stage 1, which occupies a much larger area. These calculations are for the *unscreened* interlayer, long-range potential. [From S. E. Millman and G. Kirczenow, *Phys. Rev. B: Condens. Matter* [3] **28**, 3482 (1983).]

Fig. 15, increasing the value of γ from zero to values comparable with the εZ strongly suppresses the dilute stage-1 regions of the phase diagrams. The physical origin of this effect is the large energy cost of layers with a small density of intercalant. The experimental phase diagram for Li–graphite shown in Fig. 13 clearly shows the existence of dilute stage-1 phases. This indicates that, at least for Li, and intercalant whose ionic size does not differ much from the c-axis spacing of the host carbon layers, γ is small. Alternatively, it could indicate that the extremely nonlinear form for the elastic energy overestimates the energy cost of separating the carbon layers for small values of σ. For larger intercalant atoms, the energy γ may indeed be large and the dilute stage-1 phases may be stable only at very high temperatures. More experimental study of this problem will resolve this issue.

Another modification introduced in these calculations is the effect of strongly screened potentials. The consequences of screening on long-range interactions were first discussed by Safran and Hamann,[46] who showed that, for the long-range electrostatic interactions present in intercalated graphite, a given layer interacts with its nearest-neighbor layers through the long-range, power-law interaction $V_{ij} \sim |i-j|^{-\alpha}$. However, interactions with further neighbors are much weaker because of the screening of the ionic sheets by the donated charge. Millman and Kirczenow incorporated this effect into their calculations of the phase diagrams by writing $V_{ij} = V_0 \eta_{ij} |i-j|^{-\alpha}$, where $\eta_{ij} = 1$ if there is no occupied intercalant layer between layers i and j and $\eta_{ij} = 0$ otherwise. The resulting phase diagrams are shown in Fig. 16—the strong-screening limit—for various values of γ. As expected, the strong-screening approximation eliminates the fractional stages and the entire devil's-staircase structure discussed by Bak and Bruinsma.[105] This is because these more complex structures arise from *competing* interplanar interactions, which are eliminated in the strong-screening case. The existence of fractional stages will be discussed further in the following section.

It is interesting to note that for $\gamma = 0$ the strong-screening approximation results only in stages 1 and 2 at finite temperature. The higher stages are stable only upon the introduction of finite values of γ. In general, the topology of the phase diagrams varies dramatically with small changes of the values of γ and V_0. This could be due to the extremely nonlinear forms for the density dependence of the interactions that were assumed in these calculations.

The experimental phase diagram (Fig. 13) for Li–graphite is only qualitatively reproduced by the aforementioned modifications to the lattice-gas model for staging. In particular, although the mean-field phase diagram is not symmetric about $x = \frac{1}{2}$, as predicted by the simplest model, it does show the dilute stage-1 phase which was suppressed by large values of the nonlinear

[105] P. Bak and R. Bruinsma, *Phys. Rev. Lett.* **49,** 249 (1982).

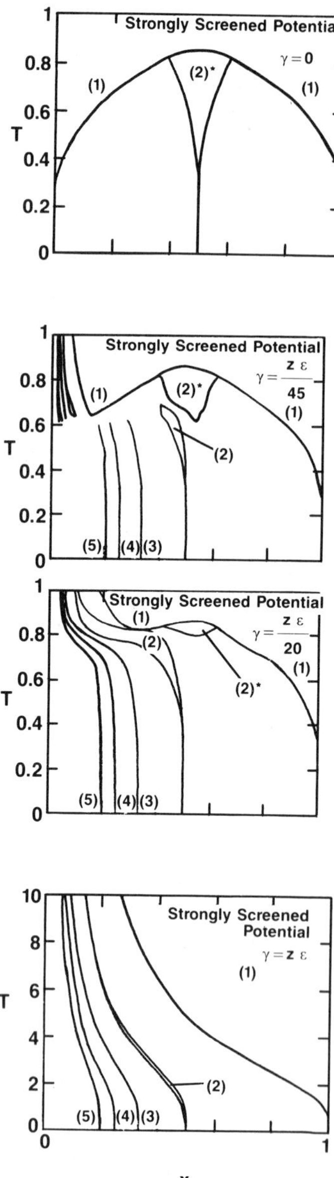

FIG. 16. The same phase diagrams as Fig. 15, for the strongly screened potential, $V_{ij} = \eta_{ij} Z\varepsilon|i-j|^{-5}/6$, where $\eta_{ij} = 1$ if there is no intercalant layer between layers i and j and is zero otherwise. [From S. E. Millman and G. Kirczenow, *Phys. Rev. B: Condens. Matter* [3] **28**, 3482 (1983).]

elastic energy γ. In addition, it shows the existence of two stage-2 regions with different in-plane densities.

DiVincenzo and Koch[80] have shown that the experimental phase diagram for Li–graphite can be understood by adding a "corrugation energy" to the simple lattice-gas model of Eq. (6.1). Although the commensurate $\sqrt{3} \times \sqrt{3}$ superlattice represents the close-packed state, there also exists a commensurate 2×2 superlattice of the intercalant layer. The energy is a local minimum near this 2×2 state. These two favorable structures lead to the existence of two stage-2 regions in the phase diagram, as shown in Fig. 17. The functional form for $U(\sigma)$ used by DiVincenzo and Koch is

$$U(\sigma) = -U_0\sigma^2 + \Delta U(\sigma) \tag{6.4}$$

where $\Delta U(\sigma)$ has a minimum at the concentration corresponding to the 2×2 superlattice, although the mean-field theory does not discriminate between disordered and long-range-ordered, in-plane, structures. The dilute (2×2) stage-2 phase is stable up to higher temperatures due to the larger in-plane entropy of this phase. The first term is minimized for the close-packed $\sqrt{3} \times \sqrt{3}$ structure, where $\sigma = 1$. In contrast, the close-packed $(\sqrt{3} \times \sqrt{3})$ stage-2 phase is unstable to coexistence with other phases due to the small entropy of this densely packed state. The model does predict some features (such as the phase labeled 1′ in Fig. 17) which are not seen experimentally. However, it does correctly explain the most striking deviations of the Li–graphite phase diagram from the predictions of the most naive lattice-gas model in terms of the in-plane structure of the Li layer.

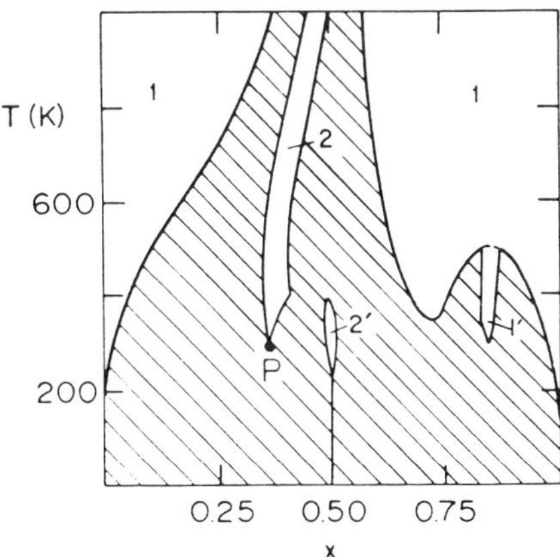

FIG. 17. Concentration–temperature phase diagram as determined from the simple, lattice-gas model augmented by the corrugation energy discussed in the text. [From D. P. DiVincenzo and T. C. Koch, *Phys. Rev. B: Condens. Matter* [3] **30**, 7092 (1984).]

Thus, while the staging phenomenon can exist in systems with a wide variety of in-plane intercalant structures, the details of the intercalant ordering and entropy are important in correlating the theoretical predictions with the experimental observations. Deviations of the in-plane energy from a simple two-body interaction show how particular values of the in-plane density can be stabilized. The lack of competing c-axis interactions due to strong screening helps explain the lack of fractional stages in most intercalation compounds. The experimental phase diagram for Li–graphite shows many of the features of the theoretical models, including the existence of dilute, stage-1 regions. These dilute, stage-1 phases may be unstable for larger intercalant atoms, where the anharmonic contributions to the elastic energy are more important.

7. Pressure-Induced Staging Transitions

In the previous section the staging transitions induced by entropy or temperature changes were discussed. In this section the effect of pressure on restaging transitions is outlined.

a. Restaging and In-Plane Densification

The temperature-induced staging transitions discussed above were all coupled (at constant total intercalant density) to changes in the in-plane density. The application of external pressure has recently been shown to drive restaging transitions in an even more dramatic manner.

The effects of the application of high pressure to intercalated graphite were first demonstrated by Clarke *et al.*[35] These experiments on KC_{24} showed that external pressure results in a densification of the in-plane structure of the intercalant. Since the experiments were reversible with pressure, the densification must be accompanied by an increase in the stage number. X-ray diffraction revealed that such restaging transitions did occur as a function of applied pressure, as shown in Fig. 18. The reversibility of these transitions implies that all the transitions occurred with no loss of intercalant, providing support for the Daumas–Herold[34] picture for staging kinetics via the motion of intercalant domains.

The original experiments on KC_{24} were extended by Wada *et al.*[36,106,107] to Cs and Rb graphite intercalation compounds. In all these cases, the high-pressure, equilibrium state consisted of a 2×2 intercalant superlattice, which would be represented by $MC_{8 \cdot n}$ for a stage-n compound. A sample with stoichiometry $MC_{12 \cdot n}$ at zero pressure evolved to a final, high-pressure phase with stoichiometry $MC_{8 \cdot n'}$, where M represents the metal atom, and n and n'

[106] N. Wada, R. Clarke, and S. A. Solin, *Synth. Met.* **2,** 27 (1980).
[107] N. Wada, *Phys. Rev. B: Condens. Matter* [3] **24,** 1065 (1981).

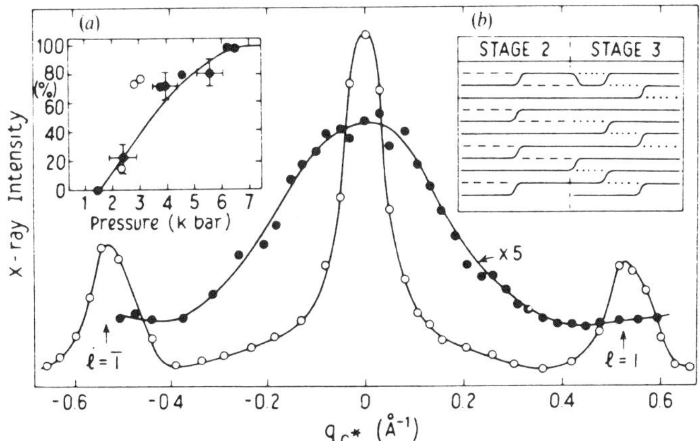

FIG. 18. X-ray intensity as a function of diffraction wave vector, q_c^*, along the $10l$ reciprocal lattice row of single crystal KC_{24} at (●) 4 kbar and (○) 11 kbar. The $10l$ scan at ambient pressure is indistinguishable from background. Insets: (a) Percentage of stage-3 phase in an HOPG sample of KC_{24}, as a function of pressure. (*b*) Deformation model for graphite layers during stage transition. [From R. Clarke, N. Wada, and S. A. Solin, *Phys. Rev. Lett.* **44**, 1616 (1980).]

are the initial and final stage numbers, respectively. To conserve the number of intercalant atoms, $n' = 3n/2$. Transitions from one stage to another are only possible for certain combinations of n and n'. For example, the transition $KC_{12 \cdot 2} \rightarrow KC_{8 \cdot 3}$ was observed, as was the transition $RbC_{12 \cdot 4} \rightarrow RbC_{8 \cdot 6}$. In the first case, the transformation was complete at a pressure $P = 6.5$ kbar, while for Rb, the sample showed pure stage-6 ordering for $P > 7.5$ kbar. For intermediate pressures, a macroscopic mixture of phases was observed. The complex behavior of K–graphite intercalation compounds at intermediate pressures has recently been studied by Kim *et al.*[42] and Sundqvist and Fischer.[108]

The observation that the 2×2 commensurate superlattice is the equilibrium, high-pressure phase of the heavy alkali metal–graphite intercalation compounds is consistent with the prediction of Miyazaki *et al.*[90] Their calculations parametrize the effects of pressure via a variable, c-axis thickness of the intercalant–graphite–intercalant sandwich. However, the same calculations show that MC_{14n} is the equilibrium stoichiometry at $P = 0$. The experimental observation of MC_{12n} can be understood since at $P = 0$ the heavy alkali metal–graphite intercalation compounds show *incommensurate* in-plane ordering.[109] The calculations of Miyazaki *et al.* included a corrugation energy, representing the graphite potential, and did not explicitly

[108] B. Sundqvist and J. E. Fischer, *Phys. Rev. B* **34**, 3532 (1986).
[109] R. Clarke, J. N. Gray, N. Homma, and M. J. Winokur, *Phys. Rev. Lett.* **47**, 1407 (1981).

consider the effects of incommensurability. In fact, the observed incommensurate phases can be thought of as domains of $\sqrt{7} \times \sqrt{7}$, commensurately ordered intercalant, separated by discommensurations. X-ray diffraction studies of CsC_{24} have been interpreted in terms of this type of incommensurate structure. Within a commensurate domain, the ordering is $\sqrt{7} \times \sqrt{7}$, corresponding to a *local* MC_{14n} stoichiometry, in agreement with the theory.

The relationship between the changes in the in-plane density and restaging transitions induced by either changes in the pressure or the temperature was elucidated by DiVincenzo *et al.*[110] They found that dilute stage-2 Li-intercalated graphite showed the same phase transformation at high T and finite P as at low T and $P = 0$. In both cases, the starting point is the 300 K state of stage-2 LiC_{16}. If the temperature is lowered, the equilibrium state is a mixture of stage-2 LiC_{12} and stage-3 LiC_{18}. Both of these phases have a commensurate $\sqrt{3} \times \sqrt{3}$ in-plane, close-packed intercalant layer. At 300 K the sample is a single phase of dilute stage 2 due to the dominance of the in-plane entropy. As discussed in Subsection 6, this effect diminishes in importance as $T \to 0$, resulting in close-packed, low-temperature states ($T \leq 250$ K). The restaging transition is required to conserve the number of Li atoms.

The same reasoning can be used to understand the transition which occurs for $P \geq 2.9$ kbar at $T = 300$ K. High pressure favors a high in-plane density (the $\sqrt{3} \times \sqrt{3}$ phase), which overcomes the higher entropy of the dilute, single stage-2 LiC_{16} phase. Again, the restaging is forced by the conservation of Li atoms.

In both cases—the temperature- and pressure-induced restaging—the entropy of the dilute stage-2 phase must be overcome. To test this crucial physical assumption, DiVincenzo *et al.* computed the slope of the phase equilibrium line dP/dT. The Clausius–Clapeyron equation yields[111]

$$\frac{dP}{dT} = \frac{\Delta S}{\Delta V} \tag{7.1}$$

where ΔS is the entropy change and ΔV the volume change for the transition. The mean-field entropy was taken to be zero in the commensurate ordered, close-packed $\sqrt{3} \times \sqrt{3}$ phases and was taken as

$$S = -\tfrac{1}{2}[\sigma \log \sigma + (1 - \sigma)\log(1 - \sigma)] \tag{7.2}$$

in the dilute, stage-2 phase. The factor of $\frac{1}{2}$ accounts for the fact that only $\frac{1}{2}$ of the layers are occupied in stage 2; S is the entropy per site. The resulting slope,

[110] D. P. DiVincenzo, C. D. Fuerst, and J. E. Fischer, *Phys. Rev. B: Condens. Matter* [3] **29B,** 1115 (1984).

[111] F. Reif, "Fundamentals of Statistical and Thermal Physics," p. 304. McGraw-Hill, New York, 1965.

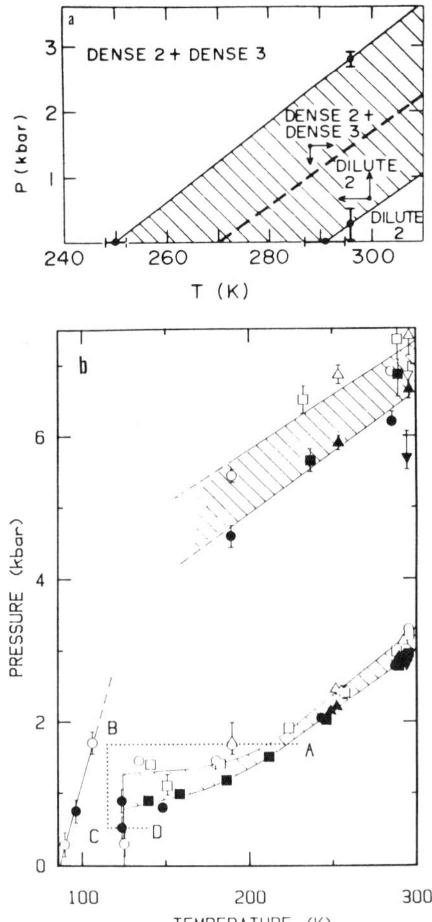

FIG. 19. (a) Pressure (P)–temperature (T) phase diagram for LiC_{16}. Dots and error bars are extremities of the hysteresis loop determined from resistivity experiments. (– – –) shows the theoretical prediction. (———) drawn parallel to (– – –) can be displaced so as to lie on the data points; the theory only predicts the slope. The stable phases are dilute stage 2 (lower right) and mixed, dense stages 2 and 3 (upper left); in the hatched region either phase is stable depending on kinetics. [From D. P. DiVincenzo, C. D. Fuerst, and J. E. Fischer, *Phys. Rev. B: Condens. Matter* [3] **29**, 1115 (1984).] (b) Pressure (P)–temperature (T) phase diagram for KC_{24} derived from resistivity anomalies. Circles, squares, and triangles are data from different samples; open and closed circles indicate passage from low P/high T to high P/low T and conversely. [From B. Sundqvist and J. E. Fischer, *Phys. Rev. B* **34**, 3532 (1986).]

using the experimentally determined value for ΔV, is shown in Fig. 19a and compares favorably with the experimental data.

DiVincenzo *et al.* note that a similar effect is *not* expected in the heavy alkali metals, where ΔV is much larger; the transition $KC_{12 \cdot 2} \rightarrow KC_{8 \cdot 3}$, which is observed at high P, is not expected to happen at positive values of T for $P = 0$. This has been confirmed in the pressure–temperature phase diagram for K–graphite recently measured by Sundqvist and Fischer[108], as shown in Fig. 19b. The transition $KC_{2\cdot12} \rightarrow KC_{3\cdot8}$ occurs in two stages at room temperature. The transition is complete at $P \sim 6$ kbar. There is no vestige of this transition at low temperature, since the slope of the phase boundary extrapolates to negative values of T. The partial restaging transition at $P \sim 3$ kbar at room temperature does extrapolate to a zero-pressure value of $T \sim 120$ K, where indeed a transition is found. However, this transition does not involve any change in stage! Sundqvist and Fischer[108] suggest that the complex in-plane transition at $T \sim 120$ K is a vestige of the high-T, high-P restaging transition. The transition and in-plane densification would continue to $T \sim 120$ K as the pressure is reduced if the strength of the host corrugation energy were large enough to induce the commensurate 2×2 structure which usually accompanies the restaging transition. At high temperatures, the entropy of the liquid phase allows an approximate 2×2 structure to form, while at low temperatures, the sample is frozen into an incommensurate structure whose density is such that the restaging transition does not occur.

These results unify staging transitions induced by either temperature or pressure. More importantly, they show that the entropy change due to densification drives the staging transition, and that it is the competition of this entropy with the lower energy of the close-packed state that allows for both concentrated and dilute staged intercalation compounds.

b. *Fractional Staging*

Although the simple lattice-gas model with long-range interactions did show fractional staging, subsequent calculations[46,40,103] showed that, if the interactions were long ranged, but restricted to nearest-neighbor pairs of intercalant layers (strong screening), the fractional layers were suppressed. However, recent observations[38,42] of fractional stage $\frac{3}{2}$ (a periodic sequence of two host layers filled by intercalant, one interlayer position empty) seem to contradict these models.

The experimental observation relies on the behavior of KC_8 under applied pressure. For $15 \leq P \leq 19$ kbar, KC_8 has been observed to transform to a mixture of stage 1 and stage $\frac{3}{2}$, as shown in Fig. 20. For $P \geq 19$ kbar, the stage-$\frac{3}{2}$ state disappears, and the sample converts to a mixed, stage-1 and stage-2 entity, with a higher in-plane stoichiometry,[112] as discussed in Subsection 7,a. After the discovery of the fractional stage by Fuerst *et al.*,[38] further experiments by Kim *et al.*[42] verified that the interplanar order in stage $\frac{3}{2}$ is

[112] J. M. Bloch, H. Katz, D. Moses, V. B. Cajpe, and J. E. Fischer, *Phys. Rev. B: Condens. Matter* [3] **31,** 6785 (1985).

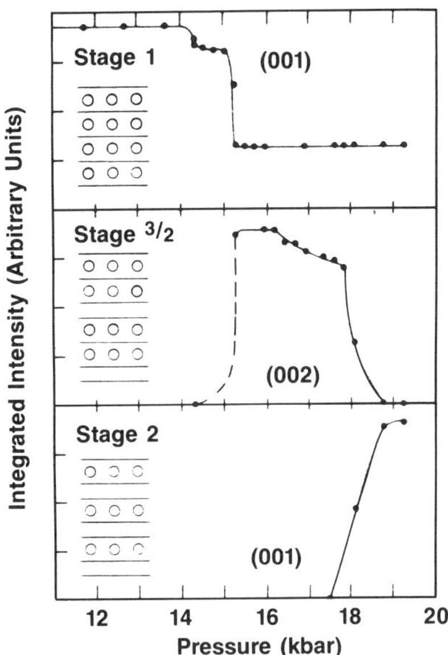

FIG. 20. Evolution of stages with pressure in KC_8. Note that all the transitions are relatively sharp. [From C. D. Fuerst, J. E. Fischer, J. D. Axe, J. B. Hastings, and D. B. McWhan, *Phys. Rev. Lett.* **50**, 780 (1983).]

fairly well correlated; the integrity of the long-range order is comparable to the other pure-stage materials observed under high pressure; it is not, however, microscopically perfectly ordered.

A model calculation by Hawrylak and Subbaswamy,[41] which considers the effect of external pressure in changing the c-axis spacings, does yield fractional stage $\frac{3}{2}$ (which those authors term $\frac{2}{3}$) for some values of the applied pressure. They considered a form for the in-plane energy $U(\sigma)$

$$U(\sigma) = -\alpha_1\sigma^{3/2} + \beta_1\sigma^2 \tag{7.3}$$

The first term accounts for the cohesive energy of the two-dimensional electron gas, while the second term includes the *bare* Coulomb repulsion of the ionized intercalant. This Coulomb term scales linearly with c_i, the spacing between the adjacent carbon layers. The in-plane screening of the ionic charge by the donated intercalant was ignored, possibly because of the large distance between the intercalant and graphite sheets. They compute the phase diagram for this nonquadratic in-plane energy. In order to couple to the c-axis spacing, they consider an expression similar to that of Dahn *et al.*[79] for the change in the intercalant–graphite thickness [see Eq. (4.2)]. In addition, they include the mean-field lattice-gas entropy [Eq. (6.1)] and the unscreened, power-law interlayer interaction. The external pressure is accounted for by a term equal

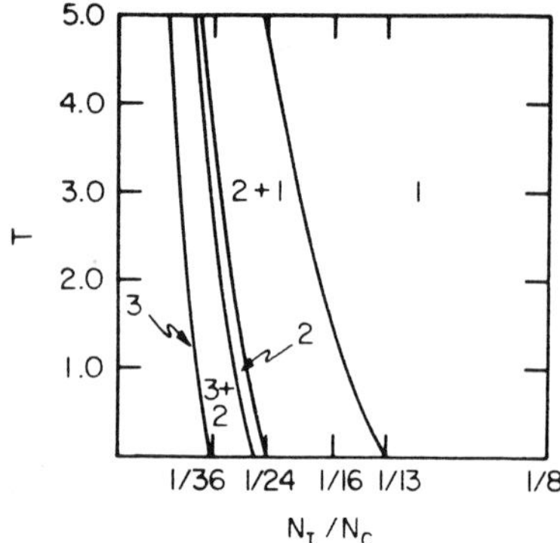

FIG. 21. Theoretical phase diagram as a function of temperature and intercalant concentration normalized to the graphite concentration (N_I/N_C) for zero applied pressure. The temperature is in units of the minimum value of the sandwich (intralayer) energy. The labels 1 + 2 and 3 + 2 denote coexisting phases. [From P. Hawrylak and K. R. Subbaswamy, *Phys. Rev. B: Condens. Matter* [3] **28**, 4851 (1983).]

to fc_i, where c_i is the separation between the ith and $(i + 1)$th carbon layers, and $f = P\sigma_C$, where σ_C is area per carbon atom.

For $f = 0$, zero applied pressure, they find regions of pure stages 1, 2, and 3, as well as coexisting phases of mixed stages for fixed intercalant concentration. It is interesting to note that in this model, stage $\frac{3}{2}$ is not stable at $P = 0$. Whether other fractional stages are stable was not investigated. The phase diagram is shown in Fig. 21. It shows very little temperature dependence and does not resemble that of Li–graphite, as shown in Fig. 13. At finite applied pressure, the phase diagram is shown in Fig. 22. In addition to the pure stages, a stage $\frac{3}{2}$ appears in coexistence with stage 1. This result is in good agreement with the observed trends in the pressure dependence of the K–graphite system.

The reason for the lack of fractional stages in this model lies in the assumed form for $U(\sigma)$, which is not symmetric about $\sigma = \frac{1}{2}$. Whether this modification to the lattice gas is sufficient to suppress all fractional stages at $P = 0$, or whether strong screening must be invoked is not yet clear. In addition, the physical origin of the fractional stage $\frac{3}{2}$ at finite values of P is not obvious. Still, this model does correctly predict the appearance of a fractional stage at finite pressure, as well as the stage 2 → mixed stage 3/stage 2 → pure stage 3 as the pressure is increased.

An alternative explanation of the existence of fractional stages with applied pressure retains the strong-screening approximation, which suppresses all fractional stages at $P = 0$. However, for finite values of the applied pressure, it is possible that the relatively short-ranged, elastic interactions of the intercalant atoms (which decay as e^{-Gz}, where G is the in-plane, intercalant

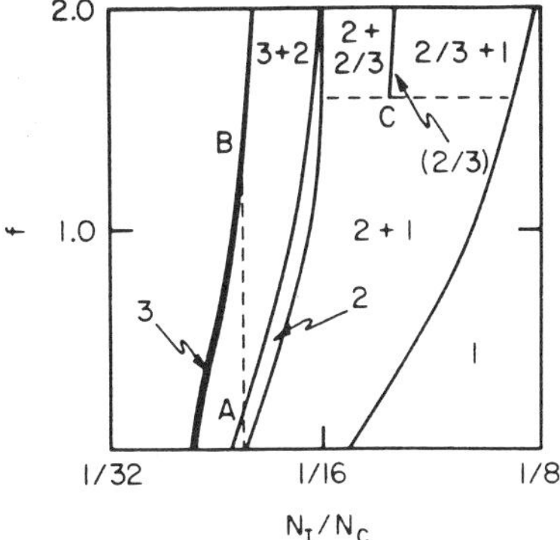

FIG. 22. Theoretical phase diagram as a function of pressure and concentration, where N_I/N_C is the ratio of intercalant to carbon atoms. The scaled pressure f corresponds to a value of ~ 12 kbar at point B. The temperature is fixed at ~ 350 K. Note that these authors denote fractional stage $\frac{3}{2}$ by $\frac{2}{3}$. [From P. Hawrylak and K. R. Subbaswamy, *Phys. Rev. B: Condens. Matter* [3] **28**, 4851 (1983).]

reciprocal lattice vector) become increasingly important. These interactions are not screened. They can provide the competition between nearest-layer and next-nearest-layer interactions which stabilize the fractional stages. This suggestion still awaits a detailed calculation.

IV. Domains and Staging Kinetics

The previous section discussed temperature- and pressure-induced restaging transitions in intercalation compounds. In the theoretical analysis, it was assumed that the intercalant layers were macroscopic (infinite) in size and that true thermal equilibrium conditions prevailed. In this section, stage disorder due to finite intercalate domain size is discussed from the point of view of both theory and experiment. The elastic interactions between domains are then treated in detail, and the effects of the resulting energy barriers are pointed out. Finally, the kinetics of staging and intercalation, as determined by recent simulations, calculations, and experiments, is reviewed.

8. Domain-Induced Random Staging

The lattice-gas models for staging presented in Section III resulted in stage disorder due to the generation of vacancies and interstitials. It was assumed that the intercalate layers in the intercalation compound extend continuously over macroscopic distances. However, there is experimental evidence that in real intercalation compounds the intercalant atoms form islands, and that every pair of host layers has an *equal* number of intercalant atoms between them.[34] Electron imaging of these islands has been reported[113–116]; a typical size is of the order of 100 Å. However, it is not certain whether these sizes are due to kinetic limitations in the intercalation process, imperfections in the host crystal, or degradation due to the electron beam. X-ray experiments have not yet reported a measurement of domain sizes. As a matter of fact, a recent[117] high-resolution x-ray study has reported coherent regions of incommensurately ordered intercalant greater than 10,000 Å! Thus, at this point, it is not at all clear whether these domains are an intrinsic phenomenon or whether they are determined by kinetics.

Notwithstanding these questions about the *origin* of domains in staged intercalation compounds, several groups have studied their effect on the integrity of staging.[8,10,11,32,118–120] As mentioned above, stage ordering is expected to be perfect (except for vacancy-induced disorder) for infinite intercalate layers. The existence[8,11,32,120] of measurable randomness in staging signifies finite domain sizes. (This kind of stage disorder refers to the existence of randomly mixed sandwiches of different stage in one compound—see Fig. 4.)

Kirczenow[10,118] has recently quantified these ideas in a calculation of the degree of randomness in staging using an extension of the lattice-gas models described above. He considers a single stack of finite-size domains. The entropy of disordering of the domains in different stacks and correlations between such domains are neglected (see the following section). In a mean-field approximation, the free energy Φ is written

$$\Phi = vU(\sigma) + \sum_i u_i v_i \sigma^2 + Tv[\sigma \log \sigma + (1-\sigma)\log(1-\sigma)] + T\sum_i v_i \log v_i \tag{8.1}$$

[113] J. M. Thomas, G. R. Millward, R. F. Schlögel, and H. P. Bohm, *Mater. Res. Bull.* **15,** 671 (1980).

[114] S. F. Flandrois, J. M. Masson, J. C. Rouillon, J. Gaultier, and C. Hauw, *Synth. Met.* **3,** 1 (1981); D. Dorignac, M. J. Lahana, R. Jagut, B. Joffrey, S. Flandrois, and C. Hauw, *in* "Intercalated Graphite" (M. S. Dresselhaus, G. Dresselhaus, J. E. Fischer, and M. S. Moran, eds.), p. 33. North-Holland Publ., Amsterdam, 1983.

[115] A. N. Berker, N. Kambe, G. Dresselhaus, and M. S. Dresselhaus, *Phys. Rev. Lett.* **49,** 249 (1982).

[116] G. Timp, M. S. Dresselhaus, P. C. Eklund, and Y. Iye, *Phys. Rev. B: Condens. Matter* [3] **26,** 2373 (1982).

[117] A. R. Kortan, A. Erbil, R. Birgeneau, and M. S. Dresselhaus, *Phys. Rev. Lett.* **49,** 1427 (1982).

[118] G. Kirczenow, *Phys. Rev. B: Condens. Matter* [3] **31,** 5376 (1985).

[119] G. Forgacs and F. Uimin, *Phys. Rev. Lett.* **52,** 6331 (1984).

[120] H. J. Kim and J. E. Fischer, *Phys. Rev. B* **33,** 4349 (1986); *Synth. Met.* **12,** 137 (1985).

Here, $U(\sigma)$ is the in-plane energy and $u_i = V_{j,j+i}$ is the interlayer repulsion discussed in Section III. The first term proportional to T is the in-plane entropy of the intercalant layer, while the last term in this equation is the entropy of randomly placing v_i sandwiches of stage i within the stack. The total number of sandwiches is $v = \sum v_i$. Note that vacancy disorder is allowed for only in an average manner, and not layer by layer. The in-plane density of intercalant is assumed to be σ, independent of the local stage index. This in contrast to the models of Section III, where it was shown that the variability of σ is strongly coupled to the stage.

The free energy is minimized subject to the constraints of (1) conservation of intercalant atoms, which introduces a chemical potential conjugate to $v\sigma$ and (2) conservation of the total number of host layers, $v_{\mathrm{H}} = \sum i v_i$. Again, this free energy is that of a single stack of finite-size domains. The fraction of stage-i units in equilibrium is f_i, defined by

$$f_i = \hat{v}_i \Big/ \sum_i \hat{v}_i = e^{c_i/T} \tag{8.2}$$

where $\hat{v}_i$ is the solution of $\partial\Phi/\partial v_i = 0$ and where

$$c_i = U(\sigma) + T[\sigma \log \sigma + (1 - \sigma)\log(1 - \sigma)] + u_i\sigma^2 - \mu\sigma - i\Psi \tag{8.3}$$

Here, μ and Ψ are the Lagrange multipliers which account for conservation of intercalant and host layers, respectively. The normalization condition, $\sum f_i = 1$, allows solution of the nonlinear system for $\{v_i\}$.

If the frequency of stage-i units is a smoothly varying function of the local stage i, an expression for f_i can be derived from an expansion about the extrema of f_i and c_i.[118] Using a saddle-point expansion, it is found that

$$f_i = f_{\bar{i}} \exp\{-[(i - \bar{i})/\tilde{\sigma}]^2/2\} \tag{8.4}$$

where $\bar{i}$ is the most probable stage index, and where

$$\tilde{\sigma}^2 = T \Big/ \left[N_0 \sigma^2 \left(\frac{\partial^2 u_i}{\partial i^2} \right)_{i=\bar{i}} \right] \tag{8.5}$$

In Eq. (8.5), N_0 is the number of sites in the layer, which is proportional to the area of the domain. As predicted, the randomness in the staging, characterized by $\tilde{\sigma}$, vanishes in the thermodynamic limit. Furthermore, this expression shows that, for a model of a single stack of domains, the randomness in the staging depends only on the interlayer interaction.

Numerical solutions of the set of nonlinear equations for the fractional probability of stage-i units f_i are shown in Fig. 23. Decreasing the range of the interlayer interaction strongly increases $\tilde{\sigma}$ and hence the staging randomness. A similar effect is obtained by decreasing the *strength* of the interlayer interaction. Thus, acceptor compounds, which generally have smaller charge

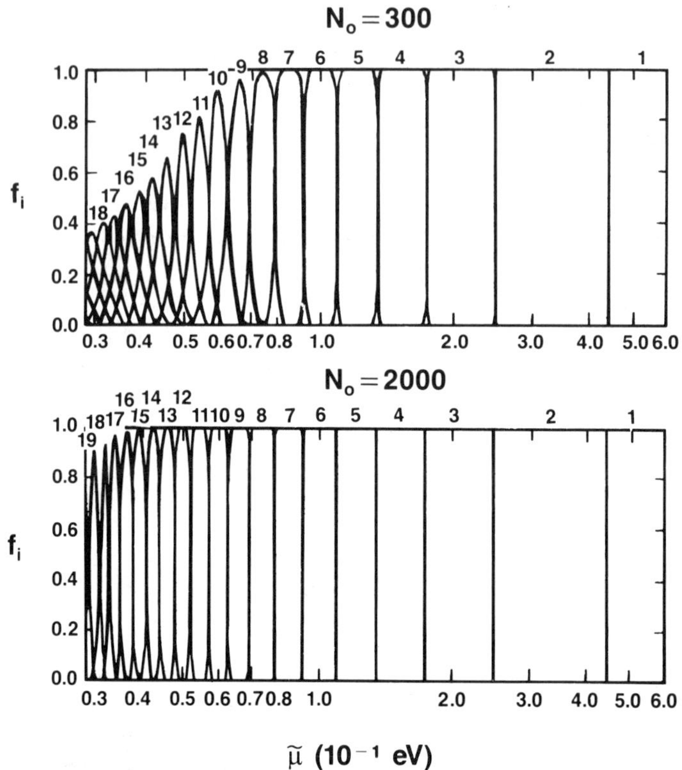

FIG. 23. Stage disorder as calculated for K–graphite. The fraction f_i of stage-i units versus the chemical potential $\tilde{\mu} = \mu - \mu_\infty$, where μ_∞ is the lowest μ at which intercalation occurs for infinite domains, is plotted. The in-plane interaction uses the parameters $\gamma = Z\varepsilon = 1$ eV and the temperature $T = 0.03$ eV. The interplanar interaction, $u_i = v_0 i^{-\alpha}$, is calculated for $v_0 = 0.3$ eV and $\alpha = 1$. The curve giving the fraction of stage i the domain is labeled by its value of i. [From G. Kirczenow, *Phys. Rev. B: Condens. Matter* [3] **31**, 5376 (1985).]

transfers than donors, should show a higher degree of random staging. It is interesting to note that this model predicts a *continuous* transition between all stages; e.g., in the stage-1 to stage-2 transition, the proportion of stage-2 units grows continuously as the chemical potential μ is increased, with a concomitant decrease in the proportion of stage-1 units.

Recent x-ray measurements have begun to address the question of random staging. In the early work of Metz and Hohlwein[32] it was already noticed that stage ordering was imperfect. Again, this effect may be related to extrinsic properties of the intercalation compound (growth kinetics, defects) which lead to finite-size domains. For example, Nishitani *et al.*[97,98] did not report any admixture of foreign stages in their study of K–graphite compounds, while

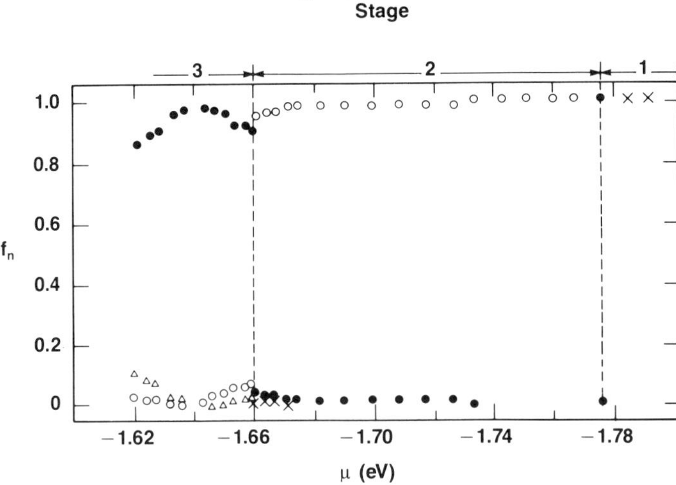

FIG. 24. Stage disorder as measured for K–graphite. The fraction of stage-i packages as a function of the measured chemical potential is shown for constant sample temperature. Arrows at the top of the figure indicate stability region of a stage-i compound and (---) indicate the staging transitions. The fraction of stage-i units is indicated by: (×), stage 1; (○), stage 2; (●), stage 3; and (△), stage 4. [From M. E. Misenheimer and H. Zabel, *Phys. Rev. Lett.* **54**, 2521 (1985).]

other studies[11,42,121] do find evidence for admixtures of stage-($n \pm 1$) units for every $n > 2$, as shown in Fig. 24. Furthermore, the stage disorder seems to increase a small amount near the stage-2 to stage-3 transitions.

The temperature and pressure dependence of randomness in stage ordering have been studied by Heiney *et al.*[122] and by Kim *et al.*,[42] respectively. These investigations show that high-stage samples are well described by a distribution of different stage units. As the temperature is raised, the randomness in this distribution increases. The change in the degree of stage order occurs where changes in the in-plane density take place as a function of either temperature or pressure. It is noted that a uniform measure of stage integrity can be obtained from a Hendricks–Teller[123] analysis of the (00l) diffraction line shapes.

Whether these observations are direct confirmations of the thermodynamic model for random staging is not yet clear. The thermodynamic model predicts

[121] C. D. Fuerst, T. C. Koch, J. E. Fischer, J. A. Axe, J. B. Hastings, and D. B. McWhan, *in* "Intercalated Graphite" (M. S. Dresselhaus, G. Dresselhaus, J. E. Fischer, and M. S. Moran, eds.), p. 39. North-Holland Publ., Amsterdam, 1983.

[122] P. A. Heiney, M. E. Huster, V. B. Cajipe, and J. E. Fischer, *J. Synth. Met.* **12**, 21 (1985).

[123] S. Hendricks and E. Teller, *J. Chem. Phys.* **10**, 147 (1942).

continuous transitions, while the *in situ* data of Misenheimer and Zabel[11] show marked discontinuities. It can be argued that these discontinuities are due to kinetic effects. However, the data show that the discontinuity in the stage order decreases with increasing stage, signifying the effects of thermal fluctuations in equilibrium. Another problem with an application of the thermodynamic model of random staging to the experiments is the prediction of only a small degree of randomness for large intercalant domains. Misenheimer and Zabel[11] argue that for the large intercalant domain sizes estimated[124] for donor compounds (5000 Å) the randomness in the staging should be small. While it is true that, even for large domain sizes, stage disorder can be large if the interlayer interactions rapidly decrease with stage, this does not seem to be the case for K–graphite compounds. The thermodynamic data of Nishitani *et al.*[97,98] indicate that the interlayer interaction falls off rather slowly with stage for stages $n < 7$. This observation was used by Kirczenow to justify a power-law falloff of the interlayer interaction $V_{ij} \sim |i - j|^{-\alpha}$, with $\alpha = 1$ for small values of n, in agreement with the Thomas–Fermi results discussed above. Thus, either the interlayer interactions are much weaker than estimated, or the domain sizes in alkali metal–graphite intercalation compounds are much smaller than previously thought, if the thermodynamic model for stage disorder is applicable.

Perhaps a more fundamental question is the intrinsic nature of this randomness and of the existence of finite-size domains. Are these domains formed as a result of the minimization of the free energy, or are they extrinsic effects due to the kinetics of the intercalation process or host sample defects? Experiments on single crystal hosts should help answer this question. If the domain size is determined by extrinsic effects such as kinetics, these same effects may be directly responsible for stage disorder (random staging). Thus the degree of random staging would be sample and perhaps time dependent. The observations[42,122] that the stage disorder is coupled to the in-plane density may also reflect kinetic constraints. An explanation of random staging based on kinetic considerations is discussed below.

9. Domain–Domain Interactions

In the previous section, randomness in stage ordering was found to be an intrinsic property of a stack of finite-size domains. The model assumed that there were no correlations between the intercalant islands. However, in real materials, the dislocations in the host lattice induced by the boundaries of these islands are responsible for correlations between domains in different stacks. These correlations are examined here in a review of the theory of the

[124] S. E. Hardcastle, M. E. Misenheimer, and H. Zabel, *Rev. Sci. Instrum.* **54,** 206 (1983).

elastic interactions between domains. The interactions between domain walls lead to energy barriers preventing the coalescence of two nearby domains, and thus have implications for the kinetics of staging and intercalation.

In the discussion of elastic interactions in Section II, it was shown that intercalant islands located in the same host plane attract each other, while islands in different planes show repulsive interactions. This conclusion was reached from simple, continuum elastic considerations. However, for a more realistic picture for the elastic interactions between intercalant islands one must take into account the discrete nature of the graphite layers. This was done first by Kirczenow,[125] who modeled the two host layers which bound the intercalate as two coupled, elastic plates. If y denotes the displacement of the center of one of the host layers from the position that it would occupy in the absence of intercalant, the elastic energy of the host in the region between the islands is written

$$E = 2L \int dx \left[\lambda \left(\frac{\partial^2 y}{\partial x^2} \right)^2 + \mu y^2 \right] \tag{9.1}$$

Here, L is the length of the domain boundary, μ characterizes the coupling between the two host layers, and λ is related to the bending modulus of the host. Typical values for graphite are $\mu = 900$ K Å^{-4}, and $\lambda = 7000$ K.[126] The stretching of the graphite layers is neglected; this is justified if the intercalant layer thickness is small compared to the host layer separation.

The energy E is minimized with respect to variations of y, consistent with the boundary conditions. Kirczenow calculates the domain-wall energy for staggered and matching domains. Staggered domains are two adjacent domains which lie in between two different pairs of host layers, while matching domains lie in between the same pair of host layers. The results are shown in Fig. 25. For matching domains, the lowest-energy state is of course represented by the zero separation configuration—consistent with the attractive in-plane elastic interactions discussed in Section II. However, the discreteness of the host layers results in a nonmonotonic behavior of the domain-wall energy as a function of the domain separation R. For $R < 8$ Å, the interaction is attractive, but for $8 < R < 16$ Å, the interaction is *repulsive*. There is an elastic energy barrier of 182 K per Å of length of domain boundary which must be overcome before the two intercalant islands can merge. This barrier can stabilize finite intercalant islands and prevent the true equilibrium state of infinite intercalant domains from occurring (at least at low temperatures).

[125] G. Kirczenow, *Phys. Rev. Lett.* **49,** 1853 (1982).

[126] P. Hawrylak, K. Subbaswamy, and G. W. Lehman, *Solid State Commun.* **51,** 787 (1984).

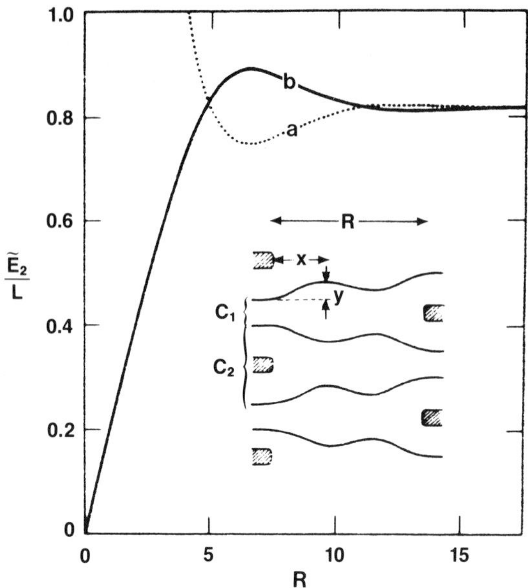

FIG. 25. Domain configuration energy $\tilde{E}_2/L$ (10^3 K per Å of length L per intercalant layer) versus domain separation R (in Å), for staggered domains shown in the inset (curve a) and matching domains (curve b). [From G. Kirczenow, *Phys. Rev. Lett.* **49**, 1853 (1982).]

For staggered domains, the interaction energy has a minimum at $R \sim 6$ Å. Although at small separations staggered domains repel, the interaction becomes attractive at large separations. The domains thus bind to each other, with a binding energy of ~ 68 K per angstrom of domain boundary.

Hawrylak *et al.*[126] extended these calculations to stages other than stage 2 and showed that the energy barrier for domain coalescence *increases* with increasing stage because the number of host layers involved in the domain-wall dislocation increases. They predict that the kinetic barrier to reaching the equilibrium state of infinite intercalate domains increases significantly with stage. These results indicate that the Daumas–Herold picture of staging (Fig. 5) may have its origins in the elastic energy barriers that prevent domain coalescence. (Of course, other transport mechanisms which do not involve island motion, but consist of the evaporation/condensation of individual atoms from one island to another, may still lead to the merger of domains.)

While the harmonic approximation used in the previous analysis is valid for small separations of the host lattice, the anharmonic terms in the interlayer potential must be considered in realistic calculations which can differentiate between different intercalants. The intercalants are characterized by their thickness, d_i. Ulloa and Kirczenow[127,128] used a Morse potential to characterize the interlayer potential in their analysis, which also included the full expression for the bending energy (not just the small-slope approximation). They found that the energy barrier is a strong function of d_i, due to the anharmonic terms in the interlayer potential. Most intercalates in graphite fall in a region where the energy barriers for domain coalescence are small—due to the large values of d_i and the resulting small interlayer interaction. However, some species fall in a region of large energy barriers. Ulloa and Kirczenow argue that this behavior should occur for Na and Tm ($d_i = 1.20$ and 1.27 Å, respectively) and that this explains the observed difficulties in the intercalation reactions of these materials. Their results imply that the slowing of intercalation kinetics by the elastic energy barriers is a very nonuniversal phenomenon and depends in some detail on the intercalant species.

10. Kinetics of Staging and Intercalation

The calculations of the energy barriers that prevent the merging of domains in intercalation compounds signify that kinetic effects may be important in understanding staging and the intercalation process in real materials. In this section, models of staging and intercalation kinetics are reviewed, along with a discussion of the small amount of experimental data on the kinetics of the staging transition. The kinetics of staging and intercalation is a problem of current research; the material reviewed here represents a first step.

The existence of finite-size domains in intercalation compounds is in itself a kinetic effect. In true equilibrium, the material would minimize all elastic strains by forming macroscopic intercalant layers. However, as the early experiments of Daumas and Herold[34] showed, the intercalation compound is most likely composed of finite-size domains. Otherwise, restaging on relatively short time scales would involve the motion of macroscopic layers—a highly unlikely process!

Recent theoretical studies have shown that, indeed, the kinetics of intercalation do result in a domain structure. The earliest study was that of Safran,[39] who used the simple lattice-gas model described in Section III as the starting point for a kinetic calculation of the evolution of a dilute stage-1 compound (with 50% in-plane vacancies) to a stage-2 material with a saturated in-plane density. The kinetic constraints, which prohibit diffusion of the intercalant atoms through the host planes, result in the Daumas–Herold

[127] S. E. Ulloa and G. Kirczenow, *Phys. Rev. Lett.* **55,** 218 (1982).

[128] S. E. Ulloa and G. Kirczenow, *Phys. Rev. B: Condens. Matter* [3] **33,** 1360 (1986).

domain structure. Because of this domain structure, the order parameter for long-range stage ordering is not unity, leading to some broadening of the structure factor describing density fluctuations.

More detailed and sophisticated calculations of the development of the domain structure in quenched intercalation compounds have been performed by Hawrylak and Subbaswamy,[129] Miyazaki and Horie,[130] and Kirczenow.[131] The simple lattice-gas model, based on the free energy of the intercalant *atoms*, was used by Hawrylak and Subbaswamy to calculate the time evolution of the in-plane density for a small number of intercalant layers. Defining $\sigma_i(\mathbf{r})$ as the local intercalant density in between layers i and $i + 1$, with $\mathbf{r}$ denoting the in-plane coordinate, the equation of motion for $\sigma_i(\mathbf{r})$ is[129]

$$\frac{\partial \sigma_i}{\partial t} = -M\nabla^2 \frac{\delta F}{\delta \sigma_i} \tag{10.1}$$

where F is the free energy and ∇^2 is the in-plane Laplacian. This expression assumes that the intercalant atoms can only diffuse within the intercalant planes with mobility M; diffusion through the host layers is prohibited.

The free energy is the sum of the free energy of the uniform layer, $F_0(\sigma_i)$, the interlayer interaction V_i, and a term which accounts for the energy cost for introducing density gradients in the in-plane structure.

$$F = \sum_i F_0(\sigma_i) + V_i + \frac{1}{A}\int d\mathbf{r}\frac{1}{2}\kappa[\nabla\sigma_i(\mathbf{r})]^2 \tag{10.2}$$

where A is the area of a plane. The first two terms in this expression are given by

$$F_0 = -\tfrac{1}{2}U_0\sigma^2 + T[\sigma\log\sigma + (1-\sigma)\log(1-\sigma)] + \gamma\sigma/(\beta+\sigma) \tag{10.3}$$

where the first term is the attraction, the second the entropy of mixing, and the third term the anharmonic elastic energy (see Section III). The interlayer interaction energy is written

$$V_i = \sum_j \tfrac{1}{2}V_0|i-j|^{-\alpha}\sigma_i\sigma_j \tag{10.4}$$

Harwlyk and Subbaswamy used the power-law form of the interaction, with $\alpha = 4$. The values $V_0/U_0 = 0.3$, $\gamma/U_0 = 1.05$, and $\beta = 0.5$ were used in these calculations.[129] The "gradient" energy is parametrized by κ, which determines the interfacial energy and the size of the islands.

[129] P. Hawrlyak and K. R. Subasswamy, *Phys. Rev. Lett.* **53,** 2098 (1984)
[130] H. Miyazaki and C. Horie, *J. Synth. Met.* **12,** 149, 155 (1985); *Phys. Rev. B* **34,** 5736 (1986).
[131] G. Kirczenow, *Phys. Rev. Lett.* **55,** 2810 (1985).

An initial configuration corresponding to a dilute stage 1 with a uniform occupancy of $\sigma_0 = 0.225$ on each layer was assumed. The system was quenched from a temperature of $T/U_0 = 1.1$ to $T/U_0 = 0.7$. The equilibrium state for this final temperature is stage 3 (see the line labeled AB in Fig. 26a). A linear stability analysis showed that fastest growing instability has a periodicity of two interlayer spacings. Thus, stage-2 regions must be established before the final stage-3 structure is reached. The evolution of the structure for six layers is shown in Fig. 26b. Harwlyak and Subbaswamy note that sometimes intermediate stages are metastable and long-lived. For a quench along the line CD in Fig. 26a, the system evolves through a fractional stage. These kinds of processes naturally lead to domain structures in staged intercalation compounds, due to the repulsive interplanar interactions.

Miyazaki and Horie[130] modified this approach to intercalation kinetics. They renormalized the in-plane length scale to that of an elementary island of intercalant, and included stochastic terms in the equation of motion. Their formulation adopts a discrete description of the island structure. Their equation of motion is similar to a discrete version of Eq. (10.1), with an additional term on the right-hand side that represents a Gaussian random force. They solve the resulting equations numerically and compute the time-dependent structure factor: $S(q, t) = \langle \sigma_q(t)\sigma_{-q}(t) \rangle$, where the wave vector q corresponds to the direction perpendicular to the layers. The system is initially at stage 3, and the chemical potential is changed so that the final state is stage 2. They observed that the stage-2 peak in the structure factor grows at the expense of the stage-3 peak. In addition, peaks corresponding to both stage 2 and stage 3 coexist at intermediate times without any significant broadening of the peaks. No fractional stages are found, even as metastable states, since the elastic interactions between islands in different columns make these fractional stages unlikely.

The kinetic models discussed heretofore examined temperature and chemical potential quenches and showed how the system evolves from one stage to another. A more ambitious goal is the modeling of the intercalation process itself. Kirczenow has calculated the kinetics of a full, three-dimensional model of the motion of intercalant islands as the host crystal is intercalated. The interlayer repulsion was taken to be a power law with an exponent $\alpha = 1$. Various terms were introduced to account for the dislocation energies of the islands and the interactions between them. Additional terms were added to represent the energetics of regions where the island intersects the host crystal surface. A guest species is placed at the surface of the host crystal; the species is in contact with a reservoir at fixed chemical potential. The movement of the elementary islands between the crystal and the reservoir is controlled by a Monte Carlo algorithm, with only nearest-neighbor in-plane hopping (of the entire island) allowed.

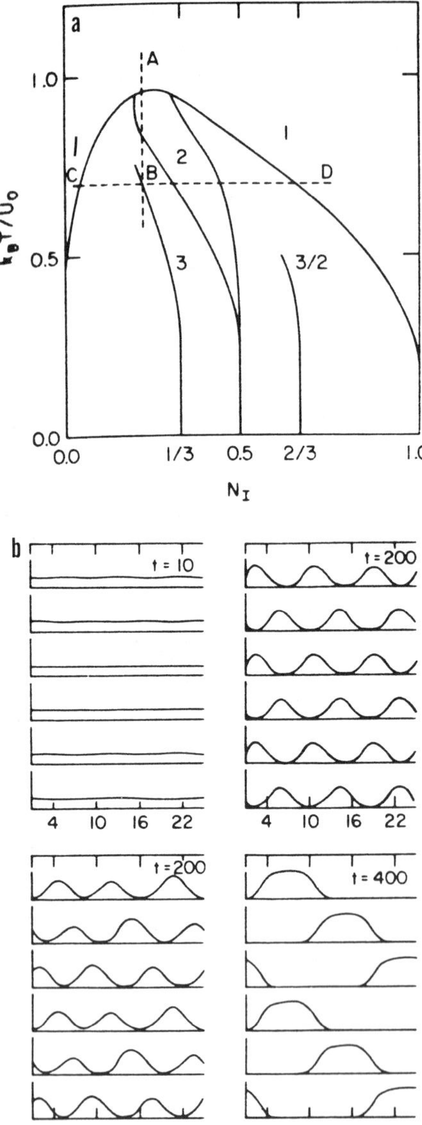

FIG. 26. (a) Equilibrium phase diagram for the free-energy model described in the text. N_I is the overall intercalant volume fraction. AB and CD represent the processes considered in the text. [From P. Hawrylak and K. R. Subbaswamy, *Phys. Rev. Letts.* **53**, 2098 (1984).] (b) Density profiles for the quenching process AB for six layers at various times. Distance is measured in units of $\kappa^{1/2}$ (~ 20 Å) and time in units of κ/MU_0 ($\sim 10^{-6}$ s). The top two figures represent the approach to stage-2 metastable configurations, while the bottom two figures show the approach to stage 3; the stage-2 metastable state has been suppressed. [From P. Hawrylak and K. R. Subbaswamy, *Phys. Rev. Lett.* **53**, 2098 (1984).]

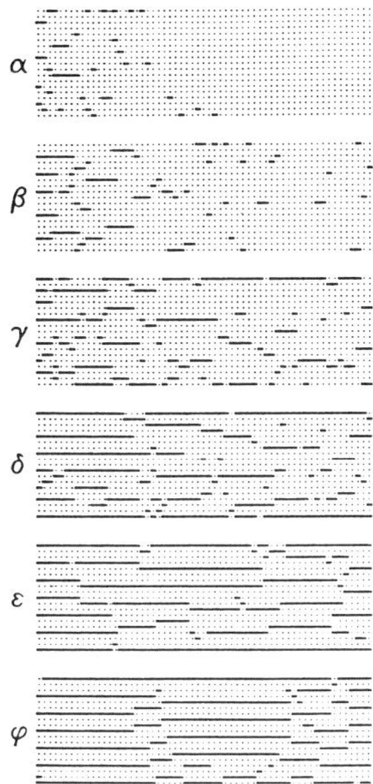

FIG. 27. Growth of stage 3 from the pristine host crystal. Frames α, β, γ, δ, ε, and ϕ are the structure after 2, 10, 80, 300, 1000, and 2505 million Monte Carlo steps, respectively. [From G. Kirczenow, *Phys. Rev. Lett.* **55**, 2810 (1985).]

The results are shown in Fig. 27. The Daumas–Herold domain structure eventually develops as the sample proceeds to its equilibrium stage-3 state. At intermediate times, the stage-3 ordering is not perfect, and non-negligible fractions of stages 2–5 are also observed. The entire surface in contact with the guest reservoir does become a single stage-3 domain at quite early times. As the stage-3 crystal is further intercalated to a final state of stage 2, considerable stage disorder is observed. The disorder is associated with the finite size of the intercalant domains.[10] In addition, kinetically induced stage disorder is observed to be more pronounced at higher stages, in agreement with the (equilibrium) model of Kirczenow[10] and with the observations of Misenheimer and Zabel.[11]

One of the outstanding questions that arises from these kinetic studies is whether equilibrium considerations (such as those discussed in Ref. 10) are at all relevant to the existence of random staging, as observed by Metz[32] and Misenheimer and Zabel.[11] Bak and Forgacz[13] have recently addressed this point. They suggest that stage disorder and randomness are due to the existence of metastable, partly disordered states that are *kinetically* induced. The staging disorder, as well as the predominance of simple (as opposed to fractional) stages is associated with kinetic hysteresis. They apparently disagree with the idea of thermally induced random staging.[10] The kinetic or equilibrium origin of random staging can be discerned by experiments which examine the temperature dependence of imperfectly staged materials. If the origin of the stage disorder is the entropy due to the random stacking of finite islands,[10] then the stage disorder should decrease as the temperature is lowered. If the origin of the stage disorder is kinetic, the frozen-in defects should persist even at low temperatures. This trend seems to be consistent with the observations of Refs. 42 and 122.

Experimental studies of the kinetics of staging transitions were performed by Nishitani *et al.*,[98] Misenheimer and Zabel,[132] and Kim and Fischer.[42] Nishitani observed that only two phases of stages n and $n-1$ coexist throughout the process and that the c-axis disorder is insignificant. These results are not in agreement with those of Ref. 11; the reasons for discrepancy may lie in the increased resolution of these latter measurements.

In their study of intercalation kinetics, Misenheimer and Zabel[132] concluded that all Bragg reflections showed resolution-limited widths. Only the heights of the $(00l)$ reflections changed with time, implying that the kinetics were nucleation controlled, rather than diffusion controlled. Recent experiments by Kim and Fischer[42] of *pressure* quenches of intercalation compounds, with the attendant restaging transitions, indicate that the interesting kinetic effects occur on a very short time scale. This suggests that constant-density quenches are nucleation controlled, rather than diffusion controlled, as assumed in the simulations discussed above. Kim and Fischer[42] suggest that some of the observed metastable states might be more closely associated with the kinetics of the in-plane ordering than with the kinetics of the staging transition itself. Domain–domain interactions may be mediated by the discommensurations between commensurate regions.[133] Further study of the relationship between stage disorder, kinetics, and temperature should determine whether the perfection of stage order is kinetically limited—and hence perhaps amenable to improvement—or whether it is an intrinsic, equilibrium state of an intercalation compound.

Acknowledgments

I am grateful to the authors cited for sharing with me their preprints and reprints and to J. E. Fischer for helpful comments. Much of my own work was the result of a collaboration with D. R. Hamann and benefitted from discussions with P. C. Hohenberg.

[132] M. E. Misenheimer and H. Zabel, *Phys. Rev. B: Condens. Matter* [3] **27,** 1443 (1983).
[133] J. E. Fischer, private communication.

SOLID STATE PHYSICS, VOLUME 40

Experimental Studies of the Structure and Dynamics of Molten Alkali and Alkaline Earth Halides

ROBERT L. McGREEVY

Clarendon Laboratory, University of Oxford, Oxford, England

I. Introduction

The liquid state of matter is largely distinguished from the solid crystalline state in that it is only ordered over a short range, whereas the crystal has long-range order. While the physics of the ordered crystalline state on the atomic scale has been studied in great depth during this century, it is only in the past two decades that improvements in experimental techniques have enabled comparable studies to be begun on liquids. There are many different types of liquids, ranging from thixotropic materials to condensed inert gases; the physical properties and complexity may vary enormously. In this article we shall be concerned with a relatively simple class of liquids—molten salts. Indeed we shall only discuss a subset of the class, molten alkali and alkaline earth halides, which may be considered as completely ionic and without molecular or complex ions. For a more general review of molten salts, including metal–molten salt mixtures and "ionic" molten alloys, the reader is referred to the article of Rovere and Tosi.[1]

Nonmolecular liquids may readily be classified in terms of the interatomic forces. There are van der Waals forces in the condensed inert gases and long-range Coulomb forces in liquid metals and molten salts. The Coulomb forces in metals are strongly screened by the conduction electrons. Our initial interest is in characterizing the structure of the short-range order and in relating this to parameters in the interatomic potential, given that the bonding within the alkali and alkaline earth group is similar and may therefore be parametrized in the same manner. We are then interested in how the atomic motions differ from those in comparable crystals with long-range ordering, and in the relation of this dynamics to the structure of the melt.

For the simple liquids mentioned above it is now known that the short-range-ordered regions have a diameter ~ 20 Å and that the ordering persists for times ~ 1 ps. The gross features of the structures are determined by packing considerations, i.e., imagining the atoms as (possibly charged) hard spheres of appropriate size, with the details being determined by the potentials. Dynamical properties, on the other hand, are more strongly potential dependent. One property of most simple liquids is the high diffusion rate, $\sim 10^{-5}$ cm^2 s^{-1}. This generally has a temperature dependence characteristic of thermal activation, $D = D_0 \exp(-E_0/k_B T)$. Diffusion constants and many other macroscopic and thermodynamic properties of molten salts have been studied in great detail, particularly for electrochemistry (see, e.g., Mamantov[2]). Such properties will not be discussed here except in relation to microscopic motions on the length and time scales noted above. We shall be more specifically concerned with collective motions of the ions, i.e., with the response of all the ions to the motion of a particular ion, and not simply with the motion of individual ions.

All liquids support long-wavelength collective mass-density oscillations, i.e., sound waves, in which the material behaves as an elastic continuum. In this limit, momentum transfer $Q = 2\pi/\lambda \rightarrow 0$, and the structure factor $S(Q)$ is related to the isothermal compressibility. It is as the wavelength of excitations is decreased that differences between liquids appear. The absence of long-range ordering means that there is no periodic dispersion relation for the single excitation spectrum as there is for phonons in the crystalline state. As $Q \rightarrow \infty$ the free-particle (gaslike) limit is approached, coincident with the disappearance of oscillations in $S(Q)$. It is the intermediate range of Q, where the structural disorder is reflected in the damped oscillatory form of $S(Q)$, as distinct from sharp Bragg peaks in the crystalline state, that is of greatest interest.

[1] M. Rovere and M. P. Tosi, *Rep. Prog. Phys.* **49,** 1001 (1986).

Molten alkali and alkaline earth halides are one of the simplest classes of binary liquids. They are expected to be totally ionic and are thus strongly characterized by the opposite charges on cation and anion. This distinguishes them from other binary liquids such as molten metal alloys (though with some similarity to alloys with a high degree of charge transfer) and mixtures of condensed inert gases. In studying the structures of binary liquids, with components α and β, we have to consider three distribution functions, i.e., $\alpha\alpha$, $\beta\beta$, and $\alpha\beta$. Similarly the single-particle dynamics are described in terms of two separate functions, α and β, and the collective dynamics by three. This is analogous to the case of a crystal with two atoms per unit cell. Here the collective dynamics may be described by two functions, acoustic and optic modes, the third vanishing due to symmetry considerations.

While the gross features of the structures will be dominated by the Coulomb forces, there will be some dependence on repulsive forces, i.e., the size and "hardness" of the ions. Because the liquids are binary there may be "optic" modes at some wavelengths similar to those in an ionic crystal. These may, however, be ill defined because of the structural disorder. If they exist, they may interact with and modify the "acoustic" modes, which will in turn be modified by diffusive motions. All of these motions will depend to some extent on the structural ordering. We shall discuss here the experimental work that has improved our understanding of these aspects of the physics of the liquid state.

II. Structure

1. Experimental Techniques

The neutron scattering cross section from a molten alkali or alkaline earth halide may be written as

$$\begin{aligned}\frac{1}{N}\frac{d\sigma}{d\Omega} &= c_+\overline{b_+^2} + c_-\overline{b_-^2} + c_+^2\bar{b}_+^2[A_{++}(Q) - 1] \\ &\quad + c_-^2\bar{b}_-^2[A_{--}(Q) - 1] + 2c_+c_-\bar{b}_+\bar{b}_-[A_{+-}(Q) - 1] \qquad (1.1)\end{aligned}$$

where N is the total number of ions, and c_α is the concentration, $\bar{b}_\alpha$ the bound-atom scattering length, and $4\pi\overline{b_\alpha^2}$ the total scattering cross section of ions of type α.

$$Q = 4\pi \sin\theta/\lambda_i \qquad (1.2)$$

[2] G. Mamantov, ed., "Molten Salts: Characterization and Analysis." Dekker, New York, 1969.

is the momentum transfer, where λ_i is the neutron wavelength and θ the scattering angle. For x-ray scattering the bs are replaced by the corresponding f values. The partial structure factors $A_{\alpha\beta}(Q)$ are related to the partial radial distribution functions $g_{\alpha\beta}(r)$ by

$$g_{\alpha\beta}(r) - 1 = \frac{1}{(2\pi)^3 \rho} \int_0^\infty 4\pi Q^2 [A_{\alpha\beta}(Q) - 1] \frac{\sin Qr}{Qr} \, dQ \tag{1.3}$$

where ρ is the number density of ions and r is the separation of ions of type α and β. This function represents, classically, the probability that there is an ion of type α at a distance r if there is one of type β at the origin. The partial structure factors defined here as $A_{\alpha\beta}(Q)$ are those conventionally used in diffraction work. A related set of partial structure factors

$$S_{\alpha\beta}(Q) = c_\alpha \delta_{\alpha\beta} + c_\alpha c_\beta [A_{\alpha\beta}(Q) - 1] \tag{1.4}$$

will be used in discussing dynamics. The nomenclature is often confused.

Since $d\sigma/d\Omega$ is a combination of all three partial structure factors, a single scattering experiment does not enable any distinction of the structural ordering of different types of ions. This combination may produce significant cancellation effects, and much of the structure is therefore hidden. Although the first x-ray diffraction measurements on molten salts[3,4] were made in the 1950s, it was a further two decades before a technique was developed for separating the partial structure factors and progress could be made. This was the neutron diffraction technique of isotopic substitution.

The neutron scattering length and cross section for a chemical species, being a property of the nucleus, may vary considerably between different isotopes. By varying the isotopic composition of a sample, the coefficients of $A_{\alpha\beta}(Q)$ in Eq. (1.1) may be altered. Performing three measurements with three different isotopic compositions, and assuming that the structure is independent of the isotopic change, the partial structure factors may then be separated by solution of the resulting set of simultaneous equations.

The first application of this technique was on a molten Cu–Sn alloy, using Cu isotopes, by Enderby *et al.*,[5] and the first on a molten salt by Page and Mika[6] on CuCl, using both Cu and Cl isotopes. Since then the work on molten salts has been restricted almost entirely to chlorides because of the lack of other suitable isotopes.

The diffractometers used for these experiments are generally of similar design. A single or multiple detector is scanned over a wide range of scattering

[3] G. Zarzycki, *J. Phys. Radium* **18,** 65 (1957).

[4] G. Zarzycki, *J. Phys. Radium* **19,** 13 (1958).

[5] J. E. Enderby, D. M. North, and P. A. Egelstaff, *Philos. Mag.* [8] **14,** 961 (1966).

[6] D. I. Page and K. Mika, *J. Phys. C* **4,** 3034 (1971).

TABLE I. NEUTRON SCATTERING PARAMETERS FOR MOLTEN CsCl[a,b]

	$Cs^{35}Cl$	CsCl	$Cs^{37}Cl$
$[^{35}Cl]$ (%)	99.4	75.53	1.8
$[^{37}Cl]$ (%)	0.6	24.47	98.2
$c_+\overline{b_+^2} + c^-\overline{b_-^2}$	1.049	0.871	0.339
$c_+^2\bar{b}_+^2$	0.076	0.076	0.076
$c_-^2\bar{b}_-^2$	0.339	0.228	0.026
$2c_+c_-\bar{b}_+\bar{b}_-$	0.320	0.262	0.089

[a] J. Locke, S. Messoloras, R. J. Stewart, R. L. McGreevy, and E. W. J. Mitchell, *Philos. Mag.* [*Part*] *B* **51**, 301 (1985).

[b] Cross sections are in units of 10^{-24} cm^2

angles, neutrons scattered at a particular angle being counted.[7] Using a typical incident neutron wavelength of 0.7 Å, a momentum-transfer range $0.5 < Q < 18$ Å^{-1} can then be covered. Oscillations in the structure factors for most (nonmolecular) molten salts do not extend to higher Q, as they do for many glasses, so this truncation is not a problem. Diffractometers on pulsed neutron sources (see, e.g., Windsor[8]), with a more intense spectrum of short-wavelength neutrons than a reactor, can extend measurements out to $Q \sim 50$ Å^{-1} if necessary.

The success of such experiments requires considerable accuracy since the differences between isotopic scattering lengths are generally small (values for Cl isotopes[9] are given in Table I). For this reason great care must be taken with methods of data analysis. Measured intensities must be corrected for scattering and absorption by furnace and sample container, for self-absorption by the sample, and for background.[10] They must be corrected for detector efficiency, inelastic (Plazcek correction) and multiple scattering, and then normalized[11] before the total scattering cross sections are obtained. Because of the small scattering-length differences, the matrix of coefficients [from Eq. (1.1)] that must be inverted in order to separate the partial structure factors is ill-conditioned; statistical errors in the data therefore become

[7] Institut Laue Langevin, "Neutron Research Facilities at the ILL High Flux Reactor." Grenoble, 1983.

[8] C. G. Windsor, "Pulsed Neutron Scattering." Taylor & Francis, London, 1981.

[9] S. W. Lovesey, "Theory of Neutron Scattering from Condensed Matter." Oxford Univ. Press (Clarendon), London and New York, 1984.

[10] P. F. J. Poncet, Ph.D. Thesis, University of Reading (1976) (unpublished).

[11] D. M. North, J. E. Enderby, and P. A. Egelstaff, *J. Phys. C* **1**, 784 (1968).

considerably amplified, and smoothing procedures must be used. This necessitates complex iterative schemes of separation, which have been discussed in detail by some authors.[10,12,13] We only comment here that some schemes may introduce "artificial" structure into the final results. Despite these difficulties, the methods of data analysis are now well established, and experimental accuracy has increased considerably since the first such experiments.

There are a number of methods for investigating the structures of binary liquids other than by neutron diffraction with isotopic substitution. The first is simply to measure a total structure factor (using either x rays or neutrons) and then to fit this with various models of the liquid structure. The large mutual cancellation of the partial structure factors within the total structure factor, due to the strong charge ordering, makes this procedure unsuitable for molten salts except in some cases where the structure is well defined.[14,15]

The second method is to alter the coefficients of the partial structure factors in Eq. (1.1), not by changing isotopes but by changing the incident radiation. For instance, three separate measurements could be made with neutrons, x-rays, and electrons, each of which will have different scattering factors for a particular ion. There are two problems with this approach. First, the different types of radiation scatter from the system in different ways, i.e., neutrons from the nuclear distribution and x rays from the electron distribution. Though the differences may be small, they become amplified during inversion of the ill-conditioned separation matrix. Indeed the differences between x-ray and neutron structure factors may be used to map the electron distribution relative to the nuclei. Second, the data correction procedures will differ between the types of radiation. This may introduce small systematic errors, which once again become amplified; use of a single type of radiation enables self-consistency checks to eliminate such errors.

A third method is the relatively new technique of energy-dispersive x-ray diffraction (EDXRD). This makes use of the fact that x-ray scattering factors change with wavelength, albeit by a small amount, notably in the vicinity of an absorption edge. With the advent of synchrotron sources it is possible to tune the wavelength close to an edge and perform three experiments on the same sample, but with different scattering factors. This method can, in principle, be used for any system with an absorption edge within the wavelength range of

[12] J. Locke, S. Messoloras, R. J. Stewart, R. L. McGreevy, and E. W. J. Mitchell, *Philos. Mag. [Part] B* **51,** 301 (1985).

[13] R. L. McGreevy and E. W. J. Mitchell, *J. Phys. C* **15,** 5537 (1982).

[14] R. Triolo and A. H. Narten, *J. Chem. Phys.* **74,** 703 (1981).

[15] Y. Takagi and T. Nakamura, *Philos. Mag. [Part] B* **51,** L43 (1985).

the source. However, the scattering length differences are very small, and the data corrections are large; EDXRD has not yet been properly evaluated in comparison to isotopic substitution, though some experiments (on solids) have been performed.[16]

The fourth method is extended x-ray absorption fine structure (EXAFS) and the related x-ray absorption near-edge structure (XANES).[17] The photoelectrons produced by x-ray absorption at an atom interfere with weak reflections from surrounding atoms, producing small oscillations above the absorption edge. These oscillations provide information on the local structure. The EXAFS part of the spectrum is dominated by two-body correlations and hence can give information on $g(r)$; however, the Q range is limited to >2.5 Å^{-1}, which biases the information content towards small r. The XANES part of the spectrum extends to lower Q but is dominated by many-atom correlations, and analysis depends strongly on theoretical models. Though the two techniques provide useful information that is unavailable due to the limitations of isotopic substitution, they are not as unambiguous in determination of structure as conventional diffraction.

2. Experimental Results

a. Alkali Chlorides

The experimentally measured scattering intensities[12] for three samples of molten CsCl with different isotopic chlorine compositions (details are given in Table I) are shown in Fig. 1. The container is made of vanadium, which is an almost totally incoherent scatterer of neutrons; the scattering pattern is therefore flat and may be readily subtracted from the sample patterns. The rise in intensity at small scattering angles is due to the incident neutron beam. The corrected total structure factors

$$F(Q) = c_+^2\bar{b}_+^2[A_{++}(Q) - 1] + c_-^2\bar{b}_-^2[A_{--}(Q) - 1] + 2c_+c_-\bar{b}_+\bar{b}_-[A_{+-}(Q) - 1] \quad (2.1)$$

are shown in Fig. 2 (as points) together with a set of total structure factors obtained from the smoothed partial structure factors (curves). In Fig. 3 the result of a direct solution for the partial structure factors (points) is compared to the iteratively smoothed solution (curves). This smoothed solution is then

[16] Y. Waseda and S. Tamaki, *Z. Phys. B* **23,** 315 (1976).
[17] T. M. Hayes and J. B. Boyce, *Solid State Phys.* **37,** 173 (1982).

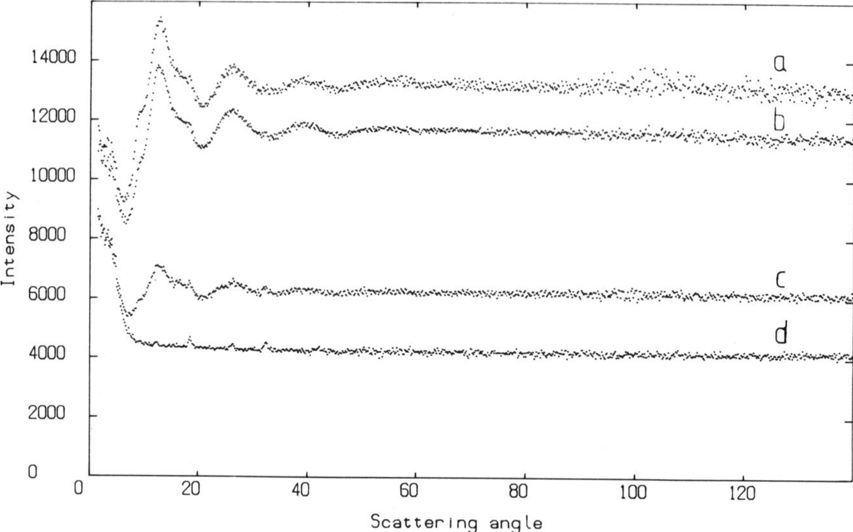

FIG. 1. Normalized scattering intensities from samples of molten (a) $Cs^{35}Cl$, (b) CsCl, (c) $Cs^{37}Cl$, and (d) the vanadium sample container. [From J. Locke, S. Messoloras, R. J. Stewart, R. L. McGreevy, and E. W. J. Mitchell, *Philos. Mag.* [*Part*] *B* **51**, 301 (1985).]

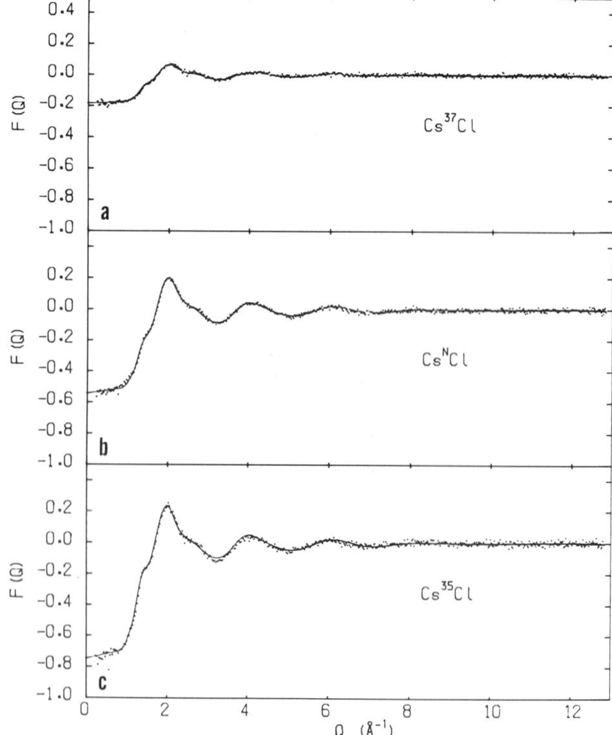

FIG. 2. Total structure factors $F(Q)$ for molten (a) $Cs^{37}Cl$, (b) CsCl, and (c) $Cs^{35}Cl$. [From J. Locke, S. Messoloras, R. J. Stewart, R. L. McGreevy, and E. W. J. Mitchell, *Philos. Mag.* [*Part*] *B* **51**, 301 (1985).]

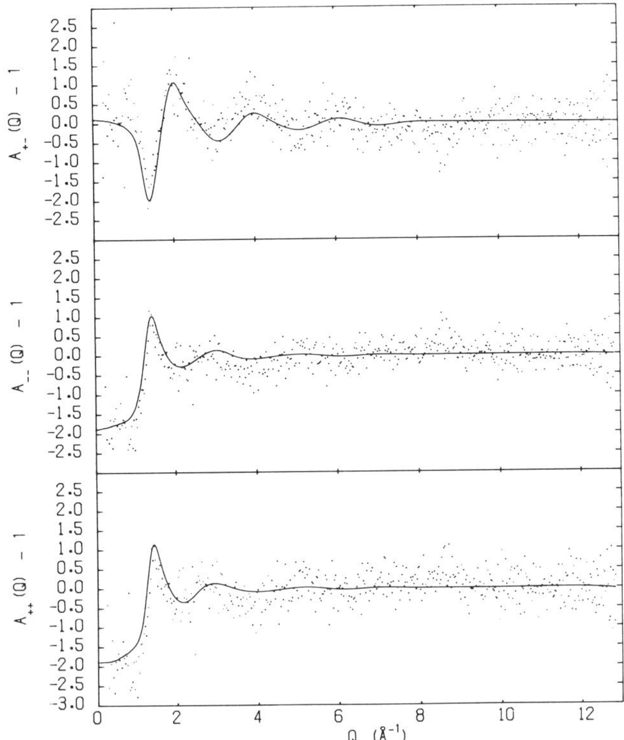

FIG. 3. Partial structure factors $A_{\alpha\beta}(Q)$ for molten CsCl. [From J. Locke, S. Messoloras, R. J. Stewart, R. L. McGreevy, and E. W. J. Mitchell, *Philos. Mag.* [*Part*]. *B* **51**, 301 (1985).]

transformed via Eq. (1.3) to give the set of partial radial distribution functions shown in Fig. 4.

These show the characteristic features of the structure of molten alkali chlorides:

(1) A well-defined first peak in $g_{+-}(r)$ with a coordination number [obtained by integration of $r^2 g_{+-}(r)$ up to the first minimum in $g_{+-}(r)$] of $\lesssim 6$.

(2) $g_{++}(r) \cong g_{--}(r)$; the first peaks in $g_{++}(r)$ and $g_{--}(r)$ occur at approximately $\sqrt{2}$ times the distance of the first peak in $g_{+-}(r)$.

(3) Charge ordering: $g_{++}(r)$ and $g_{--}(r)$ peak approximately in antiphase to $g_{+-}(r)$.

(4) Short-range order, radially over ~ 12 Å.

One of the dominant features of this structure is, then, the alternate "charge" ordering of the different ionic species. Similar ordering may occur

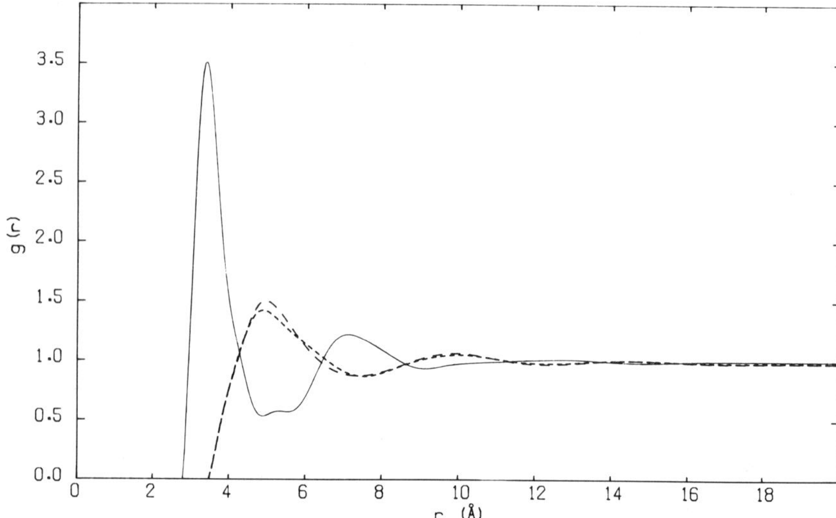

FIG. 4. Partial radial distribution functions $g_{++}(r)$ (– – –), $g_{--}(r)$ (- - -), and $g_{+-}(r)$ (———) for molten CsCl. [From J. Locke, S. Messoloras, R. J. Stewart, R. L. McGreevy, and E. W. J. Mitchell, *Philos. Mag.* [*Part*] *B* **51**, 301 (1985).]

in some molten metal alloys, e.g., Li_4Pb,[18] where the chemical properties of the two species are considerably different. In alloys such as NaCs, where the chemical properties are similar, the structure may involve clustering of the different species.[19]

In Fig. 5 we show the partial radial distribution functions for molten NaCl,[20,21] RbCl,[12,22] and CsCl,[12] presented on a distance scale normalized such that the position of the first peak in $g_{+-}(r)$, r_{+-}, is at 1. (The early results of Derrien and Dupuy[23] on molten KCl show small scale structure introduced by the method of data analysis[12] and are not therefore comparable.) It can be seen that there are strong structural similarities. The largest difference is a tendency for the height of the peaks in $g_{++}(r)$ and $g_{--}(r)$ to decrease as the cation size increases from Na^+ to Cs^+. It has been suggested[12] that this may be due to the higher polarizability of the Cs^+ ion. This is discussed further in

[18] H. Ruppersburg and H. Reiter, *J. Phys. F* **12**, 1311 (1982).

[19] M. J. Huijben, W. van der Lugt, W. A. M. Reimert, J. T. M. de Thosson and C. van Dijk, *Physica B (Amsterdam)* **97B**, 338 (1979).

[20] F. G. Edwards, J. E. Enderby, R. A. Howe, and D. I. Page, *J. Phys. C* **8**, 3483 (1975).

[21] S. Biggin and J. E. Enderby, *J. Phys. C* **15**, L305 (1982).

[22] E. W. J. Mitchell, P. F. J. Poncet, and R. J. Stewart, *Philos. Mag* [8] **34**, 721 (1976).

[23] J. Y. Derrien and J. Dupuy, *J. Phys. Orsay, Fr.* **36**, 191 (1975).

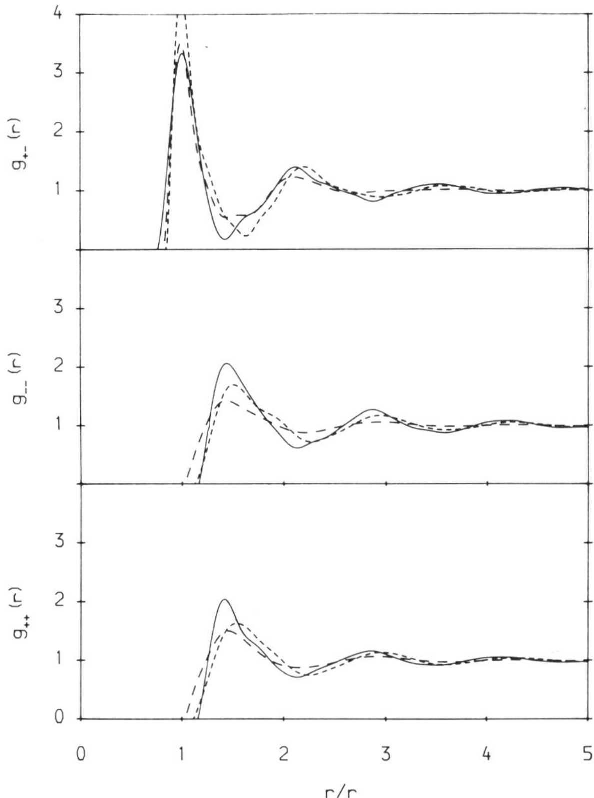

FIG. 5. Partial radial distribution functions $g_{\alpha\beta}(r)$ for molten NaCl (———) [F. G. Edwards, J. E. Enderby, R. A. Howe, and D. I. Page, *J. Phys. C* **8**, 3483 (1975); S. Biggin and J. E. Enderby, *J. Phys. C* **15**, L305 (1982)], RbCl (---) [E. W. J. Mitchell, P. F. J. Poncet, and R. J. Stewart, *Philos. Mag* [8] **34**, 721 (1976); J. Locke, S. Messoloras, R. J. Stewart, R. L. McGreevy, and E. W. J. Mitchell, *Philos. Mag. [Part] B* **51**, 301 (1985)], and CsCl (–––) (Locke *et al., ibid.*).

relation to computer simulations and ionic potentials in subsection 3. The major features of the radial distribution functions are summarized in Table II.

The coordination number of an ion in a structure is a well-defined quantity in the crystalline state, but a variety of definitions exist for liquids, and hence the significance of the concept is in some doubt. Nevertheless, it is a number that is commonly quoted and provides an idea of the type of packing within the structure. Calculation of coordination number from radial distribution involves integration of $r^2 g_{\alpha\beta}(r)$ up to a point defined by the method; the various methods have been discussed by Biggin and Enderby.[21] More

TABLE II. MAIN FEATURES OF THE PARTIAL RADIAL DISTRIBUTION FUNCTIONS OF SOME MOLTEN ALKALI CHLORIDES[a,b]

	Cation	Temperature (K)	Sum of ionic radii	Distance of closest approach	Position of first peak	Height of first peak	Coordination number
$g_{++}(r)$	Na	1148	1.9	3.19	3.9	2.05	13.0
	Rb	988	2.96	3.55	4.9	1.63	16.7
	Cs	973	3.38	3.45	4.95	1.50	15.4
$g_{--}(r)$	Na	1148	3.62	3.19	3.96	2.06	13.3
	Rb	988	3.62	3.6	4.8	1.69	17.1
	Cs	973	3.62	3.45	4.85	1.42	16.3
$g_{+-}(r)$	Na	1148	2.76	2.09	2.76	3.33	4.83
	Rb	988	3.29	2.7	3.2	4.36	7.4
	Cs	973	3.5	2.8	3.4	3.51	5.8

[a] J. Locke, S. Messoloras, R. J. Stewart, R. L. McGreevy, and E. W. J. Mitchell, *Philos. Mag.* [*Part*] *B* **51,** 301 (1985).
[b] All distances are in Å.

geometrical methods using results of computer simulations have been discussed by McGreevy *et al.*[24] and are illustrated in subsection 3. In this section all coordination numbers refer to integration to the first minimum in $g_{\alpha\beta}(r)$; they are given in Table II.

It is interesting to compare the alkali chloride structures with that of molten CuCl, as noted above the first molten salt to be investigated using isotopic substitution.[6] For this melt it is found that $g_{+-}(r)$ and $g_{--}(r)$ show the same features as for the alkali chlorides, but that $g_{++}(r)$ shows almost no structure and is similar to that for a gas. This experiment has since been repeated[25] to check these results, and it was found that there are only very weak features in $g_{++}(r)$. Recent results[26] on the structure of α-AgI suggest that this lack of order is related to the fast ion-conducting properties of these materials in the solid state. The results of Derrien and Dupuy[23] for the partial structure factors of molten AgCl show significant small-scale structure due to the method of data analysis, but it can be inferred from the total structure factors that $g_{--}(r)$ is better defined than $g_{++}(r)$.

No systematic work has yet been done on the effect of either temperature or pressure on molten alkali halide structures. However, recent shock-temperature measurements on CsI by Radousky *et al.*,[27] coupled with the theoretical calculations of Ross and Rogers,[28] suggest that at high temperature and pressure the melt undergoes a gradual change to a close-packed structure similar to that of the isoelectronic monatomic equivalent, Xe.

b. Alkaline Earth Chlorides

There are strong similarities between the structures of the molten alkali chlorides, as shown above; investigation of the small differences will require more accurate experiments. For this reason much more work has been done on the molten alkaline earth chlorides which show larger structural differences. We also include here data on molten $ZnCl_2$ and $NiCl_2$; while these are not within the alkaline earth group, the cations have similar radii to that of Mg^{2+}, and hence it is possible to examine the effect of differing electronic structure, rather than simply of ion size.

[24] R. L. McGreevy, I. Ruff, and A. Baranyai, *Phys. Chem. Liq.* **16,** 47 (1986).

[25] S. Eisenberg, J.-F. Jal, J. Dupuy, P. Chieux, and W. Knoll, *Philos. Mag. [Part] A* **46,** 195 (1982).

[26] M. A. Howe, R. L. McGreevy, and E. W. J. Mitchell, *Z. Phys. B* **62,** 15 (1985).

[27] H. B. Radousky, M. Ross, A. C. Mitchell, and W. J. Nellis, *Phys. Rev. B: Condens. Matter* [3] **31,** 1457 (1985).

[28] M. Ross and F. J. Rogers, *Phys. Rev. B: Condens. Matter* [3] **31,** 1463 (1985).

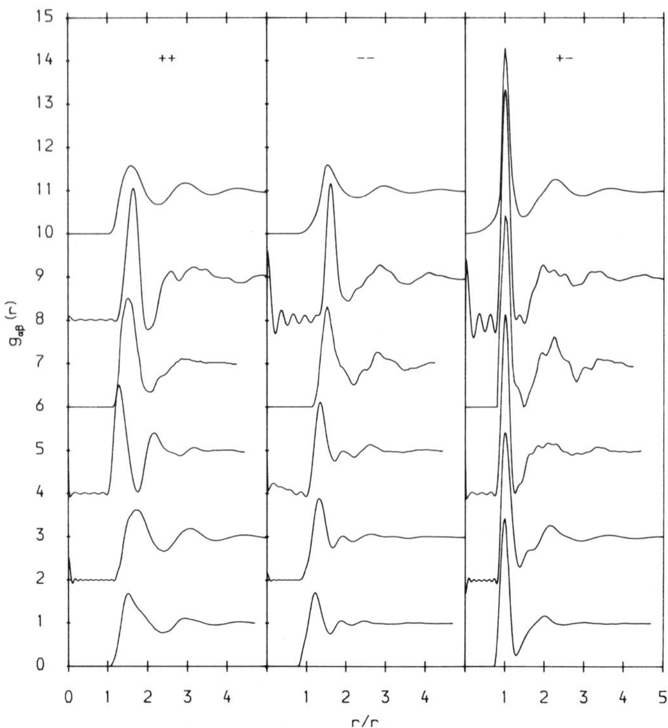

FIG. 6. Partial radial distribution functions $g_{\alpha\beta}(r)$ for (in ascending order) molten $BaCl_2$ [F. G. Edwards, J. E. Enderby, R. A. Howe, and D. I. Page, *J. Phys. C* **11**, 1053 (1978)], $SrCl_2$ [R. L. McGreevy and E. W. J. Mitchell, *J. Phys. C* **15**, 5537 (1982)], $CaCl_2$ [S. Biggin and J. E. Enderby, *J. Phys. C* **14**, 3129 (1981)], $MgCl_2$ [S. Biggin, M. Gay, and J. E. Enderby, *J. Phys. C* **17**, 977 (1984)], $ZnCl_2$ [S. Biggin and J. E. Enderby, *J. Phys. C* **14**, 3577 (1981)], and $NiCl_2$ [R. J. Newport, R. A. Howe, and N. D. Wood, *J. Phys. C* **18**, 5249 (1985).] The vertical scale is shifted by 2.0 between each curve.

The partial radial distribution functions of molten $BaCl_2$,[29] $SrCl_2$,[13] $CaCl_2$,[30] $MgCl_2$,[31] $ZnCl_2$,[32] and $NiCl_2$[33] are shown in Fig. 6. (Total structure factors have been measured for molten $MnCl_2$[31] and $CoCl_2$.[15]) As before, the distance scale has been normalized using the first peak position of $g_{+-}(r)$. This peak decreases significantly in height from the smaller cations to the larger, and the coordination number increases from ~ 4 to ~ 8. The main

[29] F. G. Edwards, J. E. Enderby, R. A. Howe, and D. I. Page, *J. Phys. C* **11,** 1053 (1978).
[30] S. Biggin and J. E. Enderby, *J. Phys. C* **14,** 3129 (1981).
[31] S. Biggin, M. Gay, and J. E. Enderby, *J. Phys. C* **17,** 977 (1984).
[32] S. Biggin and J. E. Enderby, *J. Phys. C* **14,** 3577 (1981).
[33] R. J. Newport, R. A. Howe, and N. D. Wood, *J. Phys. C.* **18,** 5249 (1985).

structural features are summarized in Table III. For the like ion distributions the first point to note is that, because of the 2:1 ratio of anions to cations, $g_{++}(r)$ and $g_{--}(r)$ must be different, unlike the alkali halides, where they are very similar. For $BaCl_2$ and $SrCl_2$, $g_{--}(r)$ is approximately isomorphous, as is $g_{++}(r)$, although the first peak in g_{Ba-Ba} is noticeably more asymmetric than that in g_{Sr-Sr}. These two salts are found overall to have similar structures in the melt; they have the same structure in the solid state below the melting point (fluorite structure). For $CaCl_2$, which has a different solid-state structure (distorted rutile), $g_{--}(r)$ is similar to that for $SrCl_2$ and $BaCl_2$, but $g_{++}(r)$ is quite different. The first peak occurs at much lower r/r_{+-} and is higher. There is then a deep minimum at the position of the first peak in g_{Ba-Ba} and g_{Sr-Sr}. Within a similar anion substructure the Ca^{2+} ions therefore occupy "nearest-neighbor" cation sites, while Sr^{2+} and Ba^{2+} ions occupy "next-nearest-neighbor" sites, as would be expected simply from the large Coulomb repulsion between the doubly charged cations. It was suggested that this close Ca–Ca grouping might break up as the temperature of the melt was increased. Day and McGreevy[34] have measured the total structure factor of molten $CaCl_2$ at 1023 and 1223 K (Fig. 7) and find that the difference (which is small because of the small coefficient of A_{++}) is consistent with such a structural change; there is also other evidence from Raman scattering[35] (discussed in Section IV) and refractive index[36] measurements. At the higher temperature $CaCl_2$ becomes isomorphous with $BaCl_2$ and $SrCl_2$.

The structure of molten $MgCl_2$ has no similarities to those of the other salts in the group. g_{++} and g_{--} peak in approximately the same position, though g_{--} is more asymmetric with a shoulder on the high r/r_{+-} side. By comparison of the relative peak positions in $g_{\alpha\beta}(r)$ and of the coordination numbers, it is possible to suggest that molten $MgCl_2$ has a layer structure in which Mg^{2+} has an approximately planar coordination with 4 Cl^-. The remaining Cl^- ions are distributed between these planes. This layering reveals itself most clearly in a "prepeak" in $A_{++}(Q)$ at $Q \cong 1.0$ Å^{-1}. This can be seen as a small peak in the total structure factor[34] for $MgCl_2$ (Fig. 7; the coefficient for A_{++} is also small for this sample). In the solid state $MgCl_2$ has a layer structure ($CdCl_2$-type), giving a small Bragg peak in the diffraction pattern at $Q = 1.07$ Å^{-1}. The total structure factor of molten $MgCl_2$ is observed to increase rapidly for $Q < 0.5$ Å^{-1} to a value much larger than that predicted by the compressibility. The origin of this feature is not understood, although it indicates

[34] S. E. Day and R. L. McGreevy, *Phys. Chem. Liq.* **15**, 129 (1985).

[35] R. A. J. Bunten, R. L. McGreevy, E. W. J. Mitchell, C. Raptis, and P. J. Walker, *J. Phys. C* **17**, 4705 (1984).

[36] Y. Iwadate, J. Mochinaga, and K. Kawamura, *J. Phys. Chem.* **85**, 3708 (1981).

TABLE III. MAIN FEATURES OF THE PARTIAL RADIAL DISTRIBUTION FUNCTIONS OF SOME MOLTEN ALKALINE EARTH CHLORIDES[a]

	Cation	Temperature (K)	Sum of ionic radii	Distance of closest approach	Position of first peak	Height of first peak	Coordination number
$g_{++}(r)$	Mg[b]	998	1.3	2.81	3.6	2.52	6.0
	Ca[c]	1093	1.98	2.7	3.6	2.5	4.2
	Sr[d]	1198	2.24	3.5	4.95	1.6	13.6
	Ba[e]	1298	2.7	3.7	4.9	1.7	14.0
	Zn[f]	600	1.48	2.8	3.8	3.1	4.7
	Ni[g]	1295	1.38	2.33	3.66	1.65	6.7
$g_{--}(r)$	Mg[b]	998	3.62	2.81	3.6	2.32	12.0
	Ca[c]	1093	3.62	2.9	3.73	2.4	7.8
	Sr[d]	1198	3.62	2.4	3.8	1.9	9.3
	Ba[e]	1298	3.62	2.5	3.86	1.7	7.0
	Zn[f]	600	3.62	3.0	3.71	3.1	8.6
	Ni[g]	1295	3.62	1.73	3.48	1.6	13.8
$g_{+-}(r)$	Mg[b]	998	2.46	1.91	2.36	4.42	4.4
	Ca[c]	1093	2.8	2.2	2.78	4.2	5.4
	Sr[d]	1198	2.93	2.4	2.9	2.9	6.9
	Ba[e]	1298	3.16	2.4	3.1	3.4	7.7
	Zn[f]	600	2.55	1.9	2.29	5.2	4.3
	Ni[g]	1295	2.5	0.0	2.32	4.4	4.9

[a] All distances are in Å.
[b] S. Biggin, M. Gay, and J. E. Enderby, *J. Phys. C* **17,** 977 (1984).
[c] S. Biggin and J. E. Enderby, *J. Phys. C* **14,** 3129 (1981).
[d] R. L. McGreevy and E. W. J. Mitchell, *J. Phys. C* **15,** 5537 (1982).
[e] F. G. Edwards, J. E. Enderby, R. A. Howe, and D. I. Page, *J. Phys. C* **11,** 1053 (1978).
[f] S. Biggin and J. E. Enderby, *J. Phys. C* **14,** 3577 (1981).
[g] R. J. Newport, R. A. Howe, and N. D. Wood, *J. Phys. C* **18,** 5249 (1985).

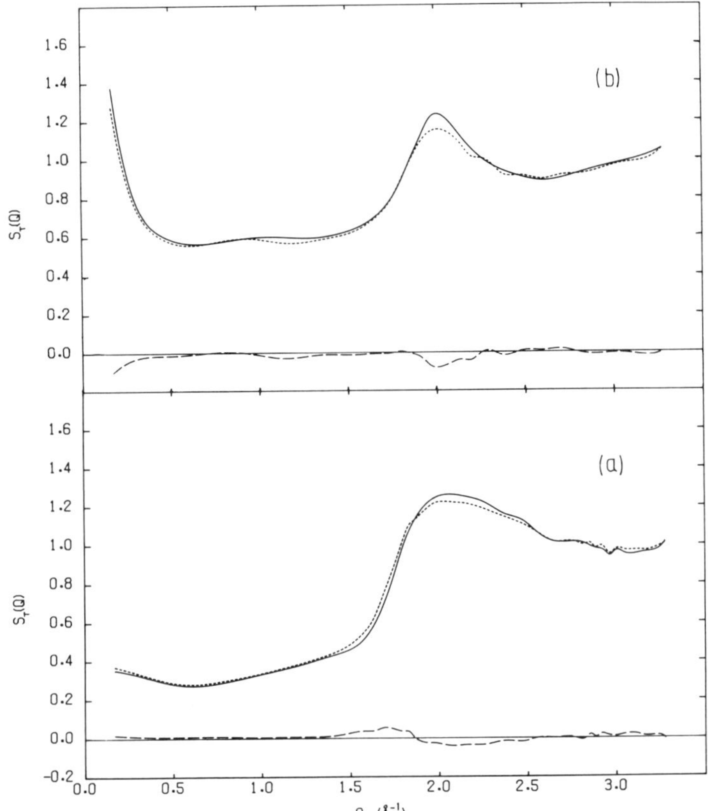

FIG. 7. Total structure factors $S_T(Q)$ for (a) molten $CaCl_2$ at 1083 K (———), 1223 K (---), and the difference (– – –) and for (b) molten $MgCl_2$ at 1023 K (———), 1223 K (---) and the difference (– – –) [S. E. Day and R. L. McGreevy, *Phys. Chem. Liq.* **15**, 129 (1986)].

some medium-range ordering in the liquid with a correlation length ~20 Å. It has been suggested[34] that it may be related to the observation of a propagating charge-density fluctuation in the Raman scattering spectrum; this is discussed in Section IV. Because the liquid is macroscopically charge neutral, the structure factor must decrease to the compressibility limit at some lower Q.

While Zn^{2+} has approximately the same ionic radius as Mg^{2+}, and g_{Zn-Cl} also has a coordination of ~4, the structure of molten $ZnCl_2$ is found to be quite different from that of $MgCl_2$. In this case the fourfold coordination is definitely tetrahedral; the strongly coordinated and closely packed tetrahedra give the liquid a very high viscosity and good glass-forming ability. The glassy state has effectively an identical structure to the liquid.[37] Ordering between the tetrahedra also gives a "prepeak" in $A_{++}(Q)$, but the structure factors have not yet been measured below $Q \sim 0.5$ Å^{-1} to see whether they show the same anomalous increase as in $MgCl_2$. X-ray diffraction[14] from molten $ZnCl_2$ has only reached $Q = 0.3$ Å^{-1}; the total structure factor confirms the tetrahedral coordination.

$MnCl_2$, although not in the same group as $MgCl_2$, has the same layer structure in the solid state, and measurement of the total (neutron) structure factor[31] suggests that it has a similar structure in the melt. $NiCl_2$ (Fig. 6),[33] on the other hand, shows more similarities to the tetrahedral coordination of $ZnCl_2$, though the coordination is not as strong, and hence there is no prepeak in $A_{++}(Q)$. The total (x-ray) structure factor[15] of molten $CoCl_2$ has also been interpreted as showing tetrahedral coordination.

The widely varying structures within this group, and the high anisotropy of some of them, cannot be explained simply in terms of ion-size effects and isotropic potentials. Possible origins of the differences will be discussed in the next section.

3. Computer Simulation and Theory

a. *Radial Distribution: Alkali Halides*

A proper review of computer simulation and theoretical studies of molten salt structures is outside the scope of this article. For more details the reader is referred to the reviews of Sangster and Dixon,[38] Parrinello and Tosi,[39] and March and Tosi.[40] However, it is useful to compare briefly the results of these calculations with experiment in order to examine the role of interionic potentials in determining molten salt structures.

The most common potential used in calculations is of the Born–Mayer–Huggins[41] form

$$V_{\alpha\beta}(r) = \frac{Z_\alpha Z_\beta e^2}{r} + a_{\alpha\beta}\exp\left(-\frac{r}{\sigma_{\alpha\beta}}\right) - \frac{C_{\alpha\beta}}{r^6} - \frac{D_{\alpha\beta}}{r^8} \tag{3.1}$$

where the first term represents the Coulomb interaction, the second the repulsive interaction due to ionic overlap, and the third and fourth terms the van der Waals interaction. A self-consistent determination of the parameters in this formula has been made by Fumi and Tosi[42] for the alkali halides.

[37] J. A. E. Desa, A. C. Wright, J. Wang, and R. N. Sinclair, *J. Non-Cryst. Solids* **51**, 57 (1982).
[38] M. J. L. Sangster and M. Dixon, *Adv. Phys.* **25**, 247 (1976).
[39] M. Parrinello and M. P. Tosi, *Riv. Nuovo Cimento Soc. Ital. Fis.* **2**, 1 (1979).
[40] N. H. March and M. P. Tosi, "Atomic Dynamics in Liquids." Macmillan, London, 1976.
[41] M.-L. Huggins and J. E. Mayer, *J. Chem. Phys.* **1**, 643 (1933).
[42] F. G. Fumi and M. P. Tosi, *J. Phys. Chem. Solids* **25**, 31 and 45 (1964).

Numerous Monte Carlo and molecular-dynamics simulations of the molten alkali halides have been performed using such "rigid-ion" potentials; for a comprehensive list see Baranyai *et al.*[43] Comparison with experiments is limited by the small amount of experimental data available, though the agreement is generally good.[44,45] Simulation results appear to be relatively insensitive to the exact form of potential used, given that the parameters are fitted to suitable properties of the appropriate crystal structure, i.e., lattice constant, compressibility, dielectric constant, etc. The largest differences between experiment and simulation are observed in the case of CsCl: A comparison of a representative simulation[45] and experiment[12] is made in Fig. 8. The simulation tends to overestimate the degree of structural ordering, giving sharper peaks in all $g_{\alpha\beta}(r)$. It has been suggested that these differences may be due to the screening effect of the polarizable Cs^+ ion. Simulations of some molten alkali halides have also been performed using polarizable ion potentials[46] in order to test this effect. However, the partial radial distribution functions for rigid and polarizable ion potentials are found to be almost identical (see Fig. 8). These polarizable potentials are all of the shell-model type[38] and are isotropic; an anisotropic potential of the deformation dipole[47] type may produce better results. This is discussed later with reference to alkaline earth halides.

All simulation results suggest that in molten alkali halides there is a general decrease in coordination of $g_{+-}(r)$ from 6 to 4 as the ratio of ionic radii decreases from ~ 1 in the case of CsF to ~ 0.3 for LiI. Geometrically this would correspond to a change from octahedral to tetrahedral coordination. This has yet to be properly tested experimentally, but the available results give a coordination for NaCl of ~ 5, while that for CsCl is ~ 6. [Direct calculation by integration of $r^2 g_{+-}(r)$ is very sensitive to errors in the position of the minimum in $g_{+-}(r)$.] Simulation of a system with a point cation gives a polymeric chain structure with a coordination of 2.[48]

Theoretical approaches to liquid structure generally involve solution of the force equation, using a two-body potential via the superposition approximation; see, e.g., March and Tosi.[40] This yields an integral equation which may be solved either analytically or numerically under a variety of further approximations. These give the various methods their different names, such as the mean spherical approximation (MSA), Percus–Yevick (PY), or hypernet-

[43] A. Baranyai, I. Ruff, and R. L. McGreevy, *J. Phys. C* **19**, 453 (1986).

[44] M. Dixon and M. J. L. Sangster, *J. Phys. C* **9**, L5 (1976).

[45] M. Dixon and M. J. L. Gillan, *Philos. Mag. [Part] B* **43**, 1099 (1981).

[46] M. Dixon and M. J. L. Sangster, *Philos. Mag.* [8] **35**, 1049 (1977).

[47] J. R. Hardy and A. M. Karo, "The Lattice Dynamics and Statics of Alkali Halide Crystals," Plenum, New York, 1979.

[48] M. Gillan, B. Larsen, M. P. Tosi, and N. H. March, *J. Phys. C* **9**, 889 (1976).

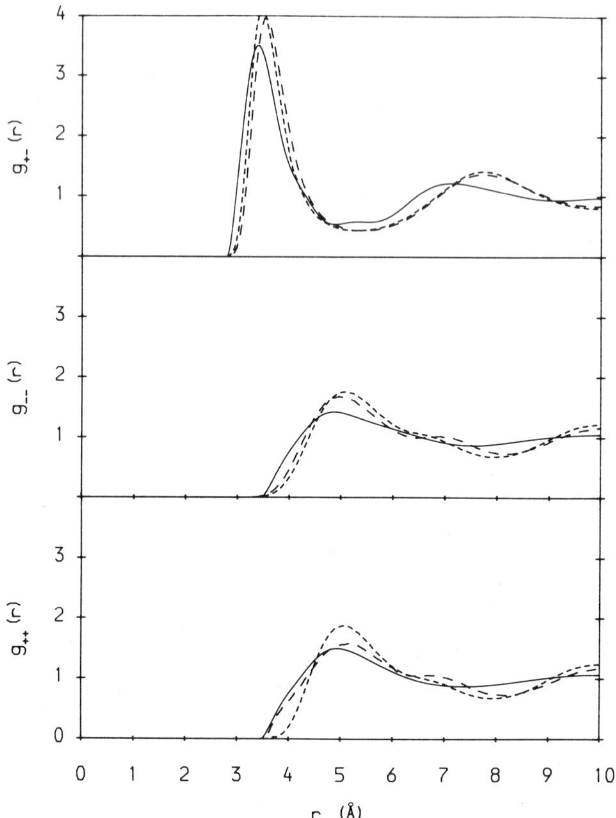

FIG. 8. Partial radial distribution functions $g_{\alpha\beta}(r)$ for molten CsCl; comparison of results from experiment (———) [J. Locke, S. Messoloras, R. J. Stewart, R. L. McGreevy, and E. W. J. Mitchell, *Philos. Mag.* [*Part*] *B* **51**, 301 (1985)] and simulations using rigid ion (---) [M. Dixon and M. J. Gillan, *Philos. Mag.* [*Part*] *B* **43**, 1099 (1981)] and polarisable ion (–––) [M. Dixon and M. J. L. Sangster, *Philos. Mag.* [8] **35**, 1049 (1977)] potentials.

ted chain (HNC). The suitability of a particular approximation depends on the type of potential being considered, a long-range Coulomb potential differing from the shorter-ranged Lennard–Jones potential. The HNC approximation is now considered to be most suitable for molten salts, and provides good agreement with simulation results.[49] Apart from theoretical interest in the validity of a particular approximation, such calculations are usually only performed in preference to simulations because of the increased computational speed.

[49] M. Abernethy, M. Dixon, and M. J. Gillan, *Philos. Mag. [Part] B* **43,** 1113 (1981).

The use of integral equations is now being extended to attempt to reverse the process and calculate the interionic potential directly from experimentally determined radial distributions.[50] Although these types of calculations are still in their infancy, improved experimental data will soon be required as input for them.

Another approach of interest is the asymptotic expansion[51,52] of $S_{\alpha\beta}(Q)$ in the limit of $Q \to 0$. The coefficients in this expansion may be related to various parameters in the long-range part of the interionic potential and other thermodynamic quantities. For instance, it is well known that the $Q = 0$ limit of the partial structure factors is related to the isothermal compressibility (K_T)[53]

$$S_{\alpha\beta}(Q = 0) = c_\alpha c_\beta \rho k_B T K_T \tag{3.2}$$

It is in principle possible, via very accurate measurement of $S_{\alpha\beta}(Q \lesssim 0.5\ \text{Å}^{-1})$, to derive quantities such as the van der Waals coefficient, though this has not yet been successfully achieved for a molten salt.

b. *Radial Distribution: Alkaline Earth Chlorides*

Though there have been numerous simulations of molten alkali halides, there have been very few of alkaline earth halides. The most comprehensive work is that of de Leeuw[54–56] on molten $SrCl_2$. His results, and those of Dixon,[57] using both rigid and polarizable ion potentials, produced reasonable agreement with experiment for $g_{+-}(r)$ and $g_{--}(r)$, but a $g_{++}(r)$ with a first peak that is twice as high and half as wide. This may be due to a screening effect of the polarizable Cl^- ion distorted by the doubly charged Sr^{2+} ion, and may be modeled empirically either by imposition of an exponential decay on the Coulomb force between cations[58] or by treating the cations as a plasma in a medium of appropriate dielectric constant.[59]

No calculations, even with such empirical potentials, have yet been successful in reproducing the structures of molten $CaCl_2$ and $MgCl_2$. In both these cases structures similar to that for $SrCl_2$ are obtained, with a larger

[50] D. Levesque, J. J. Weiss, and L. Reato, *Phys. Rev. Lett.* **54,** 451 (1985).

[51] M. Rovere, M. Parrinello, M. P. Tosi, and P. V. Giaquinta, *Philos. Mag. [Part] B* **39,** 167 (1979).

[52] M. C. Abramo, M. Parrinello, and M. P. Tosi, *J. Nonmet.* **2,** 57 (1973).

[53] J. G. Kirkwood and T. P. Buff, *J. Chem. Phys.* **19,** 774 (1951).

[54] S. W. de Leeuw, *Mol. Phys.* **36,** 103 (1978).

[55] S. W. de Leeuw, *Mol. Phys.* **36,** 765 (1978).

[56] S. W. de Leeuw, *Mol. Phys.* **37,** 489 (1979).

[57] M. Dixon, unpublished work (1983).

[58] A. Copestake, private communication.

[59] G. Pastore, P. Ballone, and M. P. Tosi, *J. Phys. C* **19,** 487 (1986).

cation separation than is actually observed. This is not surprising, since only two-body potentials are used; i.e., the potential depends on the separation of A and B, while the structures show evidence of many-body forces: i.e., the interaction of A and B affects that of B and C. The doubly charged cations are more strongly attracted to the anion than is the case for the singly charged alkali ions. In the case of a small cation such as Mg^{2+} or Ca^{2+}, there may be significant distortion of the polarizable anion, which decreases the repulsive part of the potential between that anion and a second cation. This would explain the close Ca–Ca grouping. Since the many-body interaction is related to the electronic distribution, it may have angular components which differ between cations, giving the characteristically different structures for $MgCl_2$ and $ZnCl_2$. Considerably more sophisticated potentials will be required to model such interactions; some progress may be achieved using the deformation dipole model for the polarizability.

c. *Angular Correlations and Coordination*

One of the advantages of computer simulation is that various correlations may be calculated which cannot be obtained experimentally, thus giving extra insight into the physics of a system (always remembering, however, that simulations are only models and do not, necessarily, represent reality).

For instances, $G_{+-}(r) = r^2 g_{+-}(r)$ may be decomposed into contributions from nearest neighbors, second-nearest, third-nearest, etc.[24] Examples are shown in Fig. 9. From the symmetry of the individual distributions it is obvious that for LiI the coordination is 4, there being a distinct change from fourth to fifth neighbors, while for RbCl it is $5\frac{1}{2}$; that is, the sixth neighbor is shared between first and second coordination shells. It is found in this way that all molten alkali halides have either integral or half-integral coordination numbers for unlike ions. Extra information may also be obtained on like-ion coordination. This type of analysis reveals that integration to the first minimum in $G_{+-}(r)$ produces the closest estimate to the correct coordination that is available from experimental data.

Another example of such calculations is angular correlation. There are many different approaches to these correlations ranging from the more theoretical invariant spherical harmonic approach of Nelson and coworkers,[60,61] to the standard bond-angle distributions involved in three-particle correlations. There is also the statistical geometrical approach, using Voronoi polyhedra, of Ruff and Baranyai.[62] This is of particular interest in

[60] D. R. Nelson and J. Toner, *Phys. Rev. B: Condens. Matter* [3] **24,** 363 (1981).

[61] P. J. Steinhardt, D. R. Nelson, and M. Ronchetti, *Phys. Rev. B: Condens. Matter* [3] **28,** 784 (1983).

[62] I. Ruff and A. Baranyai, *J. Chem. Phys.* **85,** 365 (1986).

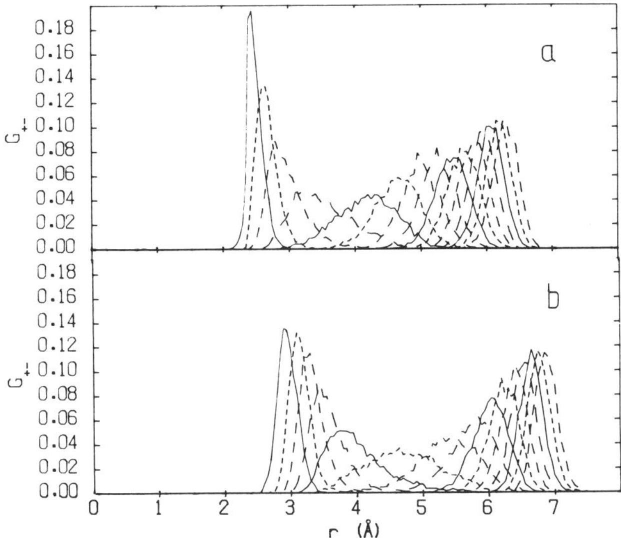

FIG. 9. Partial radial distribution functions $G_{+-}(r) = r^2 g_{+-}(r)$ for molten (a) LiI and (b) RbCl, decomposed into contributions of first, second, third neighbors, etc. [R. L. McGreevy, I. Ruff, and A. Baranyai, *Phys. Chem. Liq.* **16**, 47 (1986).]

that the polyhedra may be used to locate regions of high excess volume in the liquid, i.e., vacancies. It is found for RbCl that approximately half of the volume change on melting may be accounted for by vacancies. The excess volume is not, therefore, uniformly distributed. Further work on angular correlations is required to evaluate their general usefulness; they may have some relevance to Raman scattering spectra, as will be discussed in Section IV.

III. Dynamics: Inelastic Neutron Scattering

4. Experimental Techniques

a. Spectrometers

The principal experimental methods used in inelastic neutron spectroscopy are triple-axis spectroscopy (TAS) and the time-of-flight (TOF) technique. In TAS (see Fig. 10) the incident neutron wave vector is defined by Bragg scattering from a crystal monochromator; the neutrons are then scattered from the sample, and the scattered neutron wave vector is defined by a second monochromator (the analyzer). Because of the flexibility of choice in scattering angles and monochromator settings, this method is ideally suited

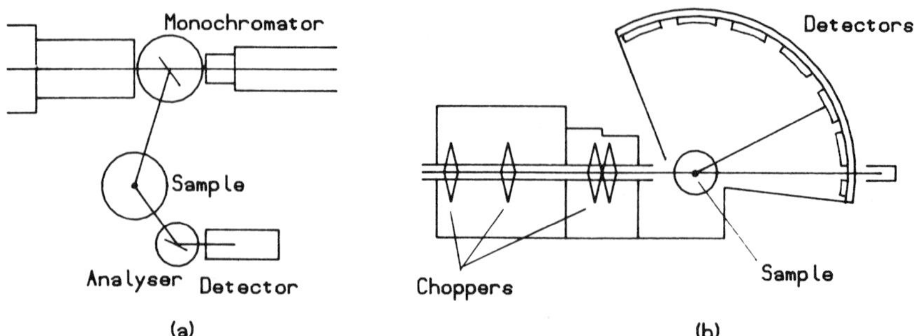

FIG. 10. Schematic representation of (a) TAS and (b) TOF inelastic neutron scattering spectrometers.

for looking at particular points in reciprocal (**Q**, ω) space, but since only one point is covered at a time, data collection rates are low, and it is most applicable to crystal work where scattering cross sections at a particular (**Q**, ω) point may be high. In TOF (Fig. 10) a short pulse of monochromatic neutrons is produced by a combination of static and/or rotating crystals or Fermi choppers. The scattered neutron energy is then calculated from the flight time to the detector relative to the arrival time of the pulse at the sample. Normally there are detectors at a number of scattering angles, and a wide range of reciprocal space is covered simultaneously. This makes the method more suitable for systems such as liquids, where the scattering cross sections may be low and spread out over reciprocal space.

Inelastic neutron scattering studies of phonon dispersion in crystals tend to take advantage of the fact that the dispersion relation is periodic in reciprocal space, and hence measurements are normally made at **Q** values in the second and third Brillouin zones, which are readily accessible using available incident neutron energies and normal scattering angles. In the case of liquids or disordered solids, where there is no long-range order and hence the dispersion relation is not periodic, the complete range of reciprocal space needs, in principle, to be covered. In practice this is not possible with a single spectrometer. The ranges covered by various spectrometers[7] are illustrated in Fig. 11, compared to the dispersion curves (100) for crystalline RbCl. (The energies of single and collective particle motions in the melt are expected to lie within a similar range.) It must be emphasized that the differential cross section depends on the incident and scattered neutron energies, and hence the intensity at a particular point in reciprocal space will differ between instruments.

Currently available instruments on nuclear reactors are not suitable to reach regions of lower Q than those illustrated, and which are of particular interest in liquids, as will be discussed later. At small energy transfers backscattering techniques are used (e.g., IN13 in Fig. 11) which prevent low-Q

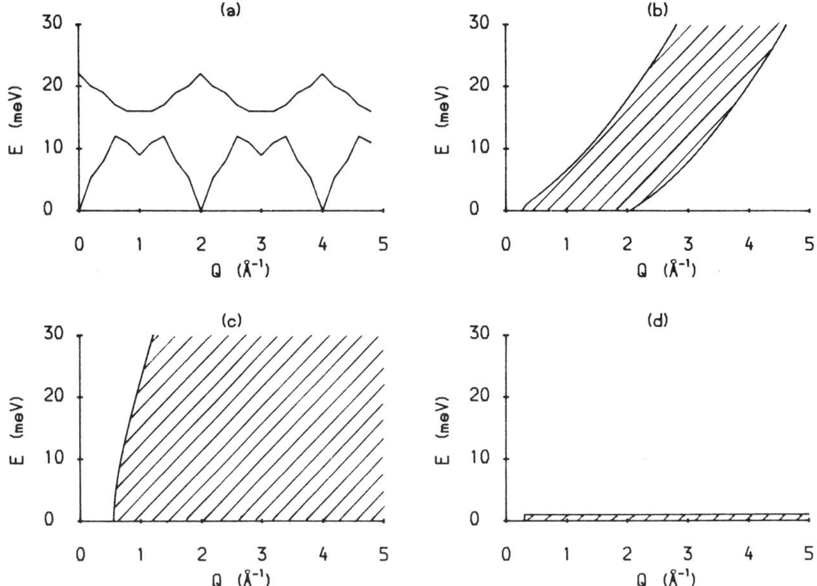

FIG. 11. (a) Dispersion curves for crystalline RbCl (100) compared to (Q, ω) ranges of inelastic neutron scattering spectrometers, (b) IN6 (TOF: $\lambda_i = 5.1$ Å), (c) IN4 (TOF: $\lambda_i = 1.0$ Å), and (d) IN13 (backscattering: $\lambda_i = 2.2$ Å) at the Institut Laue Langevin (Institut Laue Langevin, "Neutron Research Facilities at the ILL High Flux Reactor." Grenoble, 1983).

measurements. The velocities of sound in molten salts (similar to those in the crystal) are too high to fall into the available range. At higher-energy transfers it is necessary to use high incident energies to reach low Q. The flux from a thermal reactor drops off rapidly at such energies, as indeed does the scattering cross section. Many of these problems will be overcome by new instruments using pulsed neutron sources. The larger flux of high-energy neutrons available from these sources, and the flexibility of some of the TOF techniques inherent to them,[8] will enable coverage of the low-Q region at both low and high energies.

b. Correlation Functions and Isotopic Substitution

The time dependence of the relative distributions of N particles of M different species in a liquid may be described in terms of $M(M + 1)/2$ partial van Hove[63] correlation functions

$$G_{\alpha\beta}(\mathbf{r}, t) = \frac{1}{N}\left\langle \sum_{l_\alpha=1}^{N_\alpha} \sum_{j_\beta=1}^{N_\beta} \int \delta(\mathbf{r} + \mathbf{r}_{l_\alpha}(0) - \mathbf{r}')\delta(\mathbf{r}' - \mathbf{r}_{j_\beta}(t))\, d\mathbf{r}' \right\rangle \qquad (4.1)$$

[63] L. van Hove, *Phys. Rev.* **95,** 249 (1954).

where $\mathbf{r}(t)$ is a Heisenberg operator and $\langle \cdots \rangle$ denotes a thermal average. l_α indicates the lth particle of type α and

$$N = \sum_{\alpha=1}^{M} N_\alpha \tag{4.2}$$

Classically $G_{\alpha\beta}(\mathbf{r}, t)$ may be considered to represent the probability that if there is a particle of type α at the origin at time zero, then there is a particle of type β at $\mathbf{r}$ at time t.

If we define a fluctuating local particle density as

$$\rho_\alpha(\mathbf{r}, t) = \sum_{l_\alpha=1}^{N_\alpha} \rho_{l_\alpha}(\mathbf{r}, t) = \sum_{l_\alpha=1}^{N_\alpha} \delta[r - \mathbf{r}_{l_\alpha}(t)] \tag{4.3}$$

with Fourier components

$$\rho_\alpha(\mathbf{Q}, t) = \sum_{l_\alpha=1}^{N_\alpha} \rho_{l_\alpha}(\mathbf{Q}, t) = \sum_{l_\alpha=1}^{N_\alpha} \exp[i\mathbf{Q} \cdot \mathbf{r}_{l_\alpha}(t)] \tag{4.4}$$

then the space Fourier transform of the van Hove correlation function, known as the intermediate scattering function

$$S_{\alpha\beta}(\mathbf{Q}, t) = \int \exp(i\mathbf{Q} \cdot \mathbf{r}) G_{\alpha\beta}(\mathbf{r}, t)\, d\mathbf{r} \tag{4.5}$$

may be written as the autocorrelation function of these components

$$S_{\alpha\beta}(\mathbf{Q}, t) = (1/N)\langle \rho_\alpha(-\mathbf{Q}, 0)\rho_\beta(\mathbf{Q}, t)\rangle \tag{4.6}$$

The time transform

$$S_{\alpha\beta}(\mathbf{Q}, \omega) = \frac{1}{2\pi} \int \exp(-i\omega t) S_{\alpha\beta}(\mathbf{Q}, t)\, dt \tag{4.7}$$

is known as the dynamical structure factor. (There are many different definitions of such densities and correlation functions. We have used here those of McGreevy *et al.*,[64] as modified by Margaca *et al.*[65] for alkaline earth halides.)

The correlations of a single particle with itself at different times can also be distinguished

$$G_\alpha^s(\mathbf{r}, t) = \frac{1}{N} \left\langle \sum_{l_\alpha=1}^{N_\alpha} \int \delta(\mathbf{r} + \mathbf{r}_{l_\alpha}(0) - \mathbf{r}')\delta(\mathbf{r}' - \mathbf{r}_{l_\alpha}(t))\, d\mathbf{r}' \right\rangle \tag{4.8}$$

is known as the self-correlation function and has corresponding transforms.

[64] R. L. McGreevy, E. W. J. Mitchell, and F. M. A. Margaca, *J. Phys. C* **17**, 775 (1984).
[65] F. M. A. Margaca, R. L. McGreevy, and E. W. J. Mitchell, *J. Phys. C* **17**, 4725 (1984).

The differential cross section for neutron scattering may be written in terms of the various dynamical structure factors

$$\frac{1}{N}\frac{d^2\sigma}{d\Omega\, d\omega} = \frac{k}{k_0}\sum_{\alpha=1}^{M}(\overline{b_\alpha^2} - \bar{b}_\alpha^2)S_\alpha^{\mathrm{s}}(Q,\omega) + \sum_{\alpha=1}^{M}\sum_{\beta=1}^{M}\bar{b}_\alpha\bar{b}_\beta S_{\alpha\beta}(Q,\omega)$$

$$= \frac{k}{k_0}\overline{b^2}S_{\mathrm{T}}(Q,\omega) \tag{4.9}$$

where k_0 and k are the incident and scattered neutron wave vectors, $\hbar\omega$ is the neutron energy loss, and $Q = k_0 - k$ the momentum transfer. The first term is known as the incoherent scattering cross section and the second as the coherent. $4\pi\overline{b^2} = 4\pi\sum_\alpha c_\alpha \overline{b_\alpha^2}$ is the average cross section per ion and $S_{\mathrm{T}}(Q,\omega)$ is known as the total (neutron-weighted) dynamical structure factor.

As well as the components of a fluctuating particle density, we can equally well define the components of a property density

$$\rho_A(\mathbf{Q},t) = \sum_{\alpha=1}^{M} A_\alpha \rho_\alpha(\mathbf{Q},t)\Big/\sum_{\alpha=1}^{M}|A_\alpha| \tag{4.10}$$

where A_α is the "value" of the property A for particles of type α. Typical properties would be number (N), mass (m), or charge (q). The autocorrelation functions of these property densities are then

$$S_{AB}(\mathbf{Q},t) = (1/N)\langle\rho_A(-\mathbf{Q},0)\rho_B(\mathbf{Q},t)\rangle$$

$$= \sum_{\alpha=1}^{M}\sum_{\beta=1}^{M} A_\alpha B_\beta S_{\alpha\beta}(\mathbf{Q},t)\Big/\sum_{\alpha=1}^{M}\sum_{\beta=1}^{M}|A_\alpha B_\beta| \tag{4.11}$$

By analogy with particle and property densities, we may define a set of particle and property currents

$$\mathbf{j}_\alpha(\mathbf{Q},t) = \sum_{l_\alpha=1}^{N_\alpha}\mathbf{V}_{l_\alpha}\exp[i\mathbf{Q}\cdot\mathbf{r}_{l_\alpha}(t)] \tag{4.12}$$

$$\mathbf{j}_A(\mathbf{Q},t) = \sum_{\alpha=1}^{M} A_\alpha \mathbf{j}_\alpha(\mathbf{Q},t)\Big/\sum_{\alpha=1}^{M}|A_\alpha| \tag{4.13}$$

where $\mathbf{V}_{l_\alpha} = \dot{\mathbf{r}}_{l_\alpha}$. Because of the vector nature of the currents, there are both longitudinal and transverse current–current autocorrelation functions

$$C_{AB}^{\mathrm{L}}(\mathbf{Q},t) = (1/N)\langle\mathbf{Q}\cdot\mathbf{j}_A(-\mathbf{Q},0)\mathbf{Q}\cdot\mathbf{j}_B(\mathbf{Q},t)\rangle \tag{4.14}$$

$$C_{AB}^{\mathrm{T}}(\mathbf{Q},t) = (1/N)\mathrm{Tr}\langle\mathbf{Q}\times\mathbf{j}_A(-\mathbf{Q},0)\mathbf{Q}\times\mathbf{j}_B(\mathbf{Q},t)\rangle \tag{4.15}$$

The power spectrum of the longitudinal current–current autocorrelation function is then related, via the equation of continuity, to the power spectrum of the corresponding density–density autocorrelation function (the dynamical structure factor) by

$$C^{\mathrm{L}}_{AB}(Q, \omega) = \omega^2 S_{AB}(Q, \omega) \tag{4.16}$$

and hence is directly related to the differential cross section for neutron scattering via Eq. (4.9).

The moments of the dynamical structure factors

$$\langle \omega^n S_{\alpha\beta} \rangle = \int_{-\infty}^{\infty} \omega^n S_{\alpha\beta}(Q, \omega)\, d\omega \tag{4.17}$$

are of importance in discussing the dynamics of a liquid. In a classical liquid all odd moments are zero [ignoring recoil and assuming symmetrization of $S(Q, \omega)$; this is a small effect for the liquids discussed here]. The zeroth moment is the static structure factor

$$\langle \omega^0 S_{\alpha\beta} \rangle = \int_{-\infty}^{\infty} S_{\alpha\beta}(Q, \omega)\, d\omega = S_{\alpha\beta}(Q) \tag{4.18}$$

and the second moment, also the zeroth moment of the longitudinal current, is

$$\langle \omega^2 S_{\alpha\beta} \rangle = \int_{-\infty}^{\infty} \omega^2 S_{\alpha\beta}(Q, \omega)\, d\omega = c_\alpha \frac{Q^2 k_{\mathrm{B}} T}{m_\alpha} \delta_{\alpha\beta} \tag{4.19}$$

Since $\langle \omega^0 S \rangle$ depends entirely on the structure of the system, and $\langle \omega^2 S \rangle$ is independent of structure, it is immediately obvious that the shape of $S(Q, \omega)$, and not just the integral, will be structure dependent.

It can be seen that for a two-component system, such as a molten alkali or alkaline earth halide, instead of considering the system in terms of two particle densities, ρ_+ and ρ_-, we can use any two property densities. The importance of the property currents is that mass–mass and charge–charge current correlations in an ionic melt are analogous to the acoustic and optic modes in an ionic crystal. At $Q = 0$ the cross-correlation (mass–charge) is zero and the analogy is exact (the cross-correlation being zero in a perfect harmonic crystal because of symmetry). The differential cross section may be written in terms of the power spectra of the longitudinal mass (m) and charge (q) currents as

$$\begin{aligned}\frac{1}{N}\frac{d^2\sigma}{d\Omega\, d\omega} &= \frac{k}{k_0}\{(\overline{b_+^2} - \bar{b}_+^2) S_+^{\mathrm{s}}(Q, \omega) + (\overline{b_-^2} - \bar{b}_-^2) S_-^{\mathrm{s}}(Q, \omega) \\ &\quad + (1/\omega^2)[a_m^2 C^{\mathrm{L}}_{mm}(Q, \omega) + a_q^2 C^{\mathrm{L}}_{qq}(Q, \omega) + 2 a_m a_q C^{\mathrm{L}}_{mq}(Q, \omega)]\} \\ &= \frac{k}{k_0} \frac{\overline{b^2}}{\omega^2} C^{\mathrm{L}}_{\mathrm{T}}(Q, \omega)\end{aligned} \tag{4.20}$$

where

$$a_m = (c_+\bar{b}_+ + c_-\bar{b}_-)(m_+ + m_-)/(c_+m_+ + c_-m_-) \tag{4.21}$$

$$a_q = (\bar{b}_+m_- - \bar{b}_-m_+)/(c_+m_+ + c_-m_-) \tag{4.22}$$

and $C_T^L(Q, \omega)$ is the total (neutron-weighted) longitudinal current.

Strictly speaking we should, if possible, consider the dynamics at finite Q in terms of two property densities, ρ_A and ρ_B, which are independent, i.e., $\langle \omega^n S_{AB} \rangle = 0$ for all n. In the limit $Q \to \infty$ [or in practice by the time oscillations in $S_{\alpha\beta}(Q)$ have died out, i.e., at $Q \cong 10$–12 Å^{-1}] the liquid has a free-particle response and A and B become ion type, + and −. However, it is only possible to determine A and B theoretically (discussed in Subsection 6), and the process is not necessarily physically instructive. We shall therefore use ρ_m and ρ_q and pursue the analogy with crystalline acoustic and optic modes.

Because there are five terms contributing to the inelastic scattering cross section in Eq. (4.20), instead of the three in Eq. (1.1) [$S_\alpha^s(Q) = c_\alpha$ is a constant], we would in principle require five experiments with different isotopic sample compositions in order to separate all the contributions. It has already been emphasized in Subsection 1 that diffraction experiments using isotopic substitution require high accuracy and complex methods of data analysis; in inelastic experiments the available statistical accuracy is generally lower, and complete separation is simply not feasible. However, it is possible to take advantage of some properties of the partial dynamical structure factors and currents and their coefficients to achieve a substantial separation of the mass and charge currents which are of particular interest for ionic melts. This will be illustrated in the next section.

5. Experimental Results

a. Monatomic Liquids

Before considering the experimental results for molten alkali and alkaline earth halides, it is instructive to examine some results for monatomic liquids. These will fall into the categories of liquid metals (i.e., Rb[66]), condensed inert gases (i.e., Ar[67]), and quantum liquids (i.e., ^{4}He II[68]).

In the continuum limit of $Q \to 0$ (and strictly of $\omega \to 0$) $S(Q, \omega)$ has the (Landau–Plazcek) form of three Lorentzians,[69] one central (Rayleigh line)

[66] J. R. D. Copley and J. M. Rowe, *Phys. Rev. Lett.* **32**, 49 (1974).

[67] K. Skold, J. M. Rowe, and G. Ostrowski, *Phys. Rev. A* **6**, 1107 (1972).

[68] R. A. Cowley and A. D. B. Woods, *Can. J. Phys.* **49**, 177 (1971).

[69] R. D. Mountain, *Rev. Mod. Phys.* **38**, 205 (1966).

and two noncentral (Brillouin lines). The Rayleigh linewidth is determined by thermal diffusivity ($D_T Q^2$). The Brillouin line frequency is given by the sound velocity (cQ) and its width by the sound damping constant (ΓQ^2). The relative amplitudes of Rayleigh and Brillouin lines are determined by the ratio of specific heats (γ). As Q increases all lines broaden. The central linewidth becomes determined by the diffusion constant (DQ^2). The "acoustic" modes (i.e., extension of the Brillouin lines) at positive and negative frequency may broaden and overlap sufficiently that they form a single central line and cannot be resolved. For Ar this occurs at $Q \cong 0.25$ Å^{-1}, while for Rb modes it can be resolved up to $Q \cong 1.2$ Å$^{-1} = 0.8 Q_p$, where Q_p is the position of the first peak in $S(Q)$. The mode frequency is given by the sound velocity up to $\sim Q_p/3$ and has a maximum at $\sim Q_p/2$. Once the noncentral contributions can no longer be resolved, $S(Q, \omega)$ has a "quasi-Lorentzian" form which changes as Q increases into the Gaussian form of the free-particle response limit ($Q \to \infty$).

The various forms of $S(Q, \omega)$ in relation to $S(Q)$ are shown schematically in Fig. 12. In the regions where side peaks in $S(Q, \omega)$ can no longer be resolved, they may still contribute significantly to the high-energy wings of the spectrum. This is reflected in the "quasidispersion" curve of the peak in $C^L(Q, \omega) = \omega^2 S(Q, \omega)$, which has a higher energy than would be expected for purely diffusive behavior. (The term "quasidispersion" is used because $C^L(Q, \omega)$ *must* have a peak and the position is *not necessarily* related to any collective mode.) The peak in $C^L(Q, \omega)$ is an overestimate of the energy of any underlying mode [though close where the mode is resolved in $S(Q, \omega)$]; the true energy can only be determined by model fitting to experimental data. The 3-Lorentzian model of de Schepper and Cohen,[70] when applied to data for liquid Ar[71,72] and Ne,[73] predicts that a mode exists for $Q > 0.25$ Å^{-1} and, as for Rb, the dispersion is determined by the sound velocity up to $\sim Q_p/3$, and there is a maximum at $\sim Q_p/2$. Above $Q_p/2$ the mode energy decreases, and it is observed to soften (i.e., $\omega \to 0$) at $\sim 0.9\ Q_p$. There is then a dispersion gap around Q_p. At high pressures the dispersion gap in Ar becomes a minimum.[74]

In superfluid ^{4}He II,[68] because there is no dissipative component and $\gamma = 1$, there is no central line, and the acoustic modes are well defined. At higher energies there is also a broad "multiphonon" peak. The modes are resolved up to $\sim 2\ Q_p$, where their intensity decreases and they can no longer be observed. As for other liquids the frequency is given by the sound velocity up to $\sim Q_p/3$

[70] I. M. de Schepper and E. G. D. Cohen, *Phys. Rev. A* **22,** 287 (1980).

[71] I. M. de Schepper, P. Verkerk, A. A. van Well, and L. A. de Graaf, *Phys. Rev. Lett.* **80,** 974 (1983).

[72] I. M. de Schepper, P. Verkerk, L. A. de Graaf, J.-B. Suck, and J. R. D. Copley, *Phys. Rev. A* **31,** 3391 (1985).

[73] A. A. van Well and L. A. de Graaf, *Phys. Rev. A* **32,** 2384 (1985).

[74] P. Verkerk, Doctoral Thesis, Interuniversitair Reactor Institut, Delft (1985).

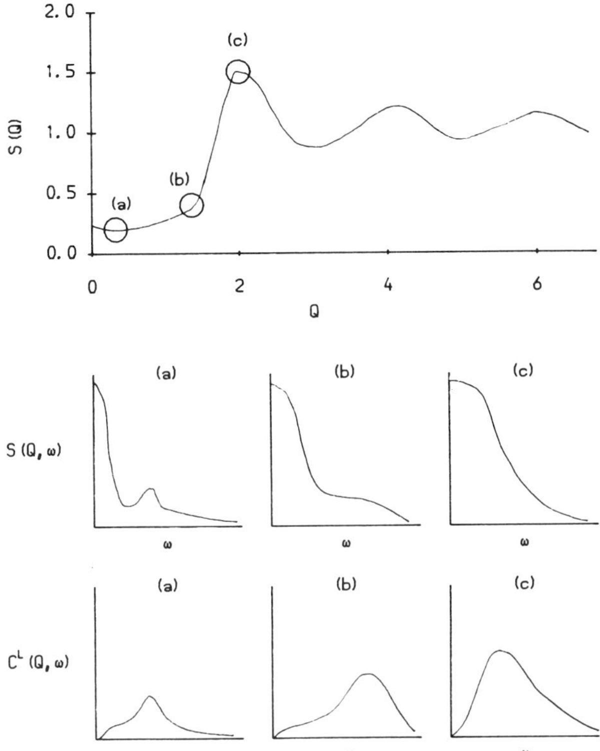

FIG. 12. Schematic representation of the various forms of $S(Q, \omega)$ and $C^L(Q, \omega)$ for a monatomic liquid corresponding to Q values in different regions below the first peak in the structure factor.

and has a maximum at $\sim Q_p/2$. There is then a minimum (the well-known "roton" minimum) at Q_p, after which the frequency increases again, initially parallel to the sound velocity. The similarity of this dispersion with that of nonquantum systems described above suggests that it is a general property of liquids. This is shown by the (albeit one-dimensional and harmonic) theory of Takeno and Goda.[75] As the range of ordering in the system increases, the minimum mode frequency at Q_p decreases until, in the limit of long-range order, the symmetric crystalline dispersion is recovered. Mode softening and the existence of a dispersion gap may be related to anharmonicity, as is the case for similar modes in crystals.

[75] S. Takeno and M. Goda, *Prog. Theor. Phys.* **47,** 970 (1972).

There is some argument[71,76–79] about the existence or significance of modes in classical liquids where they are no longer resolved as side peaks in $S(Q, \omega)$. It is generally agreed that resolved peaks correspond to "propagating" modes, though they may be heavily damped. Unresolved peaks would then correspond to overdamped modes. These may be collective in the sense that they involve correlated motions of more than one particle, but they may have little relation to phonons in crystals. In the $Q \to \infty$ limit the liquid has a free-particle response, and there are no correlated motions. These must, therefore, lose their significance at some intermediate Q. The derivation of the dispersion of overdamped modes depends on fitting of theoretical models; the prediction of mode softening (and even of the existence of overdamped modes) in liquid Ar is therefore model dependent. Perhaps surprisingly some of the experimental results for molten salts discussed in the following sections throw some light on these problems.

b. Alkali Halides

TOF spectra for molten RbCl,[64] measured using the IN6 spectrometer at the Institut Laue Langevin with an incident wavelength of 5.1 Å ($\hbar\omega_0 =$ 3.5 meV), are shown in Fig. 13. The spectra have been corrected and normalized in a similar manner to that discussed for diffraction data in Section 2. Because the spectra are measured at fixed scattering angles (θ), both ω and Q vary with time of flight (τ). The corrected and normalized intensity is then

$$I(\theta, \tau) \propto (\omega_0 - \omega)^2 S_T(Q, \omega) \tag{5.1}$$

For $\omega \ll \omega_0$ the intensity is proportional to $S_T(Q, \omega)$, while for $\omega \gg \omega_0$ it is proportional to $C_T^L(Q, \omega)$. The spectra may therefore be conveniently divided into the quasielastic part ($\hbar\omega \lesssim 2$ meV), dominated by primarily diffusive motions, and the inelastic part, ($\hbar\omega \gtrsim 5$ meV) dominated by collective motions.

The shape of the spectrum is characteristic for molten salts and many other liquids. The width of the quasielastic peak increases with $Q(\theta)$. The inelastic spectrum extends up to $\hbar\omega \cong 50$ meV and tends to peak in the region of 10–20 meV. In order to understand the origins of the different parts of the scattering intensity, it is useful to compare the spectrum of the melt with that of the polycrystalline solid. This is done for RbCl[64] in Fig. 14. At room temperature the inelastic spectrum is clearly divided into a peak at $\tau \cong 750\ \mu\text{s m}^{-1}$

[76] S. W. Lovesey, *Phys. Rev. Lett.* **53,** 401 (1984).
[77] S. W. Lovesey, *Z. Phys. B* **58,** 79 (1985).
[78] I. M. de Schepper, P. Verkerk, A. A. van Well, L. A. de Graaf, and E. G. D. Cohen, *Phys. Rev. Lett.* **54,** 158 (1985).
[79] R. L. McGreevy and E. W. J. Mitchell, *Phys. Rev. Lett.* **55,** 398 (1985).

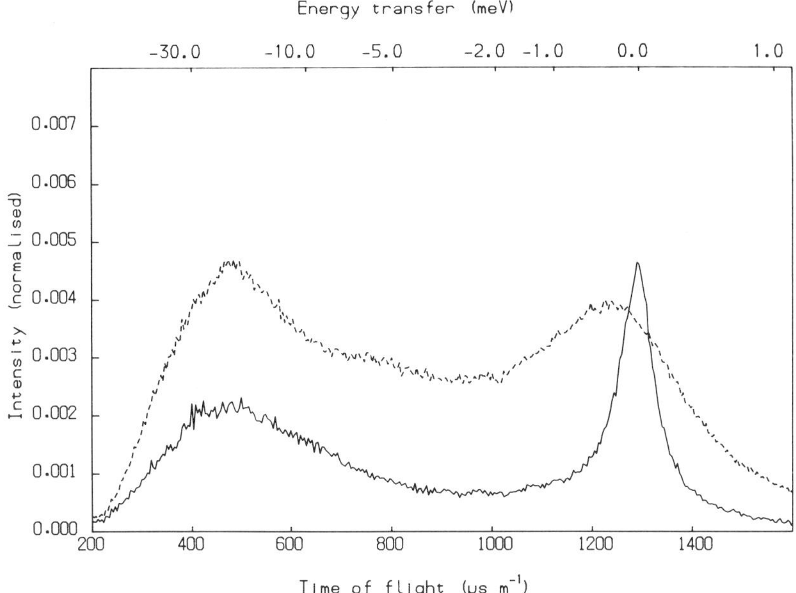

FIG. 13. Corrected and normalized inelastic neutron scattering intensities for molten RbCl at scattering angles of 35.5° (———) and 111.6° (---) [R. L. McGreevy, E. W. J. Mitchell, and F. M. A. Margaca, *J. Phys. C* **17**, 775 (1984)].

($\hbar\omega \cong 6$ meV) corresponding to acoustic modes and one at $\tau \cong 550\ \mu s\ m^{-1}$ ($\hbar\omega \cong 15$ meV) corresponding to optic modes. At low energies the elastic peak width is determined by experimental resolution. Just below the melting point the inelastic peaks have broadened considerably and increased in intensity. There is also a significant increase in intensity at high energies due to multiphonon scattering. Upon melting the largest change is in the low-energy region, where the onset of diffusion produces the broad quasielastic peak. In the inelastic region the peaks are further broadened. The optic-mode peak dominates while the acoustic-mode peak is only just resolvable. As the liquid temperature is raised, both quasielastic and inelastic peaks broaden further but retain the same basic shape.

The corrected intensities, collected at fixed scattering angles, are interpolated onto a rectangular grid of (Q, ω) points, and the functions $S_T(Q, \omega)$ and $C_T^L(Q, \omega)$ calculated. $S_T(Q, \omega)$ for molten RbCl[64] (Fig. 15), over the measured (Q, ω) range, consists of single broad peaks centred at $\omega = 0$ (no side peaks are observed). The integrated area under the peaks is simply the static structure factor and may be compared with the results of diffraction experiments. The other characteristic feature of the peak is the width, which

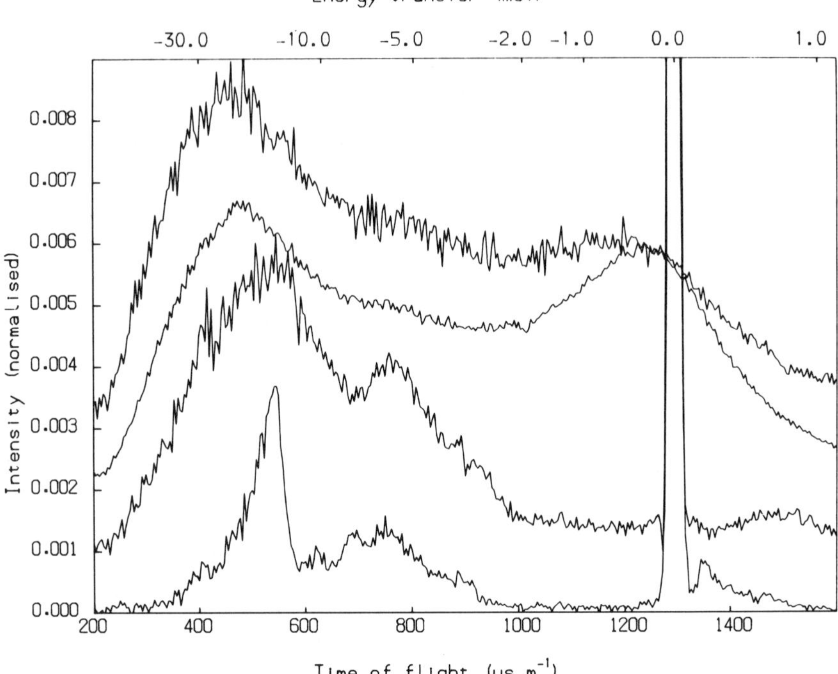

FIG. 14. Corrected and normalized inelastic neutron scattering intensities at a scattering angle of 111.6° for (in ascending order) polycrystalline RbCl at 298 and 873 K and molten RbCl at 1023 and 1173 K. The vertical scale is shifted by 0.001 between each curve [R. L. McGreevy, E. W. J. Mitchell, and F. M. A. Margaca, *J. Phys. C* **17**, 775 (1984)].

tends to increase with Q. This is shown for NaI[64] in Fig. 16. The prediction of simple diffusion is that the peak is Lorentzian with a width which should increase monotonically as $(D_+ + D_-)Q^2$, where $D_\pm$ are the diffusion constants of the two species. However, it is found that the width oscillates about this predicted value; there is a definite maximum at $Q = 1.3$ Å^{-1} and a corresponding minimum at 1.5 Å^{-1}. These features are also predicted by computer simulation,[80–82] which, however, underestimates the average diffusion rate. Such minima and maxima also occur for other molten salts.[64,65,83–85]

[80] M. Dixon, *Philos. Mag. [Part] B* **47,** 501 (1983).
[81] M. Dixon, *Philos. Mag. [Part] B* **47,** 531 (1983).
[82] M. Dixon, *Philos. Mag. [Part] B* **48,** 13 (1983).
[83] D. L. Price and J. R. D. Copley, *Phys. Rev. A* **13,** 2124 (1976).
[84] R. L. McGreevy, E. W. J. Mitchell, F. M. A. Margaca, and M. A. Howe, *J. Phys. C* **18,** 5235 (1985).
[85] R. L. McGreevy and E. W. J. Mitchell, *J. Phys. C* **18,** 1163 (1985).

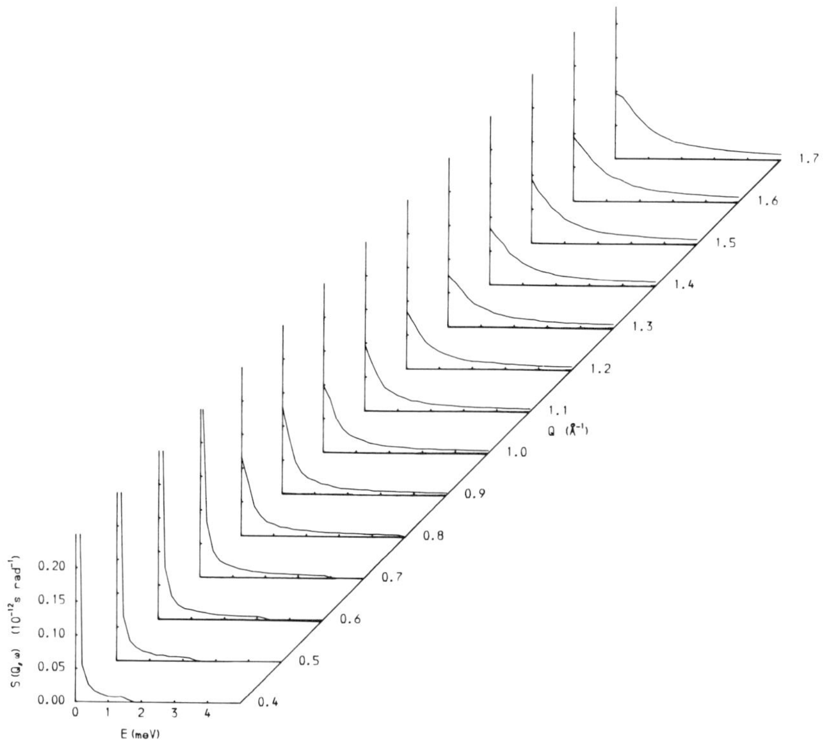

FIG. 15. Total dynamical structure factor $S_T(Q, \omega)$ for molten RbCl [R. L. McGreevy, E. W. J. Mitchell, and F. M. A. Margaca, *J. Phys. C* **17**, 775 (1984)].

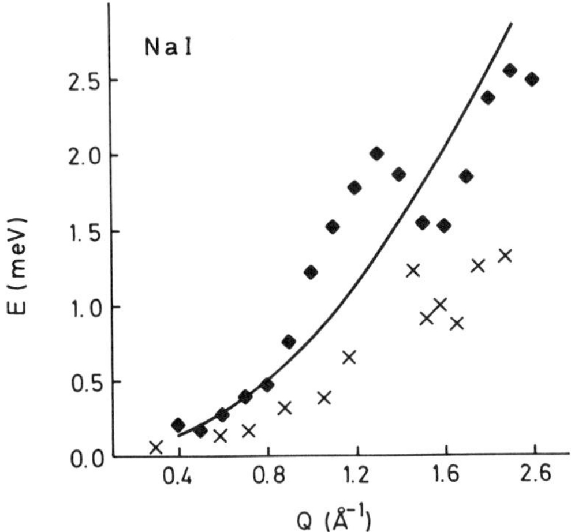

FIG. 16. Full width at half-height of $S_T(Q, \omega)$ for molten NaI [R. L. McGreevy, E. W. J. Mitchell, and F. M. A. Margaca, *J. Phys. C* **17**, 775 (1984)] (◆) compared to the predictions of simulation [M. Dixon, *Philos. Mag.* [*Part*] *B* **47**, 501 (1983)] (×) and simple diffusion (———).

For CsCl it has been found[64] that the width increases significantly more rapidly than Q^2 for $Q > 2.0\ \text{Å}^{-1}$ to values closer to the high-Q free-particle (Gaussian) limit.

There are a number of possible explanations for this behavior. The most common is the phenomenon of de Gennes[86] narrowing. If $S_\text{T}(Q, \omega)$ may be represented by a single function centered at $\omega = 0$, then in order to satisfy the requirements of both the zeroth moment (structure factor) and second moment [which increases as Q^2; see Eq. (4.19)], there will be a change in line shape and a minimum in the width at Q values corresponding to the peaks in the structure factor. Physically this corresponds to a "structural inhibition" of the diffusion mechanism. The minimum in the width of $S_\text{T}(Q, \omega)$ for NaI at $Q = 1.5\ \text{Å}^{-1}$ corresponds to the first maximum in $S_{qq}(Q)$. The first maximum in $S_{+-}(Q)$ for CsCl occurs at $Q = 2.0\ \text{Å}^{-1}$.

We know, however, that $S_\text{T}(Q, \omega)$ is a combination of five separate functions representing both single-particle and collective motions. It is possible that changes in the relative amplitudes of these functions [i.e., in $S_{\alpha\beta}(Q)$] may produce a minimum in the width of $S_\text{T}(Q, \omega)$ even though the widths of the individual functions all increase monotonically. Results from computer simulations[81] suggest that this is not the case. Indeed the width of $S_{qq}(Q, \omega)$ is found to have a high finite value (determined by the electrical conductivity and dielectric constant) at $Q \to 0$, and to have a pronounced minimum at the position of the first peak in $S_{qq}(Q)$, after which it increases again.

We also know, from comparison of scattered intensities in solid and liquid, that there are contributions to $S_\text{T}(Q, \omega)$, albeit small, which are centered at higher energies. These are the remnants of acoustic and optic modes in the solid. Softening of one of these modes may also give a maximum in the quasielastic peak width. This is discussed further in the next section. These modes do not produce resolvable side peaks in $S_\text{T}(Q, \omega)$ and may be considered to be overdamped; however, as discussed for monatomic liquids, they do contribute strongly to $C_\text{T}^\text{L}(Q, \omega)$.

The power spectrum of the total (neutron-weighted) longitudinal current, $C_\text{T}^\text{L}(Q, \omega)$, for molten NaI[64] (Fig. 17) is found to consist of single broad peaks, the width increasing with Q, with some evidence of peak splitting at lower Q values. The sharp cutoffs indicate the limited range of the experimental measurements. Computer simulation[83] produces good quantitative agreement at high Q but significantly overestimates the peak width and underestimates the height at lower Q. Assuming that $S_+^\text{s}(Q, \omega)$ and $S_-^\text{s}(Q, \omega)$ are determined largely by diffusion, and are therefore predominantly quasielastic, then $C_\text{T}^\text{L}(Q, \omega)$ contains significant contributions only from C_{mm}^L, C_{qq}^L, and C_{mq}^L, with coefficients given in Table IV. Using Eqs. (4,11), (4.16), and (4.19), it can

[86] P. G. de Gennes, *Physica (Amsterdam)* **25,** 825 (1959).

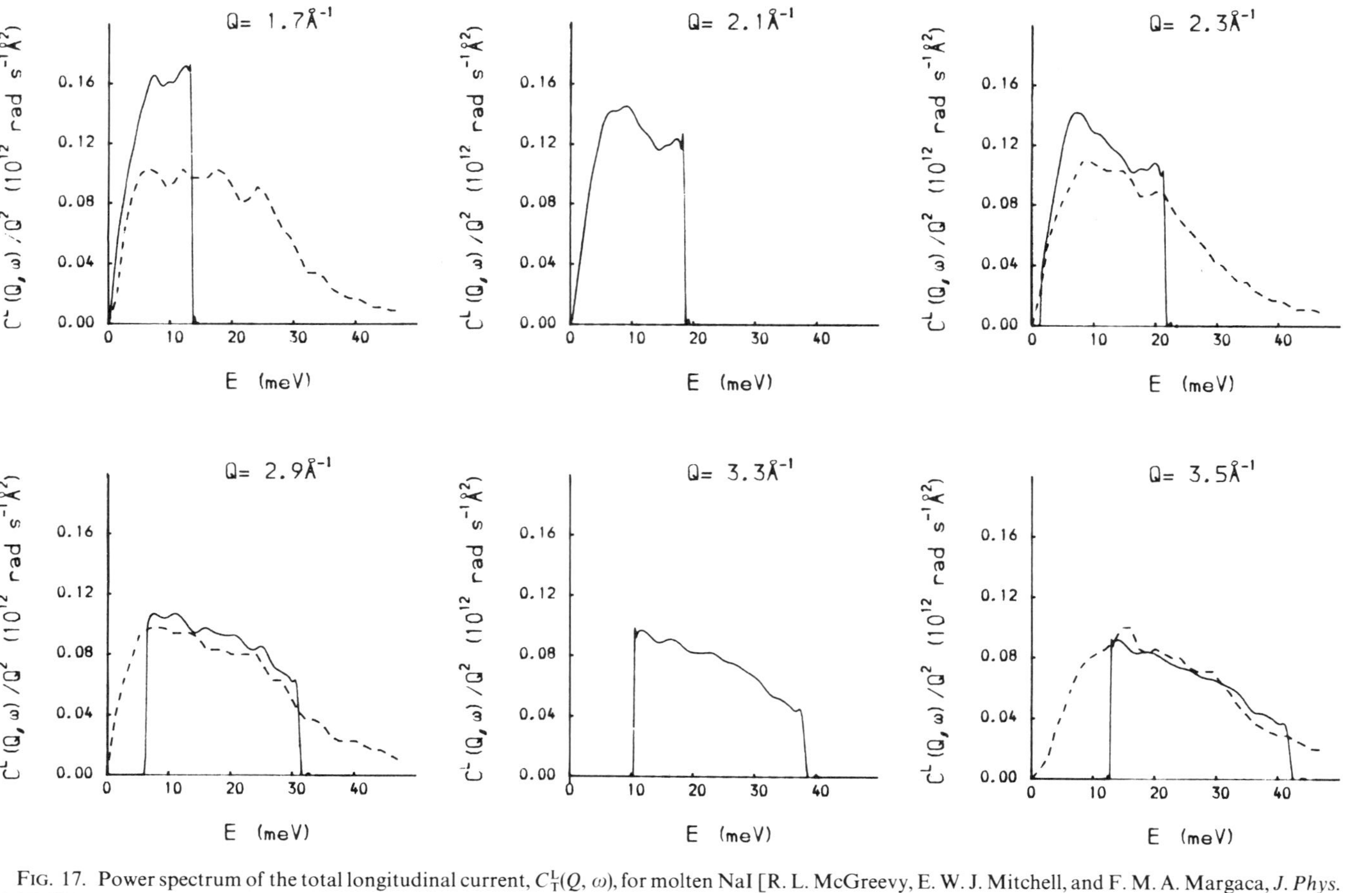

FIG. 17. Power spectrum of the total longitudinal current, $C_T^L(Q, \omega)$, for molten NaI [R. L. McGreevy, E. W. J. Mitchell, and F. M. A. Margaca, *J. Phys. C* **17**, 775 (1984)] (———) compared to the predictions of simulation [M. Dixon, *Philos. Mag.* [*Part*] *B* **47**, 501 (1983)] (– – –).

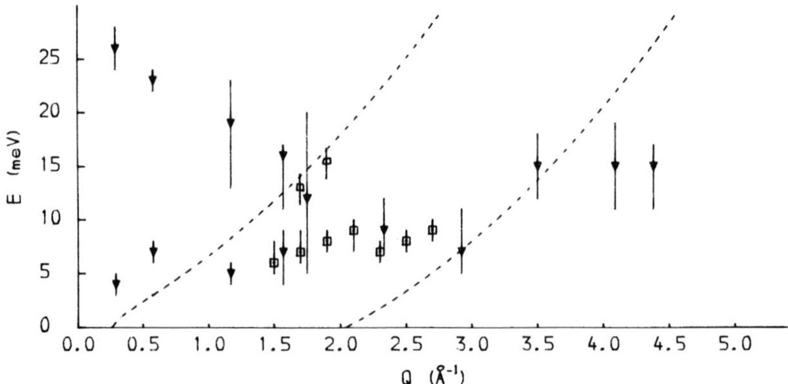

FIG. 18. Quasidispersion curve derived from the peak positions of $C_{\mathrm{T}}^{\mathrm{L}}(Q, \omega)$ for molten NaI (□) compared to the predictions of simulation (▼). The broken curves indicate the experimental (Q, ω) range [R. L. McGreevy, E. W. J. Mitchell, and F. M. A. Margaca, *J. Phys. C* **17**, 775 (1984); M. Dixon, *Philos. Mag.* [*Part*] *B* **47**, 501 (1983)].

be shown that $\langle C_{mq}^{\mathrm{L}} \rangle = 0$. If $S_{mq}(Q)$ is small, then $C_{mq}^{\mathrm{L}}(Q, \omega)$ will also be small. In NaI $a_m^2 \cong 4\, a_q^2$, so scattering is largely from C_{mm}^{L}. A quasidispersion curve may be created from the peak positions of $C_{\mathrm{T}}^{\mathrm{L}}(Q, \omega)$ as a function of Q (Fig. 18). The average peak energy of ~8 meV is similar to that of zone-edge acoustic phonons in the crystal. It is also considerably higher than would be the case if only single-particle diffusive motions existed. The small evidence of peak splitting suggests that "optic" modes, corresponding to C_{qq}^{L}, may exist at higher energy.

Since the two contributions C_{mm}^{L} and C_{qq}^{L} are not well defined or well separated in energy, it is necessary to find some way of distinguishing them. For RbBr, because the masses and neutron scattering lengths of the two ions are almost equal, $a_q \cong 0$ (see Table IV), and scattering is almost entirely from C_{mm}^{L}. The incoherent cross sections are also small, so the contributions of $S_\alpha^{\mathrm{s}}(Q, \omega)$ are reduced. For this reason RbBr was chosen for the first of these inelastic scattering experiments by Price and Copley,[83] and a complementary molecular-dynamics study by Copley and Rahman.[87] The results obtained are similar to those indicated for NaI, but with no evidence of peak splitting. Such splitting may have been observed in NaI, despite the small value of a_q, because the large mass difference between the ions increases the ratio $\langle C_{qq}^{\mathrm{L}} \rangle / \langle C_{mm}^{\mathrm{L}} \rangle$. The salt with the most favorable combination of scattering lengths for observing the contribution of C_{qq}^{L} is CsCl, for which $a_m^2/a_q^2 = 1.4$; the mass difference is also favorable. For this melt a single broad peak, with some

[87] J. R. D. Copley and A. Rahman, *Phys. Rev. A* **13,** 2276 (1976).

evidence of splitting, was again observed,[64] though with a higher average energy than for NaI, as would be expected from the larger contribution of "optic" modes.

From the results of computer simulation (Fig. 18), it appears that the positions of the peaks in $C^{\mathrm{L}}_{mm}(Q, \omega)$ and $C^{\mathrm{L}}_{qq}(Q, \omega)$ (ω_m and ω_q) only become well defined and separated in energy as $Q \to 0$, when their dispersion is similar to that of acoustic and optic modes in the crystal. At sufficiently low Q (~ 0.25 Å^{-1} for S_{mm} and ~ 0.5 Å^{-1} for S_{qq}) the modes become defined well enough to produce side peaks in the corresponding dynamical structure factors. In order to cover the appropriate range of (Q, ω) space (low Q, high ω) to observe the separation and side peaks, it is necessary to use incident neutrons of short wavelength (high energy). However, neutron fluxes from thermal reactors are lower at such wavelengths, and there is also a decrease in the inelastic scattering cross sections. Coupled with the fact that $S_{qq}(Q \to 0)$ decreases as Q^2, this makes the experiment very difficult. The attempt of Copley and Dolling[88] using KBr may have been unsuccessful because of the unfavorable value of a_q (see Table IV); however, the later experiment of McGreevy *et al.*[84] on CsCl indicated that the modes are probably broader in this region than shown by simulation. To perform such measurements successfully requires high fluxes of short-wavelength neutrons ($\lambda \sim 0.5$ Å), which should be available in the future from pulsed sources.[8]

It has already been stated that, with five contributions to the inelastic scattering cross section and with less statistical accuracy than in diffraction experiments, it is not possible to separate them all, but it is possible for some salts to achieve some separation of C^{L}_{mm} and C^{L}_{qq} by isotopic substitution. However, it is only recently that spectrometers with sufficiently high neutron fluxes have become available. For CsCl, as already indicated, $a_m^2/a_q^2 = 1.4$, while for $Cs^{37}Cl$ $a_m^2/a_q^2 = 10.6$ and scattering is predominantly from C^{L}_{mm}, as for RbBr. Assuming, as before, that single-particle motions only contribute significantly in the quasielastic region and that C^{L}_{mq} is small, a first-order difference between the spectra for the two samples should yield C^{L}_{qq}. If the cross-correlation C^{L}_{mq} is not small, then there is substantial coupling between C^{L}_{mm} and C^{L}_{qq}, and there will be no significant differences between the spectra of the two samples.

TOF spectra[84] for samples of molten $Cs^{37}Cl$, CsCl, and $Cs^{35}Cl$ are shown in Fig. 19 (coefficients a_m and a_q in Table IV). There are substantial differences between the spectra of $Cs^{37}Cl$ (mainly C^{L}_{mm}) and CsCl (C^{L}_{mm} and C^{L}_{qq}) in the inelastic region. The spectrum of $Cs^{35}Cl$ is not sufficiently different from that of CsCl in comparison with the statistical accuracy to enable second-order differences to be taken. All spectra are similar in the quasielastic region. The

[88] J. R. D. Copley and G. Dolling, *J. Phys. C* **11,** 1259 (1978).

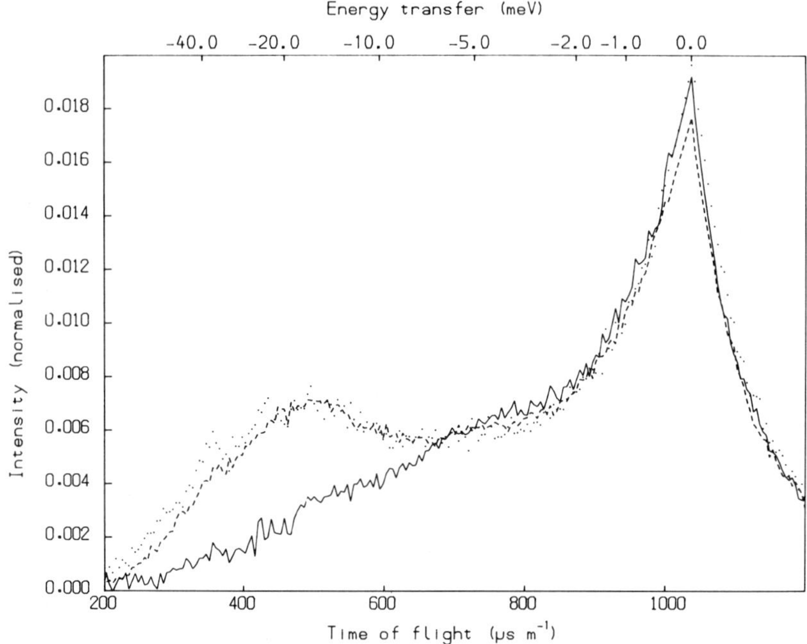

FIG. 19. Corrected and normalized inelastic neutron scattering intensities at a scattering angle of 83.8° for samples of molten $Cs^{37}Cl$ (———), CsCl (– – –) and $Cs^{35}Cl$ (· · ·) [R. L. McGreevy, E. W. J. Mitchell, F. M. A. Margaca, and M. A. Howe, *J. Phys. C* **18**, 5235 (1985)].

assumptions concerning diffusive contributions and C^{L}_{mq} are therefore justified. C^{L}_{mm} and C^{L}_{qq} derived from these spectra are shown in Fig. 20. C^{L}_{qq} is generally higher and narrower than C^{L}_{mm} and peaks at higher energy. A quasidispersion curve derived from the peak positions (Fig. 21) shows energies typical of those for acoustic and optic modes in the crystal. ω_m has a flat dispersion for $Q < 2.7$ Å^{-1}, above which it increases slightly. For $Q < 1.3$ Å^{-1} no peak is observed within the experimental region, so it must be above the cutoff; ω_m must therefore increase as Q decreases. As $Q \to 0$, ω_m is expected to be determined by the sound velocity, which is indicated. ω_q increases slightly up to $Q \cong 2.3$ Å^{-1}, has a flat dispersion up to $Q = 3.1$ Å^{-1}, and then increases slightly again. The points at $Q \cong 0.75$ Å^{-1} are derived from a total scattering experiment.[84]

The quasidispersion shows some correlation with the partial structure factors S_{mm}, S_{qq}, and S_{mq} (Fig. 21). $S_{qq}(Q)$ is a sharply peaked function, demonstrating the strong charge ordering in the liquid. $S_{mm}(Q)$, on the other hand, has a very broad peak, reflecting the weak dependence of structural

TABLE IV. COEFFICIENTS a_m AND a_q FOR SOME ALKALI AND ALKALINE EARTH HALIDES

	$[^{35}Cl]$ (%)	$[^{37}Cl]$ (%)	a_m [a]	a_q [a]
$Sr^{35}Cl_2$ [b]	99.6	0.4	2.362	−1.479
$Sr^{M}Cl_2$ [b]	52.3	47.7	1.721	−0.778
$Sr^{37}Cl_2$ [b]	5.1	94.9	1.085	−0.093
$Cs^{35}Cl$ [c]	99.4	0.6	1.715	−1.584
CsCl [c]	75.53	24.47	1.509	−1.268
$Cs^{37}Cl$ [c]	1.8	98.2	0.874	−0.268
$MgCl_2$ [d]	75.53	24.47	1.529	−0.150
$CaCl_2$ [d]	75.53	24.47	1.624	−0.586
RbCl [e]	75.53	24.47	1.669	−0.939
$Rb^{37}Cl$ [e]	7.9	92.1	1.087	−0.1
RbBr [f]	—	—	1.386	−0.014
NaI [e]	—	—	0.890	0.447

[a] Units of 10^{-12} cm.
[b] F. M. A. Margaca, R. L. McGreevy, and E. W. J. Mitchell, *J. Phys. C* **17**, 4225 (1984).
[c] R. L. McGreevy, E. W. J. Mitchell, F. M. A. Margaca, and M. A. Howe, *J. Phys. C* **18**, 5235 (1985).
[d] R. L. McGreevy and E. W. J. Mitchell, *J. Phys. C* **18**, 1163 (1985).
[e] R. L. McGreevy, E. W. J. Mitchell, and F. M. A. Margcaca, *J. Phys. C* **17**, 775 (1984).
[f] D. L. Price and J. R. D. Copley, *Phys. Rev. A* **13**, 2124 (1976).

ordering on the ionic masses. At low Q [from Eq. (3.2)]

$$S_{mm}(Q \to 0) = (c_+m_+ + c_-m_-)^2 \rho k_B T k_T/(m_+ + m_-)^2 \tag{5.2}$$

is proportional to the isothermal compressibility. Assuming, on a simple model, that

$$\omega_m^2(Q \to 0) \propto \langle \omega^2 S_{mm}(Q \to 0, \omega)\rangle / \langle S_{mm}(Q \to 0, \omega)\rangle \tag{5.3}$$

where

$$\langle \omega^2 S_{mm}\rangle = (c_+m_+ + c_-m_-)^2 Q^2 k_B T/(m_+ + m_-)^2 \tag{5.4}$$

we obtain

$$\omega_m^2 = v^2 Q^2/\gamma_m \tag{5.5}$$

where v is the isothermal velocity of sound and $1/\gamma_m$ the proportionality constant. This reduces to

$$\omega_m^2 = c^2 Q^2 \tag{5.6}$$

where c is the adiabatic velocity of sound, the expected result, if $\gamma_m = \gamma$. In the

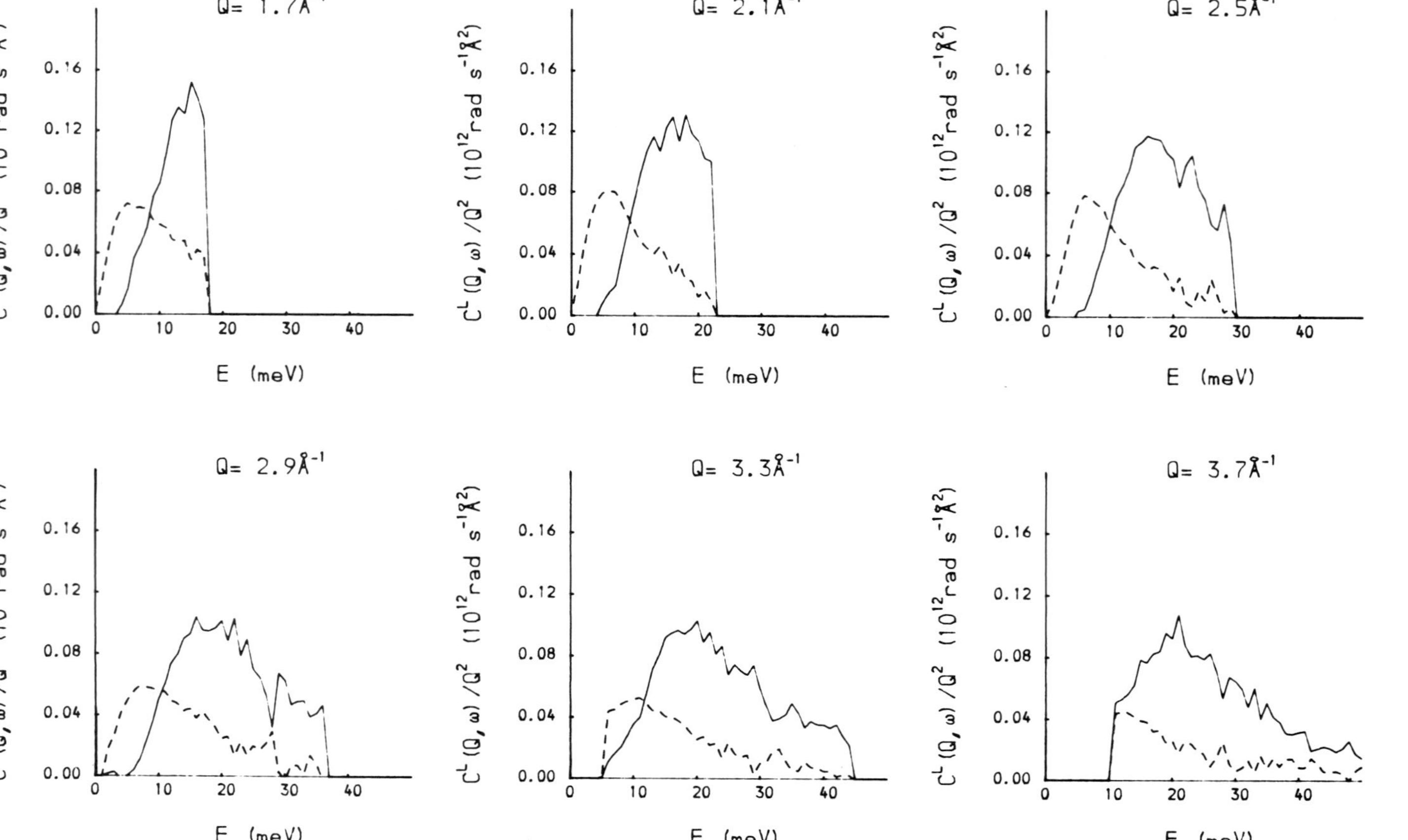

FIG. 20. Power spectra of the mass and charge longitudinal currents, $C^{L}_{mm}(Q, \omega)$ (– – –) and $C^{L}_{qq}(Q, \omega)$ (———), for molten CsCl [R. L. McGreevy, E. W. J. Mitchell, F. M. A. Margaca, and M. A. Howe, *J. Phys. C* **18**, 5235 (1985)].

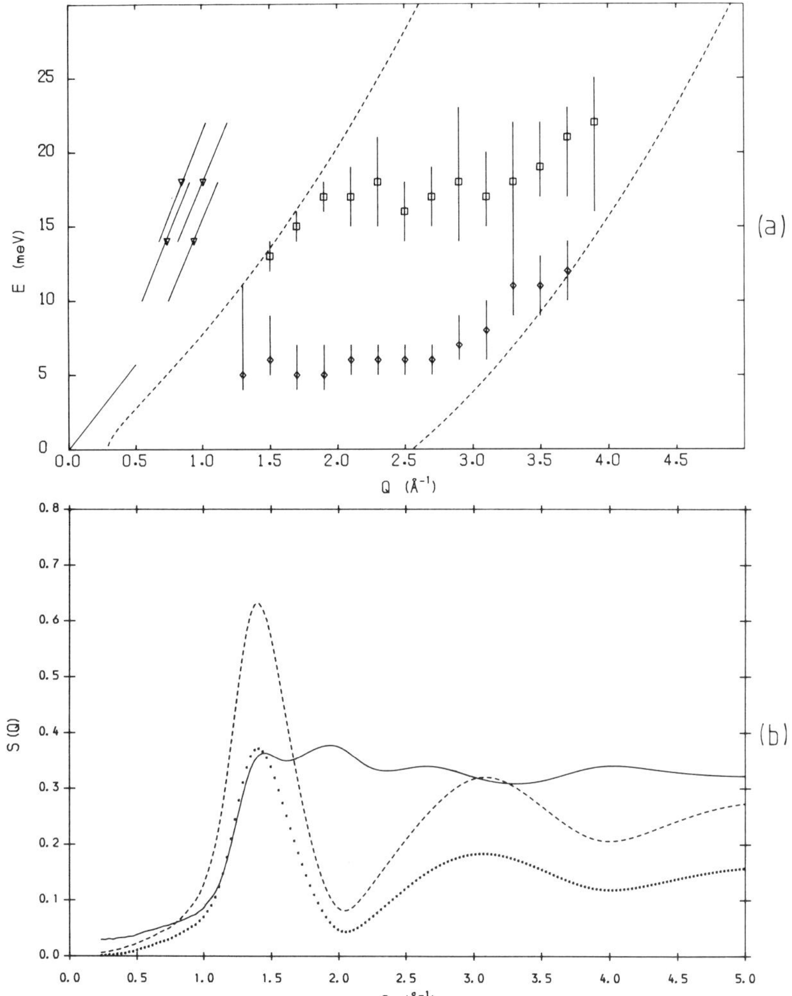

FIG. 21. (a) Quasidispersion curves derived from the peak positions of $C^{L}_{mm}(Q, \omega)$ ($\diamond$) and $C^{L}_{qq}(Q, \omega)$ ($\square$) for molten CsCl [R. L. McGreevy, E. W. J. Mitchell, F. M. A. Margaca, and M. A. Howe, *J. Phys. C* **18**, 5235 (1985)]. (———) indicates the sound velocity and (---) the experimental (Q, ω) range. The points marked (∇) are peak positions of $C^{L}_{qq}(Q, \omega)$ derived from a separate experiment. (b) Partial static structure factors $S_{mm}(Q)$ (———), $S_{qq}(Q)$ (---), and $S_{mq}(Q)$ (···) for molten CsCl [J. Locke, S. Messoloras, R. J. Stewart, R. L. McGreevy, and E. W. J. Mitchell, *Philos. Mag.* [*Part*] *B* **51**, 301 (1985)].

limit of $Q \to 0$, as stated earlier, γ determines the amplitude ratio of Rayleigh and Brillouin lines. Because the peak in $C^{\mathrm{L}}_{mm}(Q, \omega)$ for $Q < 1.3$ Å^{-1} occurs above the experimental cutoff, it may be inferred that ω_m has a maximum at $Q \cong 1$ Å$^{-1} \cong Q_{\mathrm{P}}/2$, as for monatomic liquids. In the region where $S_{mm}(Q) > S_{mm}(Q = \infty)$ (i.e., the first broad peak) ω_m has a flat dispersion. The energy here is slightly higher than expected from pure diffusion. The underlying mode may have an energy minimum or dispersion gap similar to those discussed above for monatomic liquids.

In the limit[51,52] $Q \to 0$

$$S_{qq}(Q \to 0) = \frac{Q^2 k_{\mathrm{B}} T}{\omega_{\mathrm{P}}^2} \frac{(c_+ q_+^2/m_+ + c_- q_-^2/m_-)}{(|q_+| + |q_-|)^2} \tag{5.7}$$

Since

$$\langle \omega^2 S_{qq} \rangle = Q^2 k_{\mathrm{B}} T \frac{(c_+ q_+^2/m_+ + c_- q_-^2/m_-)}{(|q_+| + |q_-|)^2} \tag{5.8}$$

on a similar model to that for ω_m, we find that

$$\omega_q^2 = \omega_{\mathrm{P}}^2/\gamma_q \tag{5.9}$$

has a nonzero value. (ω_{P} is the plasma frequency.)

$$\omega_{\mathrm{P}}^2 = (\rho e^2/\varepsilon_\infty)(c_+ q_+^2/m_+ + c_- q_-^2/m_-) \tag{5.10}$$

By analogy with the Lyddane–Sachs–Teller relation for ionic crystals, the proportionality constant is given by

$$\gamma_q = (\varepsilon_0/\varepsilon_\infty) - 1 \tag{5.11}$$

where ε_0 and ε_∞ are, respectively, the low- and high-frequency dielectric constants. From this we would estimate a value $\omega_q(Q \to 0) = 18$ meV, in good agreement with the observed values at higher Q and the trends at low Q. ω_q appears to have a minimum around the peak in $S_{qq}(Q)$ at $Q \cong 1.5$ Å^{-1}. However, $S_{mq}(Q)$ has a maximum here, and so C^{L}_{mm} and C^{L}_{qq} are coupled and ω_m and ω_q are close. In this position it may be more appropriate to consider variables other than mass and charge in order to decouple the modes. Coupling would also appear to increase as Q increases above 3.5 Å^{-1}.

Results for RbCl[64,89] obtained using first-order isotopic substitution are similar to those for CsCl, though the differences of C^{L}_{mm} and C^{L}_{qq} are smaller and the corresponding dispersion curves are closer. This would be expected from simple ideas of lattice dynamics for a binary (two atoms per unit cell) system where the difference between acoustic- and optic-mode frequencies at

[89] R. L. McGreevy and E. W. J. Mitchell, *J. Phys. C* **15**, L1001 (1982).

the Brillouin zone edge depends on the mass difference. Good separation is therefore observed in CsCl and some separation in NaI even without isotopic substitution. The separation is less for RbCl and would be expected to be very small for KCl and RbBr.

c. Alkaline Earth Halides

In Section II it was shown that the structures of molten alkali chlorides have strong similarities, but that there are considerable differences among the alkaline earth chlorides. The alkali chlorides also show strong dynamical similarities, as they do in the crystalline state. For the alkaline earth chloride crystals there are some notable differences. In particular, $SrCl_2$ (in common with alkaline earth fluorides) crystallizes in the fluorite structure and exhibits the property of superionic (or fast ion) conduction.[90] At 975 K, 223 K below the melting point, there is a partial disordering of the Cl^- sublattice, and a maximum in the specific heat. The disordered ions are highly mobile. As the temperature is increased, further disordering occurs until the ionic conductivity just below the melting point is comparable to that above. $BaCl_2$ undergoes a structural phase change at 1193 K from orthorhombic to fluorite, when it also becomes superionic. $MgCl_2$ and $CaCl_2$ do not crystallize in the fluorite structure and show no evidence of fast-ion conduction.

Margaca *et al.*[65] have measured the inelastic neutron scattering spectra of molten $Sr^{35}Cl_2$, $Sr^{M}Cl_2$, and $Sr^{37}Cl_2$ (Fig. 22; coefficients a_m and a_q in Table IV). While there are differences in the intensity of scattering in both quasielastic and inelastic regions, these are not so large as for CsCl. The total dynamical structure factors are similar to those for alkali chlorides; there is a minimum in the width corresponding to the first peak in $S_{++}(Q)$. $C_T^L(Q, \omega)$, on the other hand, shows definite evidence of peak splitting at intermediate Q for all samples (Fig. 23). Using the (second-order) differences between samples, it is found that these two peaks do not scale with a_m and a_q and are therefore not associated with mass and charge currents. Also, their dispersion (Fig. 24) is approximately linear and parallel with the same velocity as that of sound. Underlying these "additional" (A) modes are broad acoustic and optic bands which do scale with a_m and a_q and are similar to those in alkali halides. The extra modes, using isotopic differences again, have been shown to be associated with $--$ and $+-$ correlations, but not with $++$; indeed one of the modes is seen to be an extension of quasielastic broadening of the first peak in $S_{--}(Q)$. Intensity corresponding to such modes has also been observed to grow in the superionic phase of the crystal and has been associated with this

[90] W. Hayes, ed., "Crystals with the Fluorite Structure." Oxford Univ. Press (Clarendon), London and New York, 1974.

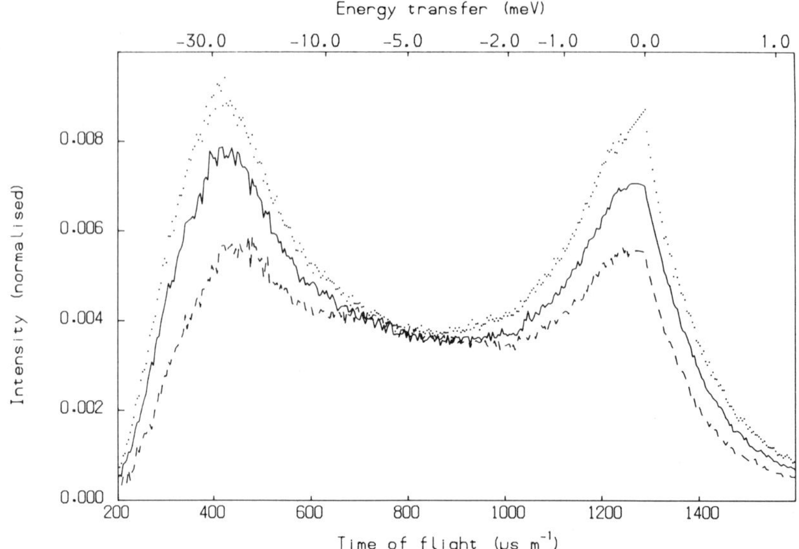

FIG. 22. Corrected and normalized inelastic neutron scattering intensities for molten $Sr^{37}Cl_2$ (– – –), $Sr^{M}Cl_2$ (———), and $Sr^{35}Cl_2$ (···) [F. M. A. Margaca, R. L. McGreevy, and E. W. J. Mitchell, *J. Phys. C* **17**, 4725 (1984)].

behavior, which involves disordering of the anion sublattice and therefore diffuse scattering from − − and + −, but not + +, correlations. The origin of these modes is not understood; they may represent modes of the vacancy/interstitial/relaxed ion complexes proposed by Catlow and Hayes[91] to account for the superionic conductivity in fluorites or alternatively some kind of cooperative ("phonon-assisted") diffusion.

TOF spectra of molten $MgCl_2$ and $CaCl_2$[85] have also been measured, though without isotopic substitution. No "additional modes" are observed in these melts, which supports the relation with superionic behavior in the crystal. However, in $MgCl_2$ (Fig. 25) there is evidence of a very weak acoustic mode which softens and disappears into the quasielastic peak. Although this mode is only directly observable in the TOF spectrum, and not with data in the form of $S_T(Q, \omega)$ or $C_T^L(Q, \omega)$, it is not an experimental artefact. As the mode softens it produces a definite maximum in the width of $S_T(Q, \omega)$ (Fig. 26), considerably above that expected for simple diffusion.

Molten $CaCl_2$ shows no direct evidence of a similar mode, though there are small maxima in the width of $S_T(Q, \omega)$, and the shapes of the TOF spectra may be interpreted as being similar to those for $MgCl_2$ but with a somewhat

[91] C. R. A. Catlow and W. Hayes, *J. Phys. C* **15**, 49 (1982).

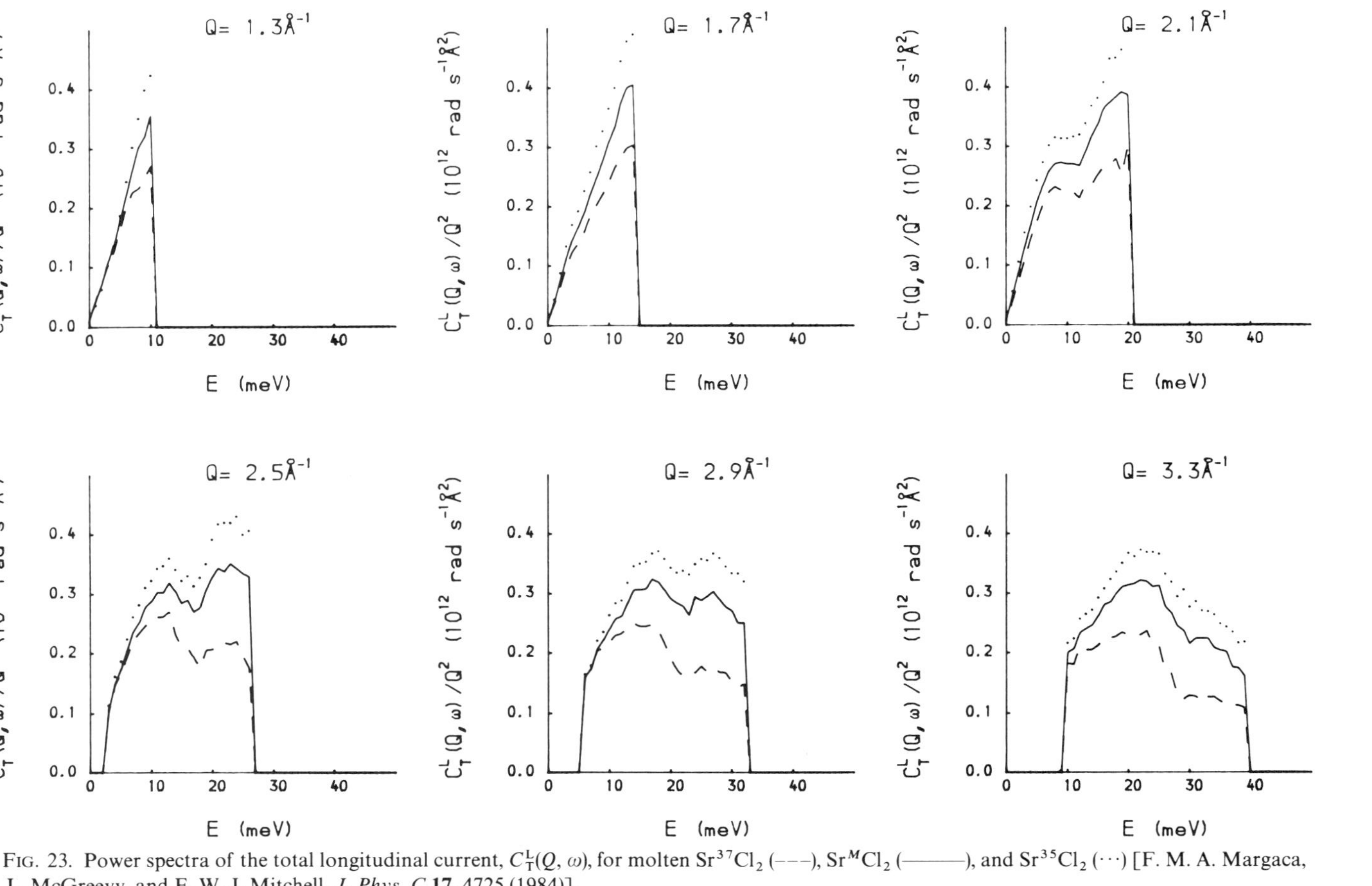

FIG. 23. Power spectra of the total longitudinal current, $C_T^L(Q, \omega)$, for molten $Sr^{37}Cl_2$ (–––), Sr^MCl_2 (———), and $Sr^{35}Cl_2$ (···) [F. M. A. Margaca, R. L. McGreevy, and E. W. J. Mitchell, *J. Phys. C* **17**, 4725 (1984)].

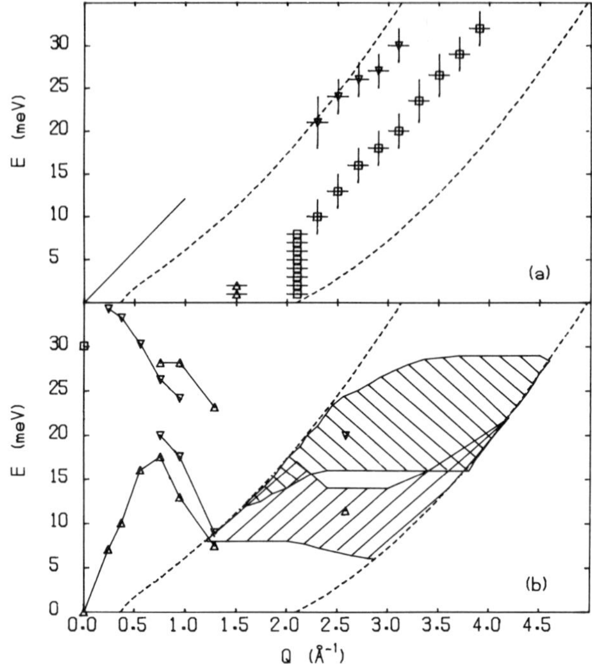

FIG. 24. Quasidispersion curves for molten $SrCl_2$. (a) "Additional" (A) modes derived from peak positions of $C_T^L(Q, \omega)$ at constant Q and ω. (———) represents the sound velocity. (b) m (////) and q (\\\\) modes derived from peak positions of $C_{mm}^L(Q, \omega)$ and $C_{qq}^L(Q, \omega)$ at constant θ compared to the predictions of simulation (m, $\triangle$; q, ∇). (– – –) indicates the experimental (Q, ω) range [F. M. A. Margaca, R. L. McGreevy, and E. W. J. Mitchell, *J. Phys. C* **17**, 4725 (1984); S. de Leeuw, *Mol. Phys.* **37**, 489 (1983)].

broader acoustic mode. In fact, maxima in the width for $CaCl_2$ are of the same magnitude as those already discussed for alkali halides. There is therefore the possibility that mode softening occurs in these salts as well, and that the maxima/minima are not due to de Gennes narrowing.

Quasidispersion curves for $MgCl_2$ and $CaCl_2$ (Fig. 27) show a higher-energy peak at 15–20 meV, largely determined by the contribution of C_{qq}^L and similar to that for alkali halides. The acoustic mode is seen to soften around $Q \cong 2$ Å^{-1}, though the method of peak determination probably overestimates this value, which should be closer to that of 1.8 Å^{-1} found from the maximum in the width of $S_T(Q, \omega)$. This is similar to the position ($\sim 0.9\ Q_P$) of mode softening proposed by de Schepper *et al.*[71] to explain the experimental results for Ar.[72]

Because of the finite width of the quasielastic peak, due to diffusion, it is not possible to say whether the mode softens completely or has a minimum. With

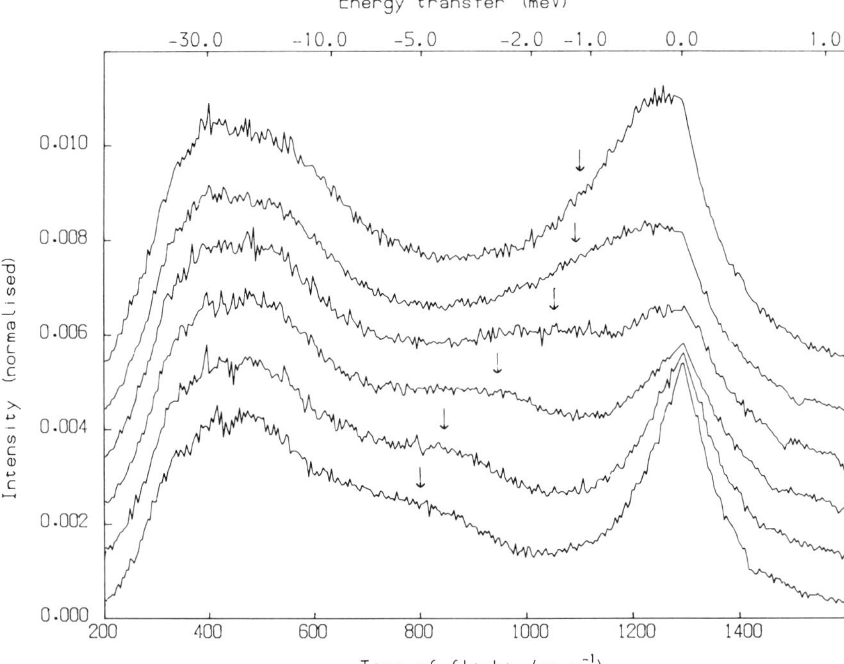

FIG. 25. Corrected and normalized inelastic neutron scattering intensities for molten $MgCl_2$ at scattering angles (in ascending order) of 71.8°, 77.8°, 84.6°, 90.6°, 97.3°, and 105.0°. The vertical scale is shifted by 0.001 between each curve. The arrows indicate the approximate position of the feature corresponding to the acoustic mode [R. L. McGreevy and E. W. J. Mitchell, *J. Phys. C* **18**, 1163 (1985)].

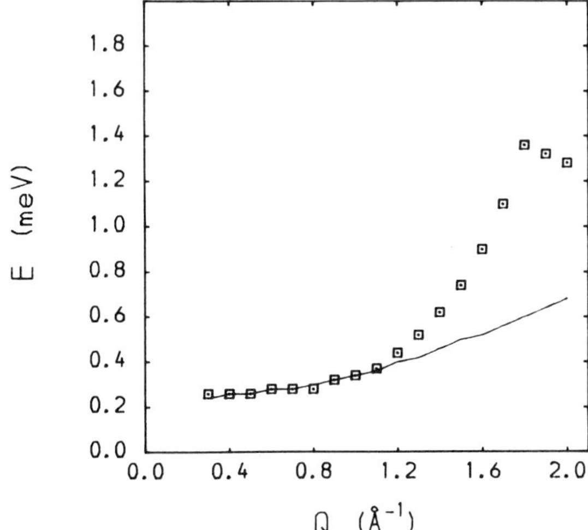

FIG. 26. Full width at half-height of $S_T(Q, \omega)$ for molten $MgCl_2$ (□) compared to the prediction of simple diffusion (———) [R. L. McGreevy and E. W. J. Mitchell, *J. Phys. C* **18**, 1163 (1985)].

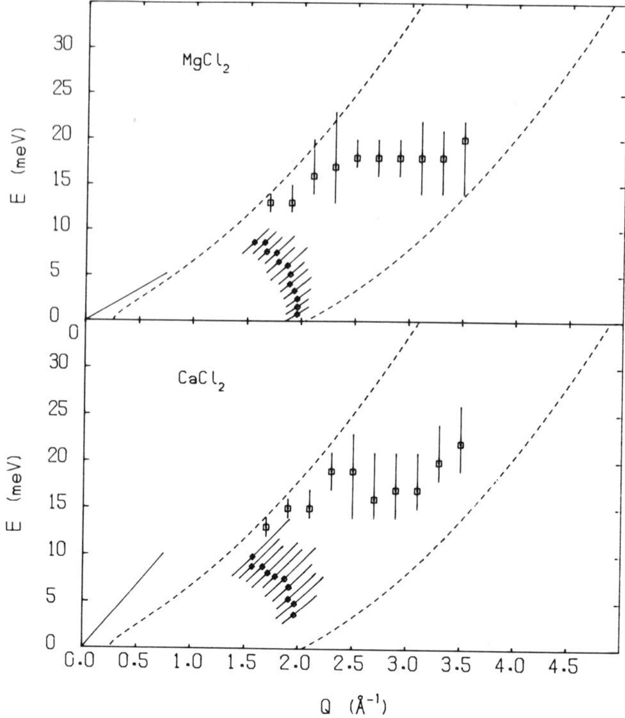

FIG. 27. Quasidispersion curves for molten $MgCl_2$ and $CaCl_2$ derived from peak positions of $C_T^L(Q, \omega)$ at constant Q (□) and $I(Q, \omega)/Q^2$ at constant θ (◇). (———) represents the sound velocity and (– – –) the experimental (Q, ω) range [R. L. McGreevy and E. W. J. Mitchell, *J. Phys. C* **18**, 1163 (1985)].

five contributions to the scattering, both coherent and incoherent, there are too many parameters to make model fitting a feasible proposition. However, recent measurements[92] on molten $MgBr_2$ and $ZnCl_2$ have shown that $MgBr_2$ behaves similarly to $MgCl_2$, but with some evidence that the mode has a nonzero minimum energy (the incoherent cross section for $MgBr_2$ is smaller than that for $MgCl_2$, which improves resolution of the mode). $ZnCl_2$, as already discussed, has an anomalously high viscosity and low diffusion constant. The quasielastic peak is therefore considerably narrower than in the other melts, and in this case the soft mode is seen as a shoulder in $S_T(Q, \omega)$ at $Q \cong 1.8$ Å^{-1} with an energy of $\cong 1$ meV. Again the mode appears not to soften completely but to have a minimum.

[92] M. Fairbanks, private communication (1985).

6. Computer Simulation and Theory

In contrast to the large number of simulations of molten alkali halide structures, there have been relatively few of dynamics, and a number of these have concentrated on hypothetical systems such as the "simple molten salt"[93] (i.e., cation and anion differing only in the sign of their charge) or on transport coefficients.[94,95] Adams *et al.*[96] have performed a detailed simulation of molten NaCl, and Copley and Rahman[87] of RbBr, using rigid ion potentials. Dixon has simulated molten NaI[80–82] and CsCl[97] using both rigid- and polarizable-ion potentials.

It has already been indicated in the previous section that these simulations agree qualitatively with experiment, although there are some obvious quantitative differences. For instance, the macroscopic diffusion constant may be underestimated, giving too small a width for $S(Q, \omega)$, and the width of peaks in $C^{L}(Q, \omega)$ may be overestimated. It has been found that, whereas the structures are relatively insensitive to the exact form of the potentials used, the dynamics are much more sensitive, as indeed is well known for dynamical calculations on crystals. In particular, the use of polarizable potentials is necessary in order to avoid the divergence of the dielectric constant that may occur with rigid-ion potentials, and this then affects the shape and dispersion of C^{L}_{qq}. There has as yet been no attempt to improve agreement with experiment by modification of potentials; however, this should probably await more precise experiments, particularly at low Q.

Adams *et al.*[96] have analyzed their simulation results for NaCl using the Mori memory function[98] formalism. [It is interesting to note that in this approach a collective mode has no significance if it is not resolved as a side peak in $S(Q, \omega)$.[99]] They find that the relaxation of the system has a short-time Gaussian and a long-time exponential behavior. A relaxation with these two characteristic behaviors has been used to analyze Raman scattering data and is discussed in Section IV.

For the alkaline earth halides the only proper dynamical simulation has been the work of de Leeuw[54,56] on molten $SrCl_2$. While simulations of the crystal, using simple rigid-ion potentials,[100] can produce superionic

[93] J. P. Hansen and I. R. McDonald, *Phys. Rev. A* **11,** 2111 (1975).

[94] G. Cicotti, G. Jacucci, and I. R. McDonald, *Phys. Rev. A* **13,** 426 (1976).

[95] G. Jacucci, I. R. McDonald, and A. Rahman, *Phys. Rev. A* **13,** 1581 (1976).

[96] E. M. Adams, I. R. McDonald, and K. Singer, *Proc. Soc. London, Ser. A* **357,** 37 (1977).

[97] M. Dixon, private communication (1983).

[98] H. Mori, *Prog. Theor. Phys.* **33,** 423 (1965).

[99] J. R. D. Copley and S. W. Lovesey, *Rep. Prog. Phys.* **38,** 461 (1975).

[100] M. Dixon and M. J. Gillan, *in* "Fast Ion Transport in Solids" (Vashista, Munday, and Shenoy, eds.). Elsevier/North-Holland, Amsterdam, 1979.

behavior, there is no evidence in either crystalline or melt simulations of the "additional modes" observed experimentally. Instead the dispersion of mass and charge currents is very similar to that in alkali halides. The absence of the A modes may be due to the limited size of the simulations; the box length of L limits any collective modes to $Q > 2\pi/L \cong 0.25$ Å^{-1}. Any interaction of diffusion and long-wavelength acoustic modes would not, therefore, be simulated. Also the vacancy/interstitial/relaxed ion clusters proposed by Catlow and Hayes[91] to account for the fast ion conduction would be too large to fit into the simulation; cooperative effects between clusters would then be on far too long a length scale. However, the simulation results do show that such effects are not *necessary* to give fast ion conduction.

Although these simulations show some quantitative differences with the experimental results, they are very valuable in interpretation. Whereas only a limited separation of the various contributions to the scattering can be made experimentally using isotopic substitution, they can all be calculated individually from simulations; indeed it was from these calculations that it was first realized that isotopic substitution was feasible. The simulations also suggested that it may be possible to observe the separation of mass and charge dispersion, and to resolve the collective modes as side peaks in $S(Q, \omega)$, by experiments at low Q, though this has not yet been achieved.

As with the simulations, there have been relatively few calculations of the theory of molten salt dynamics, largely because there are as yet no complete theories of the dynamics of simpler monatomic liquids. Most of the work that has been done[101–105] has concentrated on calculation of the moments $\langle \omega^n S_{AB} \rangle$, with A and B either mass or charge. For $n > 2$ these calculations require a knowledge of the interionic potential and radial distribution functions; they normally use the (self-consistent) results of structural simulations. The Q dependence of the fourth moment $\langle \omega^4 S_{AB} \rangle$, equal to the second moment of the longitudinal current $\langle \omega^2 C^{\mathrm{L}}_{AB} \rangle$, may be considered as a first approximation to the dispersion of the relevant mode, in the same way as the position of the peak in $C^{\mathrm{L}}_{AB}(Q, \omega)$. The approximation becomes exact, as usual, in the limit of $Q \to 0$. The second moment of the transverse current, $\langle \omega^2 C^{\mathrm{T}}_{AB} \rangle$, may also be calculated to represent transverse modes. The dispersion obtained for longitudinal modes does not agree well with the experimental results at intermediate Q because the approximation effectively ignores the effect of all

[101] M. C. Abramo, M. Parrinello, and M. P. Tosi, *J. Nonmet.* **2,** 67 (1974).
[102] P. K. Kahol, D. K. Chaturvedi, and K. N. Pathak, *J. Phys. C* **11,** 4135 (1978).
[103] G. S. Dubey, P. K. Kahol, and D. K. Chaturvedi, *J. Phys. C* **12,** L103 (1979).
[104] M. Feinstein, J. W. Halley, and P. Schofield, *J. Phys. C* **12,** 4185 (1979).
[105] M. C. Abramo, M. Parrinello, and M. P. Tosi, *J. Phys. C* **7,** 4201 (1974).

collisions and damping processes, which are obviously extremely important in such liquids.

As noted in Subsection 4, the dynamics of a two-component liquid should be considered in terms of two properties, A and B, which are, if possible, independent; i.e., $\langle \omega^n S_{AB} \rangle = 0$ for all n. These are mass and charge in the $Q \to 0$ limit and $+$ and $-$ in the $Q \to \infty$ limit. At intermediate Q there cannot be complete independence of A and B (two variables cannot represent the $3N$ coordinates of all the ions), but a good first approximation may be obtained by diagonalizing the matrix of moments $n = 0, 2, 4$. Since the second moment is naturally diagonal, this reduces to $n = 0$ (the structure factor) and $n = 4$. This has been done for NaCl by Abramo *et al.*,[105] who also have used the Mori memory function approach to calculate appropriate partial dynamical structure factors. The approximation of independence is found to be good, and the quasidispersion obtained for C^{L}_{AA} and C^{L}_{BB} shows two approximately straight and parallel branches for $Q > 1.4$ Å^{-1}, though with some small oscillations. At high Q the separation of the two branches reflects the mass difference of the ions. As yet, experimental results are too inaccurate and acquired over too narrow a range of Q and ω to consider using them for a determination of appropriate dynamical variables; however, it would be of interest to apply this approach directly to the results of a dynamical simulation.

7. Discussion

Inelastic neutron scattering measurements of the dynamical structure factors of molten alkali halides, either with or without isotopic substitution, have so far yielded little quantitative information on single-particle motion. A considerable increase in statistical accuracy will be required before the coherent and incoherent contributions can be separated. It has been found, as expected, that simple diffusion is the dominant mechanism at low Q (though this must change in the limit $Q \to 0$ to thermal diffusivity) but that it is obviously affected by structural correlations. The evidence found for collective modes suggests that the observation of maxima and minima in the width of $S_{\mathrm{T}}(Q, \omega)$ below Q_{P} are not due to de Gennes narrowing in the simple sense. They may reflect a change in the line shape of the central diffusive component, or alternatively may be due to dispersion of a noncentral collective component. While both these explanations are consistent for the results below Q_{P}, it would appear that the rise in the width of $S_{\mathrm{T}}(Q, \omega)$ for CsCl observed at $Q > Q_{\mathrm{P}}$ to values well above the diffusion prediction indicates a change in the mechanism that determines the linewidth from diffusion to Brownian (gaslike) motion.

This may be interpreted approximately by saying that, for alkali halides, the ions undergo primarily gaslike (uncorrelated) or collective oscillations within a cage of nearest neighbors, and then diffuse by hopping to a vacant next-neighbor site. Diffusive hops must be predominantly longer than the nearest-neighbor distance. The ion cannot hop to a nearest-neighbor site since this would be energetically unfavorable; such sites are normally occupied by ions of the opposite charge. In the alkaline earth halides the situation is more complex because of the differing numbers of cations and anions and corresponding differences in $g_{++}(r)$ and $g_{--}(r)$. Many-body interactions in these liquids will also significantly distort the diffusive behavior. Cooperative diffusion may occur, as is probably the case in $SrCl_2$.

The use of isotopic substitution when applied to measurements of the current correlations in molten alkali halides has proved invaluable. It has been shown that mass and charge are appropriate (if not exact) variables for a description of dynamics in the range $Q < 4$ Å^{-1} and that the analogy with acoustic and optic modes in the crystal holds. Although no modes that are resolvable as side peaks in $S(Q, \omega)$ have yet been observed, the quasidispersion of C^{L}_{mm} and C^{L}_{qq} shows sufficient similarity to the crystalline dispersion that it may be concluded that collective motions do exist in some sense, even though they are highly overdamped and do not propagate.

Isotopic substitution has proved its more general usefulness in the work on $SrCl_2$, where first it has been shown that second-order substitution is feasible and may yield qualitatively useful information, and second that the method may be used to derive this information rather than simply to show consistency with a model. The technique should now be applied to, for instance, $MgCl_2$, in order to enhance the observation of mode softening and determine the nature (i.e., acoustic or not) of the mode. The optic-mode dispersion may be simultaneously derived. At present it is only feasible to apply isotopic substitution within the intermediate Q range (1–4 Å^{-1}) that has been predominantly discussed. However, measurements at lower and higher Q are important for our understanding of dynamics and its relation to structure; improvements in spectrometers and the use of pulsed neutron sources should begin to provide this information soon.

The direct measurement of a soft mode in molten $MgCl_2$ is definite evidence that there are noncentral contributions to $S_T(Q, \omega)$ even when the are not resolved as a side peak. It seems likely that the mode does not soften completely and have a dispersion gap, but that there is a low-energy minimum. However, this neither confirms nor disproves the prediction of a dispersion gap for Ar.[71] Since the dispersion observed[92] for $ZnCl_2$, where the mode is resolved almost, is similar to that of $MgCl_2$, it suggests that the resolution of the side peak may not be of fundamental importance. The criteria for resolution, and also for the direct observation of a soft mode, have not yet been determined. A possible relation with features of the Raman scattering spectrum is discussed in the next section.

IV. Dynamics: Light Scattering

8. Experimental Techniques

a. Scattering Cross Sections

The differential cross section for light scattering from a fluid may be written[106] as the power spectrum of the polarizability density autocorrelation function

$$\frac{d^2\sigma}{d\Omega\, d\omega} = \frac{k_S^4}{2\pi}\int dt\, e^{-i\omega t}\int d\mathbf{r}\, e^{i\mathbf{Q}\cdot\mathbf{r}}\langle[\boldsymbol{\varepsilon}_I\cdot\alpha(\mathbf{R}\cdot\mathbf{t}')\cdot\boldsymbol{\varepsilon}_S]$$
$$\times\,[\boldsymbol{\varepsilon}_I\cdot\alpha(\mathbf{R}+\mathbf{r}, t+t')\cdot\boldsymbol{\varepsilon}_S]\rangle \tag{8.1}$$

where $\mathbf{k}_S$ is the wave vector of the scattered light, $\boldsymbol{\varepsilon}_I$ and $\boldsymbol{\varepsilon}_S$ are the polarization vectors of incident and scattered electric fields, and $\alpha\,(\mathbf{r}, t)$ is the polarizability density tensor. In discussions of light scattering from crystalline solids[107] the components of α may be expanded in a Taylor series about the equilibrium state (indicated by 0) using a set of normal coordinates $\mathbf{Q}_k$.

$$\alpha_{ij} = (\alpha_{ij})_0 + \sum_k\left(\frac{d\alpha_{ij}}{dQ_{k_0}}\right)Q_k + \sum_{k,l}\left(\frac{d^2\alpha_{ij}}{dQ_k\, dQ_l}\right)_0 Q_kQ_l + \cdots \tag{8.2}$$

The correlations of terms involving $\langle(\alpha_{ij})(\alpha_{ij})_0\rangle$ give rise to Rayleigh–Brillouin (RB) scattering (long-wavelength density fluctuations); those terms $\langle(\alpha'_{ij})_0(\alpha'_{ij})_0\rangle$ (the prime indicating differentiation with respect to a normal coordinate) give first-order Raman scattering (one phonon), and the terms $\langle(\alpha''_{ij})_0(\alpha''_{ij})_0\rangle$ give second-order Raman scattering (two phonon), etc.

In a disordered system there is no long-range order, and hence the normal coordinates become the displacements around equilibrium atomic positions. Shuker and Gammon[108] have shown that in this case the Raman scattering terms contain contributions from all momentum transfers Q. In liquids there are no "equilibrium" positions of the atoms due to diffusion, and it is more

[106] S. Bratos and G. Tarjus, *Phys. Mod. Mater.* **2,** 571 (1980).
[107] W. Hayes and R. Loudon, "Scattering of Light by Crystals." Wiley, New York, 1978.
[108] R. Shuker and R. W. Gammon, *Phys. Rev. Lett.* **25,** 222 (1970).

appropriate to consider the normal coordinates to be density fluctuations[109] (e.g., mass or charge for ionic melts) of all wavelengths. The Raman scattering spectrum is therefore related, in a complex manner, to the partial dynamical structure factors $S_{\alpha\beta}(Q, \omega)$ at all Q. The different mechanisms by which the polarizability may be modulated, and hence give rise to Raman scattering, are discussed in the next section.

The RB spectrum, on the other hand, is simply proportional to the sum of $S_{\alpha\beta}(Q, \omega)$, appropriately weighted with the mean ionic polarizabilities $\alpha_{\pm}$, in the low-Q limit. (For 90° scattering with light of wavelength ~5000 Å, $Q \cong 2.5 \times 10^{-3}$ Å^{-1}.) The differential cross section may be written in terms of the mass- and charge-density fluctuations as

$$\frac{d^2\sigma_{\mathrm{RB}}}{d\Omega\, d\omega} \propto \alpha_m^2 S_{mm}(Q, \omega) + \alpha_q^2 S_{qq}(Q, \omega) + 2\alpha_m \alpha_q S_{mq}(Q, \omega) \tag{8.3}$$

with α_m and α_q defined in the same manner as a_m and a_q in Eqs. (4.21 and 4.22). The integrated cross section is then related to the low-Q limits of the partial structure factors. Since $S_{mm}(Q \to 0) \propto \rho k_{\mathrm{B}} T K_{\mathrm{T}}$ [Eq. (5.2)], while $S_{qq}(Q \to 0)$ and $S_{mq}(Q \to 0)$ decrease as Q^2 [Eq. (5.7)], the contribution from charge-density fluctuations is a factor of 10^5 smaller than that from mass-density fluctuations. The spectrum will, therefore, be similar in form to the RB spectra of monatomic fluids.

b. *Raman Scattering*

The technique of Raman scattering is familiar to most chemists and physicists. Although the scattering mechanism is the same (fluctuations in atomic polarizability due to motions of other atoms), the applications may be divided broadly into two categories: The first is the investigation of collective excitations in crystalline materials, e.g., phonons, and the second is the vibrational spectroscopy of molecules, which may be in solids, liquids, or gases. In the former the energy transfers are usually smaller than $\bar{\nu} = 500$ cm^{-1} ($2\pi c\bar{\nu} = \hbar\omega$; 1 meV = 8 cm^{-1}), while in the latter they are larger. The spectra discussed here will fall into the first category, there being no significant scattering observable from molten alkali or alkaline earth halides at $\bar{\nu} > 500$ cm^{-1}.

The major difficulty in measuring the Raman spectrum of molten salts at low $\bar{\nu}$ is the large amount of stray light that may be scattered from many sources, including sample containment, liquid surface, fine particles in the liquid, etc. For this reason many molecular spectroscopists do not even extend

[109] J. Giergl, K. R. Subbaswamy, and P. C. Eklund, *Phys. Rev. B: Condens. Matter* [3] **29**, 3490 (1984).

their measurements below 500 cm^{-1}, although useful information on the liquid may be obtained at lower $\bar{\nu}$.

A typical experimental setup for Raman scattering[110] from molten salts is shown in Fig. 28. Plane-polarized light from an Ar ion laser (multimode), usually $\lambda = 4880$ Å, is scattered through 90°. By a combination of a half-wave plate in the incident beam and an analyzing polaroid in the scattered beam, spectra may be measured with the electric vectors of the incident and scattered waves either parallel (polarized: I_{xx} or I_{VV}) or perpendicular (depolarized: I_{yx} or I_{VH}). Using such an arrangement with a high-resolution double monochromator, Giergl *et al.*[109] have been able to extend measurements down to $\bar{\nu} \cong 10$ cm^{-1}, while by concentration on details of furnace design[111] and sample containment, Raptis and co-workers[110,112] have reached 4 cm^{-1}. Most alkali and alkaline earth halides, if sufficiently pure, may be contained in optically polished silica cells without problems of rapid corrosion. For fluorides, however, nickel or platinum containers[109,113] with small slits for incident and scattered beams must be used, the melt being held in under its own surface tension.

c. *Brillouin Scattering*

Since RB scattering is typically confined to $\bar{\nu} < 1$ cm^{-1}, different experimental methods are used than in Raman scattering. A typical setup[114] is shown in Fig. 29. The natural width of the laser line is reduced by using a single-mode configuration. A Fabry–Perot etalon provides the high-$\bar{\nu}$ resolution required. Stray light is again a major problem, and spatial filtering is used to reduce it. Polarized and depolarized spectra may be measured; sample containment is as for Raman scattering experiments.

9. Experimental Results

a. *Raman Scattering: Alkali Halides*

Polarized and depolarized Raman scattering spectra of molten alkali chlorides[109,110,115,116] and iodides[109,110,112,116,117] are shown in Figs. 30 and 31 (spectra for the bromides[110,116] are similar; comparable fluoride

[110] E. W. J. Mitchell and C. Raptis, *J. Phys. C* **16,** 2973 (1983).
[111] C. Raptis, *J. Phys. E* **16,** 749 (1983).
[112] C. Raptis, R. A. J. Bunten, and E. W. J. Mitchell, *J. Phys. C* **16,** 5351 (1983).
[113] A. S. Quist, *Appl. Spectrosc.* **25,** 80 (1971).
[114] W. Yao, H. Z. Cummins, and R. H. Bruce, *Phys. Rev. B: Condens. Matter* [3] **24,** 424 (1981).
[115] J. H. R. Clarke and L. V. Woodcock, *J. Chem. Phys.* **57,** 1006 (1972).
[116] C. Raptis and E. W. J. Mitchell, *J. Phys. C* (1987) (in press).
[117] J. Giergl, P. C. Eklund, and K. R. Subbaswamy, *Solid State Commun.* **40,** 139 (1981).

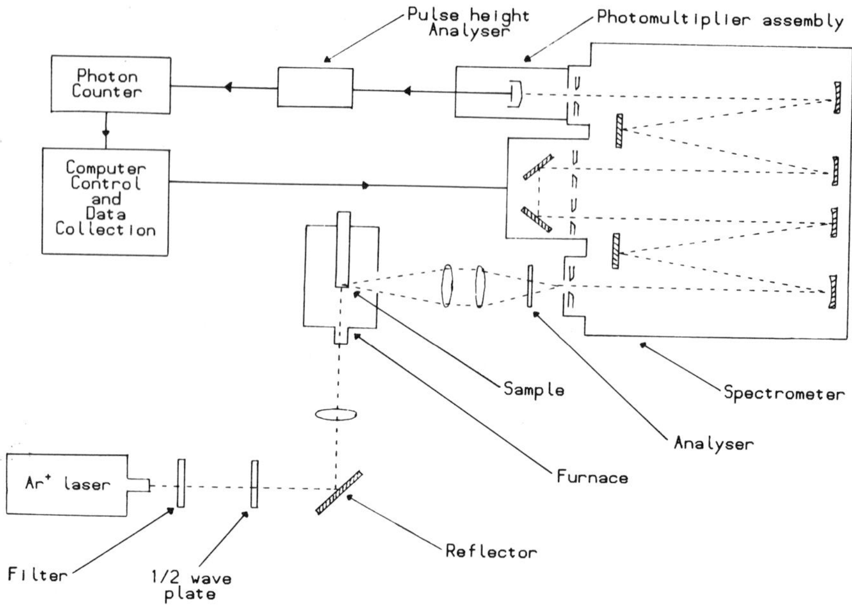

FIG. 28. Schematic representation of a typical experimental arrangement for Raman scattering from molten salts [E. W. J. Mitchell and C. Raptis, *J. Phys. C* **16**, 2973 (1983)].

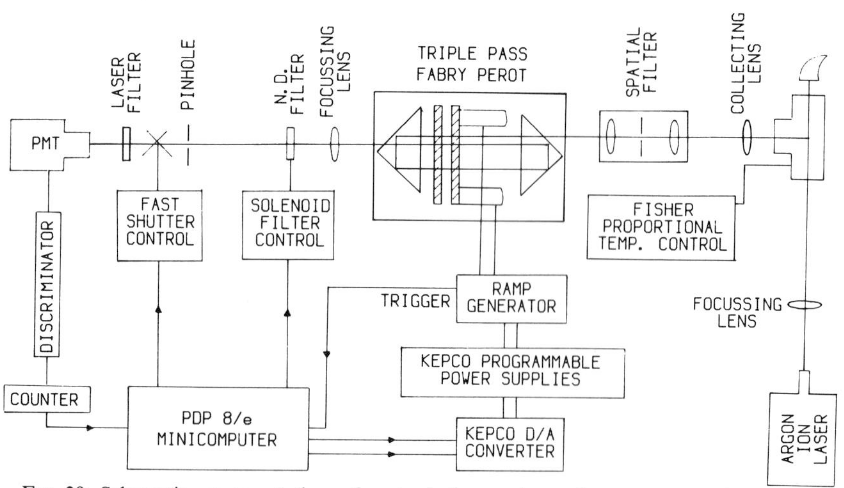

FIG. 29. Schematic representation of a typical experimental arrangement for Brillouin scattering from molten salts [W. Yao, H. Z. Cummins, and R. H. Bruce, *Phys. Rev. B: Condens. Matter* **24**, 424 (1981)].

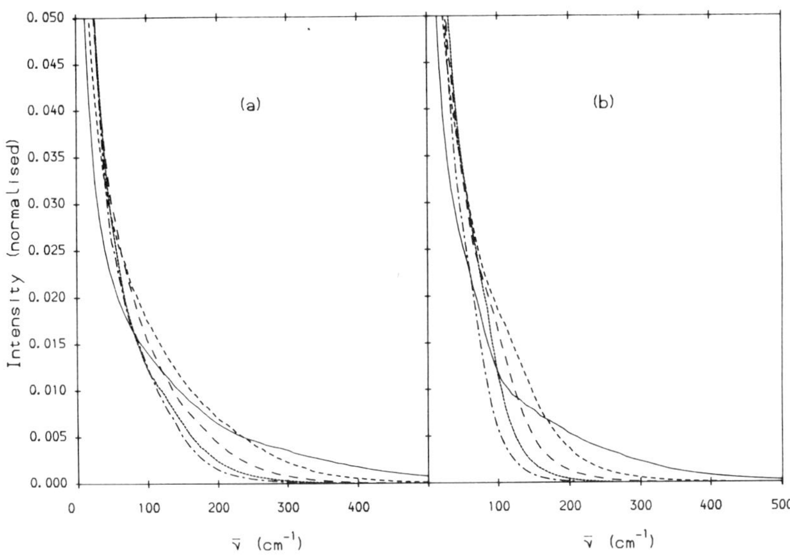

FIG. 30. Polarized Raman scattering intensities $I_{xx}(\bar{\nu})$ for molten alkali (a) chlorides [J. Giergl, K. R. Subbaswamy, and P. C. Eklund, *Phys. Rev. B: Condens. Matter* [3] **29**, 3490 (1984); E. W. J. Mitchell and C. Raptis, *J. Phys. C* **16**, 2973 (1983); J. H. R. Clarke and L. V. Woodcock, *J. Chem. Phys.* **57**, 1006 (1972); C. Raptis and E. W. J. Mitchell, *J. Phys. C* (1987) (in press)] and (b) iodides [Giergl *et al.*, *ibid.*; Mitchell and Raptis, *ibid.*; Raptis and Mitchell, *ibid.*; C. Raptis, R. A. J. Bunten, and E. W. J. Mitchell, *J. Phys. C* **16**, 5351 (1983); J. Giergl, P. C. Eklund, and K. R. Subbaswamy, *Solid State Commun.* **40**, 139 (1981)]. Cations are Li^+ (———), Na^+ (---), K^+ (– – –), Rb^+ (···), and Cs^+ (—·—). The intensities are normalized using the integrated intensity.

spectra have not been measured). The general features are (1) a broad central peak of half-width $\sim 10\ cm^{-1}$, (2) a long exponential tail extending up to $\bar{\nu} \cong 500\ cm^{-1}$, and (3) a weak shoulder or shoulders at intermediate $\bar{\nu}$ (in some cases, e.g., CsCl, no shoulder is visible). The total extent of the spectra scales with the difference in ionic masses, as does the density of states for comparable ionic crystals.

The scattered intensity from the liquid is an order of magnitude larger than that from the crystalline state just below the melting point.[109,110] Though difficult to measure, Raptis and co-workers[110,112,116,118] have reported scattering cross sections for some of these melts. For CsCl the integrated polarized cross section is $(1/N)\, d\sigma/d\Omega \cong 2.5 \times 10^{-29}\ cm^2\ sr^{-1}$. This is of the same order of magnitude as that expected from the contribution of S_{mm} to RB

[118] C. Raptis, *Appl. Phys.* **59,** 1644 (1986).

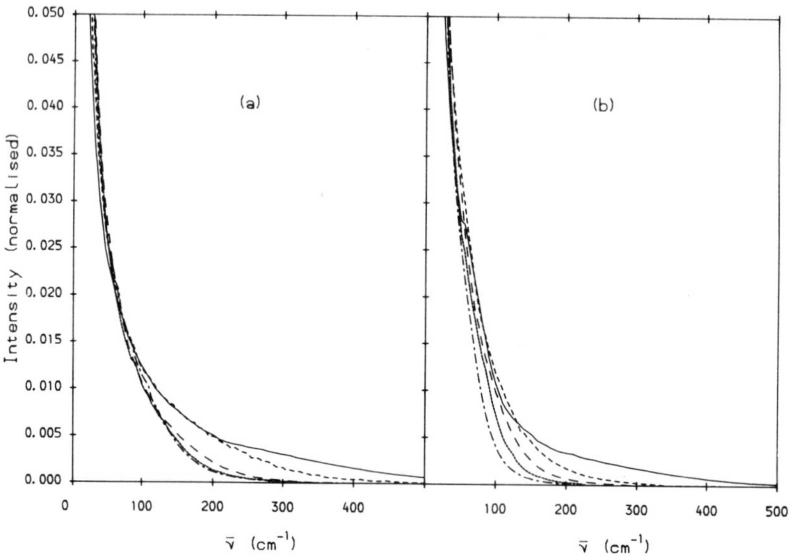

FIG. 31. Depolarized Raman scattering intensities $I_{yx}(\bar{\nu})$ for molten alkali (a) chlorides [J. Giergl, K. R. Subbaswamy, and P. C. Eklund, *Phys. Rev. B* **29**, 3490 (1984); E. W. J. Mitchell and C. Raptis, *J. Phys. C* **16**, 2973 (1983); J. H. R. Clarke and L. V. Woodcock, *J. Chem. Phys.* **57**, 1006 (1972); C. Raptis and E. W. J. Mitchell, *J. Phys. C* (1987) (in press)] and (b) iodides [Giergl *et al.*, *ibid.*; Mitchell and Raptis, *ibid.*; Raptis and Mitchell, *ibid.*; C. Raptis, R. A. J. Bunten, and E. W. J. Mitchell, *J. Phys. C* **16**, 5351 (1983); J. Giergl, P. C. Eklund, and K. R. Subbaswamy, *Solid State Commun.* **40**, 139 (1981)].

scattering; i.e., it is $\sim 10^5$ times higher than that expected from the S_{qq} contribution, which is not, therefore, the origin of the scattering.

The intensity tends to increase as the ion sizes and polarizabilities increase from LiCl to CsI, as might be expected. Giergl *et al.*[109] report that the scattering from LiBr decreases with increasing temperature, although Raptis[119] finds it to increase; this discrepancy may be due to fogging of the sample cell walls at the higher temperatures. Fairbanks[120] has made careful measurements on CsCl (the least reactive and most stable melt), which show that the scattered intensity increases approximately with the square of the ionic number density. However, it is found that the line shape does not change (within experimental accuracy) over the few hundred degrees above the melting point for which it has been measured.

While polarized and depolarized spectra have similar shapes, they are not simply proportional. The depolarization ratio, $\rho_d(\bar{\nu}) = I_{yx}(\bar{\nu})/I_{xx}(\bar{\nu})$, for

[119] C. Raptis, private communication (1986).

[120] M. Fairbanks, private communication (1986).

molten alkali chlorides and iodides is shown in Fig. 32. For a macroscopically isotropic scatterer such as a liquid, the maximum possible value[107] for ρ_d is 0.75. It is found that ρ_d decreases as $\bar{\nu}$ increases, reaches a minimum value, and then either remains approximately constant or increases. The most striking trend is the large variation in the average ρ_d from a high value for Cs^+ salts to a low value for Na^+ salts.

From inelastic neutron scattering experiments we know that the appropriate density and current correlations in these molten salts show contributions from (1) low-frequency diffusive motions with characteristic time scales ~ 1 ps, (2) high-frequency "Brownian" (gaslike) motions ~ 0.1 ps, and (3) "acoustic" and "optic" modes with intermediate time scales. These time scales are obviously reflected in (1) the central peak at low $\bar{\nu}$, (2) the high-$\bar{\nu}$ tail, and (3) the intermediate-$\bar{\nu}$ shoulders. The spectra are similar to those from neutron scattering in that the collective terms are overdamped and no side peaks are observed. However, the quantitative relation of the light scattering spectrum to the dynamical structure factors is much more complex, since there are a

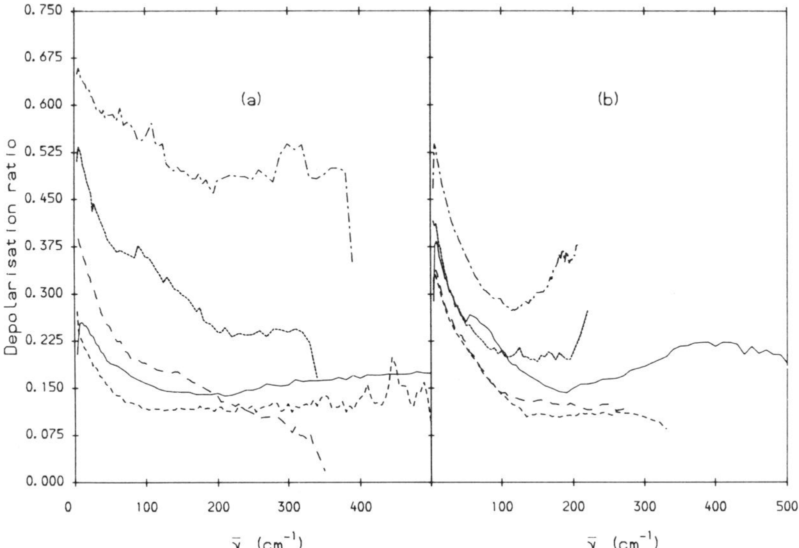

FIG. 32. Depolarization ratio $\rho_d(\bar{\nu}) = I_{yx}(\bar{\nu})/I_{xx}(\bar{\nu})$ for molten alkali (a) chlorides [J. Giergl, K. R. Subbaswamy, and P. C. Eklund, *Phys. Rev.* [*Part*] *B* **29**, 3490 (1984); E. W. J. Mitchell and C. Raptis, *J. Phys. C* **16**, 2973 (1983); J. H. R. Clarke and L. V. Woodcock, *J. Chem. Phys.* **57**, 1006 (1972); C. Raptis and E. W. J. Mitchell, *J. Phys. C* (1986) (submitted for publication)] and (b) iodides [Giergl *et al.*, *ibid.*; Mitchell and Raptis, *ibid.*, Raptis and Mitchell, *ibid.*; C. Raptis, R. A. J. Bunten, and E. W. J. Mitchell, *J. Phys. C* **16**, 5351 (1983); J. Giergl, P. C. Eklund, and K. R. Subbaswamy, *Solid State Commun.* **40**, 139 (1981)].

large number of possible scattering mechanisms, and fluctuations at all Q may contribute.

The interpretation of the Raman spectra of glasses by Shuker and Gammon[108] is that they may be used to produce an effective density of states

$$g(\bar{\nu}) = \bar{\nu} I(\bar{\nu})/[n(\bar{\nu}) + 1] \tag{9.1}$$

where $n(\bar{\nu})$ is the Bose–Einstein population factor. $g(\bar{\nu})$ for the molten alkali halides[109,110] normally consists of a single broad peak, similar in shape to $C^{L}(Q, \omega)$ and peaking in the region $100 < \bar{\nu} < 200\ \mathrm{cm}^{-1}$. In the case of LiI the peak is split. Although the use of $g(\bar{\nu})$ is not entirely appropriate for liquids, since the lack of equilibrium atomic positions due to diffusion introduces additional low- and high-frequency scattering, it is found that the peak positions coincide quite closely with the Debye temperatures of the corresponding crystals.[110] This would suggest that the density-of-states argument is not entirely inappropriate either, at least at intermediate $\bar{\nu}$.

McTague *et al.*[121] have measured the Raman spectra of gaseous Ar and Kr. At low densities they may be described by a single exponential, $I(\bar{\nu}) = I_0 \exp(-\bar{\nu}/\Delta)$. The spectra are almost completely polarized; i.e., $\rho_d \cong 0.75$. At higher densities and in the liquid two exponential regions are observed.[121] Fleury *et al.*[122] have found (empirically) that the high-$\bar{\nu}$ exponents show a universal dependence on density and temperature

$$\Delta = 3(k_B T/m\sigma^2)^{1/2}[1 + (2\sigma^3\rho^2)^2] \tag{9.2}$$

where σ is the appropriate atomic radius (i.e., Lennard–Jones length parameter). The first term in this expression has been explained by Mahan[123] on the assumption that isolated binary collisions (IBC) are the dominant process in the (hard-sphere) fluid, with scattering via the dipole–induced-dipole (DID) mechanism. The density dependence then reflects deviations from hard-sphere potentials. McTague *et al.*[121] also assume the DID mechanism but use the theory of Stephen,[124] which considers scattering to be due to pairs of excitations (second-order Raman). Taking only pair interactions to be important, and factoring the resultant four-particle correlation function into two-particle correlations, leads to terms in the cross section of the form $S(-Q, \omega' - \omega)\, S(Q, \omega')$. This gives better agreement for the high-$\bar{\nu}$ behavior than the IBC model. Deviations at low $\bar{\nu}$ are interpreted as indicating the increased importance of three- and four-body correlations. Gelbart[125] has

[121] J. P. McTague, P. A. Fleury, and D. B. Dupre, *Phys. Rev.* **188,** 303 (1969).
[122] P. A. Fleury, J. M. Worlock, and H. L. Carter, *Phys. Rev. Lett.* **30,** 591 (1973).
[123] G. D. Mahan, *Phys. Lett. A* **44,** 287 (1973).
[124] M. J. Stephen, *Phys. Rev.* **187,** 279 (1969).
[125] W. Gelbart, *Adv. Chem. Phys.* **26,** 1 (1974).

shown that the IBC and excitation pair approaches are equivalent in the low-density limit, since here $S(Q, \omega)$ is dominated by binary collisions.

Giergl *et al.*[109] have applied the excitation pair approach to KCl, again assuming DID interactions, using $S(Q, \omega)$ from the simple molten salt simulation of Hansen and McDonald.[93] This also produces good agreement at high $\bar{\nu}$ but deviations at low $\bar{\nu}$. Clarke and Woodcock[115] have pointed out that the DID mechanism produces complete depolarization, while in KCl ρ_d is considerably smaller than 0.75. They suggest that in an ionic liquid the predominant scattering mechanism will be local charge-density fluctuations, termed charge-induced anisotropy (CIA). Since this interaction is longer ranged than DID, many-particle correlations will contribute. Calculations on this basis again produce good agreement in the high-$\bar{\nu}$ exponential tail, but deviate at low $\bar{\nu}$. While this mechanism may qualitatively explain the low depolarization ratios observed, they have not been predicted quantitatively.

Giergl *et al.*[109] have also examined the hard-sphere IBC, DID approach of Mahan,[123] $\Delta \propto (k_B T/m\sigma^2)^{1/2}$, in relation to the exponents of the high-$\bar{\nu}$ tails in the molten salt spectra. By using the reduced mass for m and the Debye screening length for σ they obtain $\Delta \propto \omega_p$, the plasma frequency. They find that this agrees well with experiment with a proportionality constant of 0.35. Raptis and Mitchell[116] have used the mean ionic radius of the ions for σ and obtain a proportionality constant of 0.375.

At high densities overlap of the electron distributions of the ions may occur, leading to changes in polarizability. Like DID this electron overlap (EO) interaction will be short ranged. Fairbanks *et al.*[126] have proposed an explanation for the variation in the average depolarization ratio, dominated by the low-$\bar{\nu}$ part of the scattering, based on an EO scattering mechanism. Defining the (symmetrized) time Fourier transform of the scattered intensity[35] as

$$I(t) = \int_0^\infty \frac{I(\omega)}{k_S^4} e^{-h\omega/2k_B T} \cos \omega t \, dt \tag{9.3}$$

then the time-dependent depolarization ratio is

$$\rho_d(t) = I_{yx}(t)/I_{xx}(t) \tag{9.4}$$

[not equal to the transform of $\rho_d(\bar{\nu})$]. At $t = 0$ we have

$$\rho_0 = \rho_d(t = 0) = I_{yx}(t = 0)/I_{xx}(t = 0) \tag{9.5}$$

which is approximately equal to the ratio of the integrated depolarized and polarized scattered intensities. (Spectra are only measured for $\bar{\nu} > 4$ cm^{-1}.

[126] M. Fairbanks, R. L. McGreevy, and E. W. J. Mitchell, *J. Phys. C* **19**, L53 (1986).

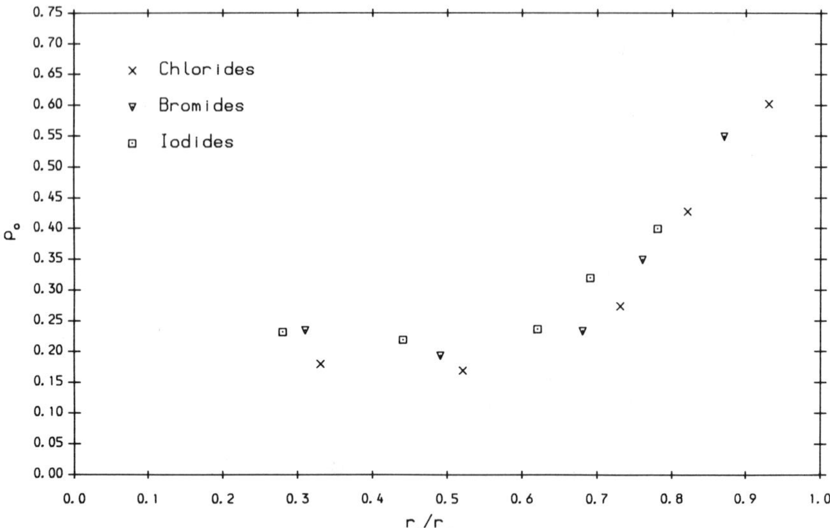

FIG. 33. Depolarization ratio $\rho_0 = I_{yx}(t = 0)/I_{xx}(t = 0)$ for molten alkali halides [M. Fairbanks, R. L. McGreevy, and E. W. J. Mitchell, *J. Phys. C* **19**, L53 (1986)].

Extrapolation to $\bar{v} = 0$ has been discussed by Bunten *et al.*[35,127] The extension of the low-$\bar{v}$ Raman scattering to $\bar{v} = 0$ has been confirmed by the Brillouin scattering results of Qiu *et al.*,[128] discussed in subsection 11.) Values of ρ_0 for the alkali halides are plotted as a function of the ratio of ionic radii, r_+/r_- (polarizabilities scale with r^3) in Fig. 33. There are obvious trends within the group; $\rho_0 \rightarrow 0.75$ as $r_+/r_- \rightarrow 1$ (equivalent to a monatomic liquid) and has a minimum value of ~ 0.15 at $r_+/r_- \cong 0.5$.

This trend is explained on the basis that EO is only significant for pairs of unlike ions. (It is found from the radial distribution functions that on average only one ion is close enough to the repulsive core of another ion to cause significant distortion.) If the ions differ only in their charge (i.e., have the same ionic radius, polarizability, etc.), then the distortion will produce equal and opposite dipole moments on the two ions. The net dipole moment is then zero; so the trace of the polarizability tensor α is zero and $\rho_0 = 0.75$. If the ions differ in radius/polarizability, then the dipole moments will not cancel, the trace of α is nonzero, and ρ_0 is less than 0.75. In the limit of ions of significantly different sizes, the dipolar term will dominate and $\rho_0 \rightarrow 0$. However, in this limit, approached for Li^+ salts, there will be significant overlap of the larger (like)

[127] R. A. J. Bunten, R. L. McGreevy, E. W. J. Mitchell, and C. Raptis, *J. Phys. C* **19**, 2925 (1986).

[128] S. L Qiu, R. A. J. Bunten, M. Dutta, E. W. J. Mitchell, and H. Z. Cummins, *Phys. Rev. B: Condens. Matter* [3] **31**, 2456 (1985).

ions. Overlap of like ions gives $\rho_0 = 0.75$; so the average value of ρ_0 will begin to rise again at small radius ratios, as is indeed observed for Li^+ salts. (This EO argument would also hold for the depolarization ratio of inert gas mixtures.)

The EO explanation works well for $\rho_d(t = 0)$, which is dominated by low-$\bar{\nu}$ scattering. DID has been used to explain high-$\bar{\nu}$ scattering in monatomic liquids and both DID and CIA in ionic melts. A combination of all three mechanisms is probably necessary for a general description of the Raman spectra of molten alkali halides. None of the calculations has yet explained the presence of shoulders at intermediate $\bar{\nu}$. It is observed experimentally that these are most predominant for systems with a large ionic mass ratio, e.g., LiI. This suggests a connection with zone-edge excitations in the crystal. Even if highly overdamped, they would involve many-body correlations which have not yet been properly treated in relation to any of the mechanisms. It is interesting to speculate on whether the observation of noncentral shoulders is related to the symmetry of the modes, via some short-range equivalent of the group-theory selection rules for Raman scattering from crystals. Symmetries would differ between tetrahedrally coordinated LiI and octahedrally coordinated CsCl. Calculations of angular correlations, discussed in Section II, may be of use in this respect. However, these are small contributions to the scattering for alkali halides. In the alkaline earth halides they are considerably more important, as will be discussed in the next section.

b. Raman Scattering: Alkaline Earth Halides

The polarized and depolarized Raman spectra of molten alkaline earth chlorides[35,129] and bromides[127] are shown in Figs. 34 and 35. While low- and high-$\bar{\nu}$ behaviors are similar to those for alkali halides, the shoulders at intermediate $\bar{\nu}$ are significantly stronger, in the case of $MgCl_2$ and $MgBr_2$ being clearly resolved as a side peak. Collective (many-body) effects are obviously more important in these melts, as has already been suggested by both neutron diffraction data in Section II and inelastic neutron scattering in Section III.

The side peaks are seen to be completely polarized, indicating a definite symmetry to the collective motion. The average depolarization ratio shows some of the same trends as for the alkali halides,[126] being higher for salts with similar ionic radius, e.g., $BaCl_2$, and lower for those with different radii, e.g., $MgBr_2$. In general the values are higher than for alkali halides with the same radius ratio, and this is explained on the EO model[126] as being due to the large amount of anion–anion overlap in the structures. There will also be deviations due to the many-body correlations.

[129] C.-H. Huang and M. H. Brooker, *Chem. Phys. Lett.* **43,** 180 (1976).

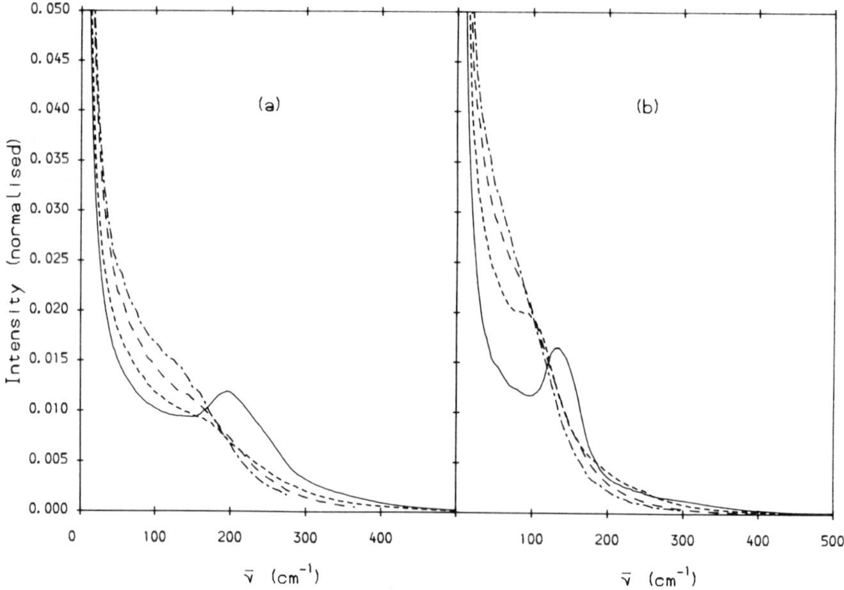

FIG. 34. Polarized Raman scattering intensities $I_{xx}(\bar{\nu})$ for molten alkaline earth (a) chlorides [R. A. J. Bunten, R. L. McGreevy, E. W. J. Mitchell, C. Raptis, and P. J. Walker, *J. Phys. C* **17**, 4705 (1984); C.-H. Huang and M. H. Brooker, *Chem. Phys. Lett.* **43**, 180 (1976)] and (b) bromides [R. A. J. Bunten, R. L. McGreevy, E. W. J. Mitchell, and C. Raptis, *J. Phys. C* **19**, 2925 (1986)]. Cations are Mg^{2+} (———), Ca^{2+} (- - -), Sr^{2+} (— — —) and Ba^{2+} (—·—).

The scattering intensities are similar to those for the alkali halides. They are found to increase with temperature, while the line shape remains approximately the same, except in the case of $CaCl_2$.[35] This is the only melt for which a significant change in line shape has been observed, and involves broadening of the central peak. It has been suggested[34] that this is due to the change in cation structure discussed in Section II. We discuss later a model in which the width of the central line is related to the electrical conductivity and dielectric constant. While the conductivity[130] shows no anomalous behavior, the refractive index measurements of Iwadate *et al.*[36] show a temperature coefficient for $CaCl_2$ that is of the opposite sign and an order of magnitude larger than for other molten salts. Both these pieces of evidence support the idea of a structural change in molten $CaCl_2$ over ~200 K above the melting point.

The Raman spectra of both glassy and molten $ZnCl_2$[131] show a well-resolved side peak with additional small shoulders. This may be deconvoluted

[130] J. G. Janz, "Molten Salts Handbook." Academic Press, New York, 1967.

[131] F. Aliotta, G. Maisano, P. Migliardo, C. Vasi, F. Wanderlingh, G. Pedro Smith, and R. Triolo, *J. Chem. Phys.* **75**, 613 (1981).

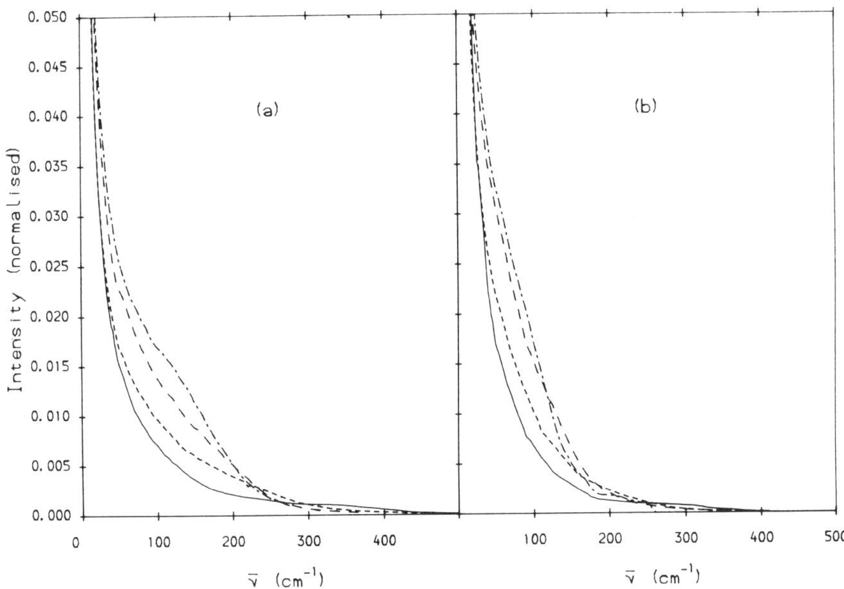

FIG. 35. Depolarized Raman scattering intensities $I_{yx}(\bar{\nu})$ for molten alkaline earth (a) chlorides [R. A. J. Bunten, R. L. McGreevy, E. W. J. Mitchell, C. Raptis, and P. J. Walker, *J. Phys. C* **17**, 4705 (1984); C.-H. Huang and M. H. Brooker, *Chem. Phys. Lett* **43**, 180 (1976)] and (b) bromides [R. A. J. Bunten, R. L. McGreevy, E. W. J. Mitchell, and C. Raptis, *J. Phys. C* **19**, 2925 (1986)].

into a set of Gaussians which are in agreement with the modes expected for a set of $ZnCl_2$ tetrahedra edge connected and randomly oriented (an approximate description of the structure; see Section II). On a similar basis (commonly used in Raman spectroscopy of molecules or complex ions), Huang and Brooker[129] have deconvoluted the spectra for $MgCl_2$ to indicate the presence of tetrahedral $MgCl_4{}^{2-}$ complexes. Neutron diffraction results (Section II) have shown that $MgCl_2$, unlike $ZnCl_2$, does not have a tetrahedral coordination. The Gaussian analysis for the $MgCl_2$ spectrum may not be entirely appropriate because of the high mobility of the ions (the mobility in $ZnCl_2$ is low); in this case vibrational bands will not necessarily be Gaussian or symmetric, and it is difficult to separate them from the low-$\bar{\nu}$ diffusion and high-$\bar{\nu}$ collision backgrounds. It is interesting to note that addition of alkali halide to the $MgCl_2$ melt produces a significant change in the spectrum.[132] The alkali-rich compositions show vibrational bands that are much better defined than in the pure $MgCl_2$ melt and very similar in shape to those of $ZnCl_2$, suggesting that the additional alkali interferes with the Mg–Cl–Mg many-body interactions postulated to explain the layer structure (Section II),

[132] M. H. Brooker and C.-H. Huang, *Can. J. Chem.* **58,** 168 (1980).

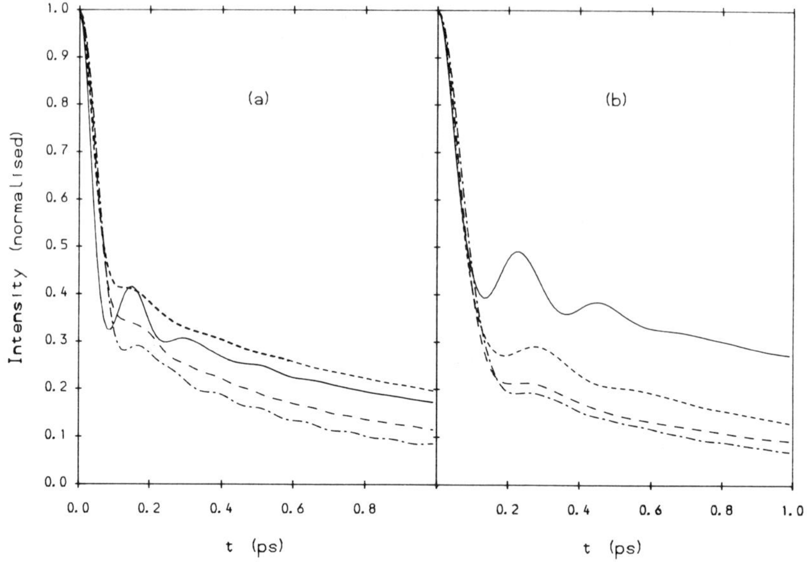

FIG. 36. Polarized Raman scattering intensities $I_{xx}(t)$ for molten alkaline earth (a) chlorides [R. A. J. Bunten, R. L. McGreevy, E. W. J. Mitchell, C. Raptis, and P. J. Walker, *J. Phys. C* **17**, 4705 (1984); C.-H. Huang and M. H. Brooker, *Chem. Phys. Lett.* **43**, 180 (1976)] and (b) bromides [R. A. J. Bunten, R. L. McGreevy, E. W. J. Mitchell and C. Raptis, *J. Phys. C* **19**, 2925 (1986)].

leading to the tetrahedral coordination expected for isotropic two-body interactions.

Bunten *et al.*[35] have developed a different method of analysis. The time transforms $I(t)$ of the spectra, defined in Eq. (9.3), are shown in Figs. 36 and 37. The exponential tail in $I(\bar{\nu})$ is reflected in the rapid decay of $I(t)$ at short times, and the sharp central peak in $I(\bar{\nu})$ in the long-time tail of $I(t)$. The noncentral contributions to $I(\bar{\nu})$ are seen as a damped oscillatory modulation of $I(t)$. Bunten *et al.*[35] have used a semiempirical model for the form of $I(t)$

$$\frac{I(t)}{I(t=0)} = \frac{a_1\tau_1^2}{t^2+\tau_1^2} + a_2 \exp\{-[(t^2+\tau_2^2)^{1/2} - \tau_2]/\tau_3\}$$
$$+ (1 - a_1 - a_2)\exp\{-[(t^2+\tau_4^2)^{1/2} - \tau_4]/\tau_5\}\cos\omega_0 t \quad (9.6)$$

The first term represents the transform of the exponential $\bar{\nu}$ tail and, therefore, corresponds to second-order Raman scattering, whether from pairs of excitations or binary collisions. The second and third terms are equivalent to the Rayleigh and Brillouin lines in the RB spectrum. The second term is a dissipative component corresponding to the central $\bar{\nu}$ peak, and the third a damped oscillatory component corresponding to the intermediate-$\bar{\nu}$ shoulder or side peak. This would be (disorder-allowed) first-order Raman scattering.

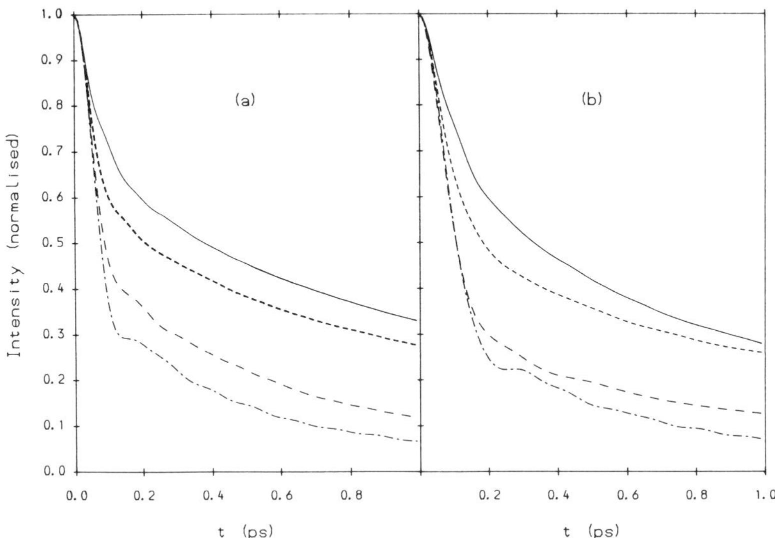

FIG. 37. Depolarized Raman scattering intensities $I_{yx}(t)$ for molten alkaline earth (a) chlorides [R. A. J. Bunten, R. J. McGreevy, E. W. J. Mitchell, C. Raptis, and P. J. Walker, *J. Phys. C* **17**, 4705 (1984); C.-H. Huang and M. H. Brooker, *Chem. Phys. Lett.* **43**, 180 (1976)] and (b) bromides [R. A. J. Bunten, R. L. McGreevy, E. W. J. Mitchell, and C. Raptis, *J. Phys. C* **19**, 2925 (1986)].

The quasiexponential damping has the same short-time Gaussian (free-particle) and long-time exponential (diffusive) behavior as the relaxation used by Adams *et al.*[96] to fit molecular dynamics results for molten NaCl, as noted in Subsection 6. The functional form was suggested by Egelstaff and Schofield[133] because it had the correct limits and also an analytic transform (modified Bessel function of the second kind, K_1). It has also been used to describe $S(Q, \omega)$ for monatomic fluids.[79]

Examples of fits of this model, transformed back to $\bar{\nu}$, are shown in Fig. 38. It is found that, while in certain cases some of the parameters may be zero, it is in general necessary to use all eight parameters to adequately fit the polarized and depolarized spectra of all the molten alkaline earth chlorides and bromides. The time constants τ_1 (0.05–0.15 ps; see Table V) have been compared to the binary collision prediction, $\tau_1 = (m\sigma^2/k_B T)^{1/2}/3$, using the average mass for m and the average ionic radius for σ. Reasonable agreement is found for the chlorides and is improved by including the empirical density dependence found for inert gases [Eq. (9.2)]. However, values of τ_1 for the bromides are almost identical to those of the corresponding chlorides,

[133] P. A. Egelstaff and P. Schofield, *Nucl. Sci. Eng.* **12,** 260 (1962).

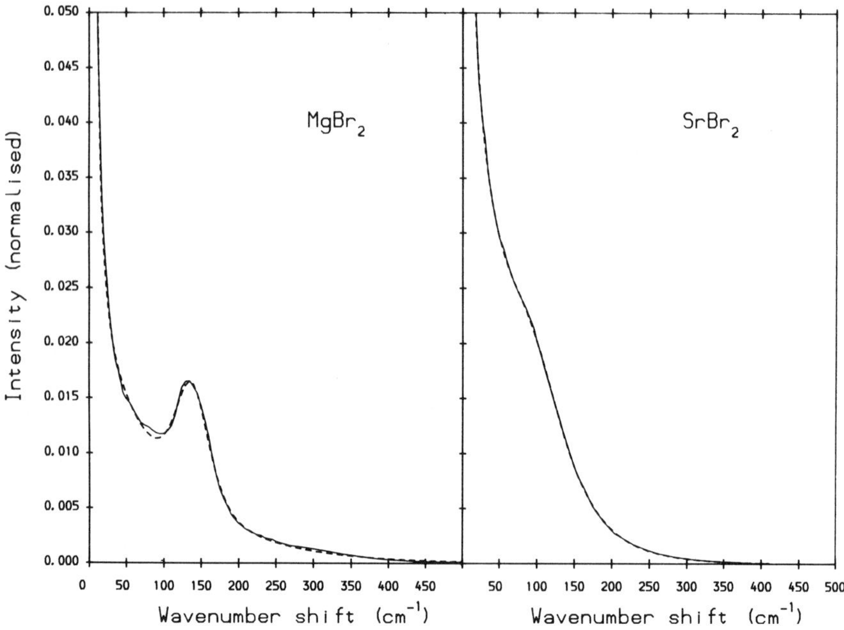

FIG. 38. Fits of the semiempirical model of Bunten *et al.* [*J. Phys. C* **17**, 4705 (1984)] (---) to the Raman scattering intensities [R. A. J. Bunten, R. L. McGreevy, E. W. J. Mitchell, and C. Raptis, *J. Phys. C* **19**, 2925 (1986)] (———) $I_{xx}(\bar{\nu})$ for molten $MgBr_2$ and $I_{yx}(\bar{\nu})$ for molten $SrBr_2$.

indicating that they are largely independent of the cation. The binary collision prediction for the bromides is, therefore, poor.

The values of the individual time constants τ_2, τ_3, τ_4, and τ_5 will depend on the precise form of the damping function used, and their individual significance is not known; so average exponential damping times

$$\tau_{23} = (\tau_3^2 + 2\tau_2\tau_3)^{1/2} \tag{9.7}$$

$$\tau_{45} = (\tau_5^2 + 2\tau_4\tau_5)^{1/2} \tag{9.8}$$

have been calculated. These are of order 1–2 ps for τ_{23} and 0.1–0.3 ps for τ_{45} (see Table V). τ_{23} obviously represents a diffusive-type relaxation. Bunten *et al.*[35] have suggested that the process causing the largest modulation of the polarizability at low frequencies will be antiphase motions of oppositely charged ions, i.e., charge currents. τ_{23} will then be related to the electrical conductivity and dielectric constant

$$\tau_{23} = \varepsilon_0/\sigma_0 \tag{9.9}$$

TABLE V. RELAXATION TIMES DERIVED FROM RAMAN SPECTRA OF MOLTEN ALKALINE EARTH CHLORIDES[a] AND BROMIDES[b]

Spectrum	T (K)	τ_1 (ps)	τ_{23} (ps)	τ_{45} (ps)	ω_0 (10^{-12} rad s^{-1})
$MgCl_2$ I_{xx}	1018	0.056	1.319	0.108	38.41
$MgBr_2$	1023	0.055	1.903	0.305	26.03
$CaCl_2$	1073	0.072	1.469	0.079	30.26
$CaBr_2$	1043	0.067	1.280	0.193	19.85
$SrCl_2$	1184	0.085	1.047	0.096	27.08
$SrBr_2$	951	0.091	1.256	0.139	17.15
$BaCl_2$	1255	0.112	1.048	0.110	22.96
$BaBr_2$	1148	0.094	1.088	0.177	15.83
$MgCl_2$ I_{yx}	1018	0.064	1.433	—	—
$MgBr_2$	1023	0.095	1.176	—	—
$CaCl_2$	1073	0.053	1.447	—	—
$CaBr_2$	1043	0.081	1.713	—	—
$SrCl_2$	1184	0.097	0.842	0.087	18.42
$SrBr_2$	951	0.097	1.379	0.125	12.55
$BaCl_2$	1255	0.153	0.782	0.105	18.32
$BaBr_2$	1148	0.102	0.761	0.212	14.63

[a] R. A. J. Bunten, R. L. McGreevy, E. W. J. Mitchell, C. Raptis, and P. J. Walker, *J. Phys. C* **17,** 4705 (1984).

[b] R. A. J. Bunten, R. L. McGreevy, E. W. J. Mitchell, and C. Raptis, *J. Phys. C* **19,** 2925 (1986).

[This also gives the limiting width of $S_{qq}(Q \to 0, \omega)$.] Although static dielectric constants have not been measured, the values derived from this equation, using the experimentally determined values of τ_{23} and σ_0, are of the expected magnitude.

The time constants τ_{45} are intermediate between predicted binary collision times and diffusive relaxation times. They represent the lifetime of the collective mode. If τ_{45} is sufficiently long compared to the period $2\pi/\omega_0$, then the mode will be resolved as a side peak and may be considered to propagate. Otherwise it is overdamped. The criterion for the existence of a side peak is found to be $\tau_{45} > 1.25\pi/\omega_0$; i.e., the relaxation time is greater than about half the period (exactly one-half would be expected for a Gaussian peak).

τ_{23} and τ_{45} only vary by a factor of 2 between all the salts, which may be fitted with the same model for the line shape. The longest relaxation time derived is 1.5 ps. The resolution of a side peak is not, therefore, evidence for the existence of either complex ions or quasimolecules in these liquids.

Values of ω_0 have been found[127] to have an approximately linear dependence on the reciprocal of the reduced mass, $1/\mu$, rather than ω_0^2, as might be expected. This may, however, include some dependence on ionic radii or polarizabilities, which tend to increase with the ionic masses. Bunten *et al.*[35] have noted that the values of ω_0 are of the magnitude expected for TO modes (non-Raman active) in the crystal; in the case of $SrCl_2$, ω_0 is very close to the zone-center TO frequency. They originally suggested that ω_0 may represent a long-wavelength ($Q \to 0$) TO mode in the melt, first-order Raman scattering being allowed by the disorder. They later noted, however, that an alternative explanation was possible, bearing in mind the result of Shuker and Gammon[108] that all Q, and not simply $Q = 0$, may contribute in the disordered case. ω_0 may then represent an average frequency over the TO band. In cases such as $MgCl_2$, where the crystalline dispersion is very flat, the band will be narrow and the side peak well defined. For $SrCl_2$ and $BaCl_2$, where the band shows some dispersion, it will be broad and the side peak is not resolved. The fact that the side peaks in Mg^{2+} and Ca^{2+} salts are almost totally polarized suggests that they are dominated by a narrow range of Q, probably at a zone edge rather than zone center.

In $MgCl_2$ and $MgBr_2$ spectra (both polarized and depolarized) there is evidence of weak additional scattering at $\bar{\nu} \cong 350$ and $300\ \text{cm}^{-1}$, respectively. This may be due to an LO band, which is more weakly coupled to the (transverse) light wave. The observation of this feature, together with the resolution of a side peak for the TO mode, is evidence of longer-range ordering in Mg^{2+} salts than in the other alkali and alkaline earth halides. This correlates with the observation of a rise in $S_T(Q)$ at low Q for $MgCl_2$ (Section II). Ordering with a sufficiently long correlation length may allow charge-density fluctuations to propagate within that length. It may also be noted that it is only in melts with a resolved side peak in the Raman spectrum that acoustic-mode softening is directly observed (Section III). This may also be related to the range of structural ordering.

The model of Bunten *et al.*,[35] while very successful for the alkaline earth halides, is not generally so successful for the alkali halides. While many of the spectra may be adequately fitted using only the first two terms in Eq. (9.6) (i.e., no collective-mode term), those with stronger shoulders (e.g., LiI, NaI) are not well fitted. It many be that the TO band cannot be adequately represented by a single frequency, as is obviously the case for LiI. Relaxation times for those spectra which may be fitted are comparable to those for the alkaline earths. The model is most useful in that it demonstrates that the occurrence of a side peak in the spectrum does not indicate a difference in the scattering process. The relaxation times derived reflect the processes in the liquid but give no direct information on the scattering mechanism. Until this is better understood, the differences between polarized and depolarized spectra will remain unclear.

c. Brillouin Scattering

While the Raman spectra of a large number of alkali and alkaline earth halides have been measured, there have been very few Brillouin scattering studies. Torell and Gunilla Knape[134] have measured RB spectra of molten NaCl and KCl, but with low resolution, and only sound velocities (in agreement with values obtained by ultrasonic measurements) were determined. High-resolution experiments, comparable to those on monatomic liquids,[135] have been made on molten KCl and CsCl by Qiu *et al.*[128] and on some alkali iodides and alkaline earth chlorides by Li *et al.*[136] (quantitative results from the latter experiments are not yet available).

RB spectra of molten CsCl (free spectral range FSR = 0.178 cm^{-1}) at a series of temperatures[128] are shown in Fig. 39 (VT polarization = $VV + VH$ = $xx + yx$). These are convolutions of three sets of Rayleigh and Brillouin lines; the Stokes Brillouin line corresponding to the central Rayleigh line is indicated by an arrow. The RB spectrum has the same three-Lorentzian form expected for monatomic liquids,[69] i.e., $S_{mm}(Q \to 0, \omega) \equiv S(Q \to O, \omega)$ (see Subsection 8); this has then to be convoluted with the experimental resolution function before fitting to the data. Estimates of the Rayleigh linewidth and intensity are unreliable due to stray light; it is not possible to prepare samples of the purity required for this. The Brillouin peak position ($\sim$0.15 cm^{-1}) and temperature shift are found to agree well with ultrasonic measurements. However, the Brillouin linewidth ($\sim$0.004 cm^{-1}) is much narrower than expected and indicates a bulk viscosity that is an order of magnitude smaller than in the low-frequency ultrasonic region. The linewidth only increases slightly with temperature, while the intensity increases slightly and then decreases again.

The Brillouin lines are found to be completely polarized, and the Rayleigh line is largely polarized (this is difficult to estimate because of the stray light component). Measurements with a wider FSR (up to 20 cm^{-1}) showed no evidence of Rayleigh wing features associated with rotational motion, indicating (as expected from all other experimental results) that there are no complex ions or quasimolecules in these melts. However, these measurements did show a flat, highly depolarized, background. The integrated cross section for this background agreed well with that estimated by Mitchell and Raptis[110] for the low-$\bar{\nu}$ Raman scattering, showing that this scattering does extend smoothly to $\bar{\nu} = 0$.

The RB spectra of molten KCl are compared in Fig. 40 to those of the crystal (Q parallel to the 110 axis) just below the melting point.[128] The

[134] L. M. Torell and H. E. Gunilla Knape, *Z. Naturforsch., A* **34A,** 899 (1979).
[135] V. Ghaem-Maghami and A. D. May, *Phys. Rev. A* **22,** 692 (1980).
[136] W. Li, R. L. McGreevy and H. Z. Cummins, private communication (1985).

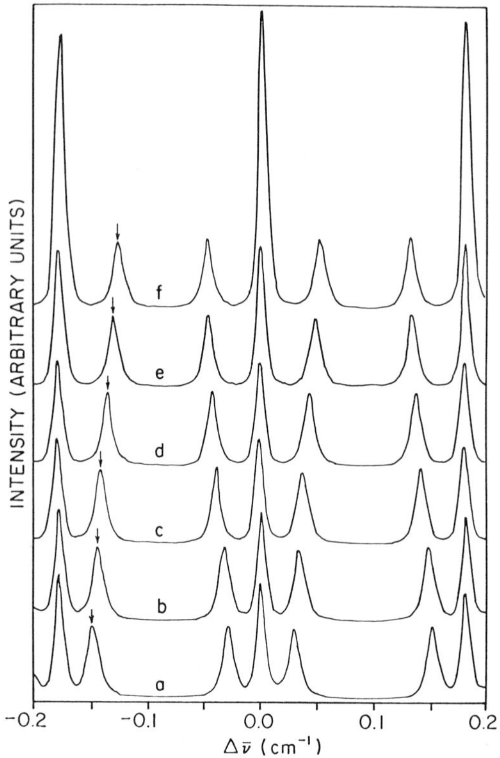

FIG. 39. Brillouin scattering intensities $I_{VT}(\bar{\nu})$ for molten CsCl at (a) 932, (b) 970, (c) 1023, (d) 1074, (e) 1122, and (f) 1152 K. The arrow indicates the Stokes Brillouin line associated with the central Rayleigh line [S. L. Qiu, R. A. J. Bunten, M. Dutta, E. W. J. Mitchell, and H. Z. Cummins, *Phys. Rev. B: Condens. Matter* [3] **31**, 2456 (1985)].

Brillouin line shift upon melting is as expected from the change in density and compressibility. However, it is found that the linewidth decreases by a factor of ~3, and the intensity increases by a factor of ~10. These changes, together with the narrowness of the lines compared to ultrasonic data, are not understood. Possibly the more localized low-frequency modes in the disordered liquid do not scatter long-wavelength sound modes as well as the more extended modes found in the crystal.

RB spectra for the molten alkali iodides and alkaline earth chlorides[136] are similar to those for the alkali chlorides. Measurements in the alkaline earths are considerably more difficult because their highly hygroscopic nature makes preparation of pure samples much harder. However, it is found that Brillouin line intensities are low in comparison to the alkali halides, although no comparison with crystals has been made. It is worth noting that the Brillouin

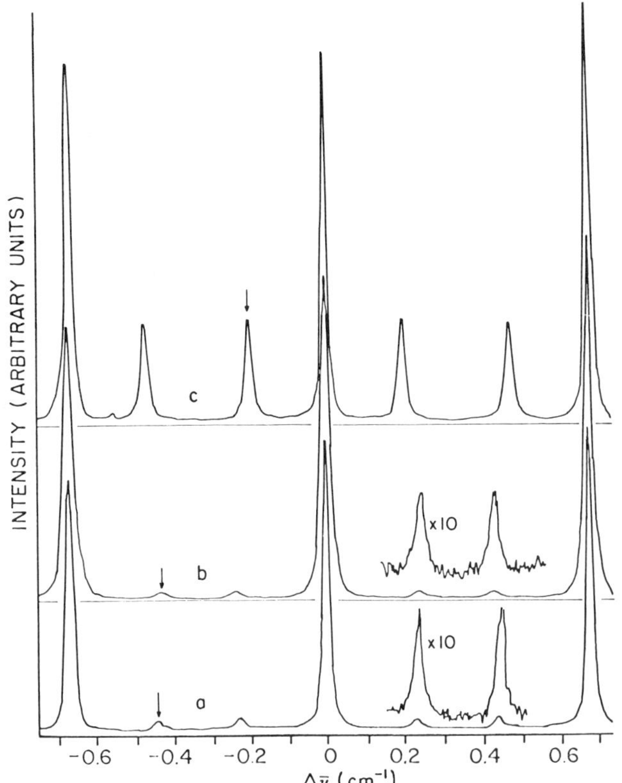

FIG. 40. Brillouin scattering intensities $I_{VT}(\bar{\nu})$ for crystalline KCl(110) at (a) 980 and (b) 1043 K and (c) molten KCl at 1083 K. The arrow indicates the Stokes Brillouin line associated with the central Rayleigh line [S. L. Qiu, R. A. J. Bunten, M. Dutta, E. W. J. Mitchell, and H. Z. Cummins, *Phys. Rev. B: Condens. Matter* [3] **31**, 2456 (1985)].

peak position for molten $MgCl_2$ is in agreement with the ultrasonic value (interestingly there is no shift with temperature). $S_T(Q)$ must therefore decrease from the high value observed at $Q \cong 0.15$ Å^{-1} (Section II) to the predicted compressibility limit before $Q \cong 10^{-3}$ Å^{-1}

d. Infrared Absorption and Reflection

Infrared absorption and reflection are commonly used experimental techniques for the study of optic modes in crystals. Little comparable work has been done on molten salts. Barker[137] has measured near-infrared absorption

[137] A. J. Barker, *J. Phys. C* **5**, 2276 (1972).

spectra of molten LiF, NaCl, and KBr in both high-temperature crystalline and molten phases. From the small changes observed he concludes that three-phonon processes occur similarly in solid and liquid, and that the high-frequency limit of the vibrational density of states remains approximately the same across the melting transition. Mead[138] has measured the far-infrared reflection and emission spectra of high-temperature crystalline and molten LiF. Using a Kramers–Kronig method of analysis, he determines $\omega_{TO} \cong 300$ cm^{-1} (37 meV) and $\omega_{TO} \cong 420$ cm^{-1} (53 meV) for the liquid, compared to the solid values of 271 and 430 cm^{-1}, respectively. This close correspondence, together with the results of Barker, is in accord with the evidence for the existence and dispersion of optic modes in the melt from inelastic neutron and Raman scattering discussed earlier.

V. Conclusions

The experimental work discussed in the previous sections has provided a good basis for understanding the structures and dynamics of molten alkali and alkaline earth halides. The alkali halides are found to have short-range order radially over 10–12 Å. Charge ordering is the dominant feature. The partial radial distribution functions are predictable using a simple repulsive and Coulomb potential, although there is some evidence that ionic polarizability is important for larger cations such as Cs^+. CsCl shows octahedral coordination similar to that of the NaCl crystal structure, while there are indications that melts with a smaller cation such as Li^+ will show tetrahedral coordination more similar to the zinc blende or wurtzite structures.

The molten alkaline earth halides show considerably more structural variation, though charge ordering again dominates. Coordination varies from fourfold to eightfold. There is strong evidence of the importance of many-body forces caused by distortion of the polarizable anions by small, doubly charged, cations such as Mg^{2+}. These forces are not simply isotropic and are dependent on electronic structure, giving characteristically different local ordering for $MgCl_2$ and $ZnCl_2$. In $CaCl_2$ the many-body interaction leads to a change in cation structure with temperature, while in $MgCl_2$ there is evidence of medium-range ordering with a correlation length $\sim$20 Å. Such many-body forces have not yet been successfully modeled.

While the dynamical structural factors of the molten alkali halides are largely determined by single-particle, diffusive motions, there is strong evidence from inelastic neutron scattering experiments of small contributions from collective motions. These are the remnants of acoustic and optic modes

[138] D. G. Mead, *J. Phys. C* 7, 445 (1974).

in the crystal. It can be shown by the use of isotopic substitution that they are equivalent to mass and charge-current fluctuations in the melt. Though the modes are overdamped, their energies are similar to those in the crystal. There is some relation between their dispersion and the corresponding structure factors, although its precise form is not yet clear. While the neutron scattering cross section is only related to longitudinal currents, Raman scattering experiments show possible evidence of transverse optic modes.

The dynamics of the alkaline earth halides, while basically similar to that of the alkali halides, is more complex. Molten $SrCl_2$ shows evidence of "additional modes" that are associated with fast ion conduction in the solid state. Acoustic-mode softening is observed in $MgCl_2$, $MgBr_2$, and $ZnCl_2$; in $ZnCl_2$ the soft mode almost propagates. Propagating transverse optic modes are clearly visible in the Raman spectra of these melts. Such features have been associated with medium-range ordering due to many-body interactions.

Though the term "mode" has been used, it should be emphasized that the actual atomic motions are not necessarily similar to those associated with phonons in low-temperature crystals. However, in high-temperature crystals, where there are strong phonon–phonon interactions due to anharmonicity, the appropriate collective modes will not be extended single phonons but localized phonon "clouds," which have more in common with modes in the liquid than in the low-temperature crystal. In those cases where liquid modes have been said to propagate, this does not imply infinite extension but simply that the correlation length is longer than the wavelength (or equivalently that the relaxation time is longer than the period). It must also be emphasized conversely that the modes should not be considered as vibrations of quasimolecules or complex ions. There is no evidence in these systems that an ion remains correlated with any other ion or group of ions for times significantly longer than the average diffusional relaxation time, of order 1 ps.

Now that a general picture of the behavior of these liquids is emerging, further work will be necessary in order to clarify some of the details. In the absence of any progress on the theoretical description of the Raman spectra, there is little work to suggest in this field. Possible areas for investigation using neutron scattering will include:

(1) Alkali halide structures. Quantitatively more accurate experiments, particularly at low Q, to distinguish between different potentials.

(2) Alkaline earth halide structures. Development of potentials to model many-body interactions, possibly using inversion techniques from the experimentally determined structure factors.

(3) Alkali halide dynamics. Quantitatively more accurate experiments in the intermediate-Q range to enable second-order isotopic substitution to be used to determine the importance of mass–charge coupling and to enable

some separation of the incoherent scattering. Measurements at lower and higher Q, probably using pulsed neutron sources.

(4) Alkaline earth halide dynamics. Further measurements within the group, particularly using isotopic substitution. Comparison with the dynamics of corresponding fast ion conducting crystals.

The experimental techniques described here should now also be applied, where possible, to studies of other binary liquids in order to determine which features are due to the binary nature of the system and which to the opposing charges on cation and anion. While extensive work has been done on the structures of molten binary alloys, there has been hardly any work on their dynamics; similarly there has been no work on binary mixtures of condensed inert gases. Unlike molten salts these systems can be produced with a wide variety of relative concentrations of the two species. In this respect metal–molten salt solutions are of interest; they cover the range from the (electronic) conducting liquid metal to the insulating molten salt. Though some structural work has been done (see Rovere and Tosi[1]), there have been no dynamical studies.

The work on molten salts also contributes to our understanding of the wider field of disordered systems. As well as monatomic and polyatomic fluids, this includes amorphous materials and glasses. There are also systems showing partial disorder, ranging from the low-disorder limit of crystals with defects or impurities to highly disordered fast ion conductors such as α-AgI. Acoustic modes observed in molten salts show similarities in dispersion to those in monatomic liquids such as argon, despite coupling to charge fluctuations. Observation of acoustic-mode softening has shown that there are collective contributions centered at nonzero frequency even when they are not resolvable as a side peak in $S(Q, \omega)$, a subject of recent argument. Work on the dynamics of some binary liquids, particularly chemically ordered molten alloys, may offer some hope of determining criteria for the resolution of side peaks.

Collective-mode behavior in molten salts also shows strong similarities to that in binary metallic glasses.[139,140] This suggests that the onset of diffusion at the melting point does not fundamentally alter the collective dynamics, but just introduces further damping. The observation of both structural and dynamical similarities between fast ion conductors and their melts indicates that the phenomenon of fast ion conduction may well be considered as a partial melting. By a study of the structure and dynamics of liquids that show interesting properties in the solid state, such as glass-forming ability or fast ion

[139] J. Hafner. *J. Phys. C* **16,** 5773 (1983).

[140] J.-B. Suck, H. Rudin, H.-J. Guntherodt, and H. Beck, *Phys. Rev. Lett.* **50,** 49 (1983).

conduction, rather than simply studying the solid, we may obtain a better understanding of these properties. This approach may also be applied to the melting transition in ordinary crystals. The work described here has shown some of the experimental techniques that may be used.

ACKNOWLEDGMENTS

The author wishes to thank Professor E. W. J. Mitchell for his encouragement and for critical reading of the manuscript. He also wishes to thank Mr. M. A. Howe for proofreading and Miss M. S. Avery for typing.

Author Index

Numbers in parentheses are reference numbers and indicate that an author's work is referred to although his name is not cited in the text.

C

D

H

I

J

K

L

M

Y

Z

Subject Index

B

C

G

H

I

J

K

L

M

N

O

P

Q

R